New Insights into Abiotic and Biotic Stress in Plants

New Insights into Abiotic and Biotic Stress in Plants

Editor: Roger Wyatt

STATES
ACADEMIC PRESS
www.statesacademicpress.com

States Academic Press,
109 South 5th Street,
Brooklyn, NY 11249, USA

Visit us on the World Wide Web at:
www.statesacademicpress.com

ISBN: 978-1-63989-378-2 (Hardback)

Cataloging-in-Publication Data

New insights into abiotic and biotic stress in plants / edited by Roger Wyatt.
p. cm.
Includes bibliographical references and index.
ISBN 978-1-63989-378-2
1. Plants--Effect of stress on. 2. Plant physiology. 3. Crops--Effect of stress on.
4. Crops--Physiology. I. Wyatt, Roger.
QK754 .N49 2022
581.7--dc23

Table of Contents

Preface ... VII

Chapter 1 **Role of ABA in Arabidopsis Salt, Drought and Desiccation Tolerance** ... 1
V. C. Dilukshi Fernando and Dana F. Schroeder

Chapter 2 **Consequences of Water Deficit on Metabolism of Legumes** ... 19
Allan Klynger da Silva Lobato, Gélia Dinah Monteiro Viana, Gleberson Guillen Piccinin, Milton Hélio Lima da Silva, Gabriel Mascarenhas Maciel and Douglas José Marques

Chapter 3 **Evaluation of TMV Lesion Formation and Timing of Signal Transduction during Induction of Systemic Acquired Resistance (SAR) in Tobacco with a Computer-Assisted Method** ... 34
Zoltán Á. Nagy, György Kátay, Gábor Gullner and Attila L. Ádám

Chapter 4 **Abiotic and Biotic Elicitors–Role in Secondary Metabolites Production through In Vitro Culture of Medicinal Plants** ... 44
Poornananda M. Naik and Jameel M. Al-Khayri

Chapter 5 **Signalling Crosstalk of Plant Defence Responses to Xylem-Invading Pathogens** ... 75
Sabina Berne and Branka Javornik

Chapter 6 **Molecules and Methods for the Control of Biotic Stress Especially the Insect Pests — Present Scenario and Future Perspective** ... 104
Santosh Kumar Upadhyay and Sudhir P. Singh

Chapter 7 **Drought Adaptation in Millets** ... 128
Zerihun Tadele

Chapter 8 **Antioxidant Enzyme Activities and Abiotic Stress Tolerance Relationship in Vegetable Crops** ... 152
Sebnem Kusvuran, Sevinç Kiran and S. Sebnem Ellialtioglu

Chapter 9 **Abiotic Stress Alleviation with Brassinosteroids in Plant Roots** ... 178
Sevgi Marakli and Nermin Gozukirmizi

Chapter 10 **Stress-Induced Proteins in Recalcitrant Seeds during Deep Dormancy and Early Germination** ... 200
Marina I. Azarkovich

Chapter 11 **Plant Evolution in Response to Abiotic and Biotic Stressors at "Rear-edge" Range Boundaries** **219**
Gunbharpur S. Gill, Riston Haugen, Jesse Larson, Jason Olsen and David H. Siemens

Permissions

List of Contributors

Index

Preface

This book was inspired by the evolution of our times; to answer the curiosity of inquisitive minds. Many developments have occurred across the globe in the recent past which has transformed the progress in the field.

The immobility of plants forces them to endure and adapt to a variety of environmental stresses. These stresses are categorized into biotic and abiotic stresses. Biotic stresses are caused by living organisms such as nematodes, bacteria, fungi, weeds and arachnids. They exhibit a parasitic behavior and deprive plants of their nutrients. Abiotic stresses include desertification, soil salination, scarcity of water and extreme weather. Some of the common adaptive measures employed by plants include increased reactive oxygen scavenging activity, reduced photosynthesis, stomatal closure, increased root length and reduced leaf growth. Plants may also produce toxic compounds like phytoalexins and reactive oxygen species to combat pathogens. This book covers in detail some existent theories and innovative concepts revolving around abiotic and biotic stress in plants. It includes some of the vital pieces of work being conducted across the world, on various topics related to biotic and abiotic stresses. Those with an interest in the field of abiotic and biotic stress in plants would find this book helpful.

This book was developed from a mere concept to drafts to chapters and finally compiled together as a complete text to benefit the readers across all nations. To ensure the quality of the content we instilled two significant steps in our procedure. The first was to appoint an editorial team that would verify the data and statistics provided in the book and also select the most appropriate and valuable contributions from the plentiful contributions we received from authors worldwide. The next step was to appoint an expert of the topic as the Editor-in-Chief, who would head the project and finally make the necessary amendments and modifications to make the text reader-friendly. I was then commissioned to examine all the material to present the topics in the most comprehensible and productive format.

I would like to take this opportunity to thank all the contributing authors who were supportive enough to contribute their time and knowledge to this project. I also wish to convey my regards to my family who have been extremely supportive during the entire project.

Editor

1

Role of ABA in Arabidopsis Salt, Drought, and Desiccation Tolerance

V. C. Dilukshi Fernando and Dana F. Schroeder

Abstract

The ability of plants to respond to environmental stimuli is essential to plant survival. Abscisic acid (ABA) is a phytohormone with roles at various stages of plant development. ABA also plays a major role in mediating physiological responses to environmental stresses such as salt, osmotic, and cold stress. Plant responses to environmental stress have been widely studied in the model plant *Arabidopsis thaliana* and ABA signaling mechanisms elucidated. In general, the adaptive responses of plants to various stress conditions can be either ABA-dependent or ABA-independent. Here we focus on the role of ABA in stress signaling and abiotic stress tolerance. We describe the intrinsic mechanisms that confer stress tolerance via ABA, as well as how ABA-regulated gene products play a role in salt and drought tolerance at different stages of the life cycle. In addition, the contribution of ABA to regulation of stomatal aperture and therefore desiccation tolerance will be discussed. Understanding ABA signaling mechanisms in abiotic stress provides avenues for improving plant performance.

Keywords: ABA signaling, salt, drought, dessiccation, Arabidopsis

1. Introduction

Due to their sessile nature, plants cannot avoid environmental stresses, thus they have evolved mechanisms to overcome the detrimental effects of stress. For example, plant endogenous developmental programs are modified such that structural and metabolic changes assist to overcome adverse environmental conditions such as salinity and drought. Failure to adapt to adverse environmental conditions can significantly reduce yield by impacting plant development and productivity. Abiotic stress conditions initiate a number of molecular, biochemical, and physiological changes at both the cellular and whole plant levels [1]. One major biochemical change in response to stress is elevation of abscisic acid (ABA) levels, which in turn triggers expression of a cascade of stress-responsive genes [2]. Cellular ABA levels are induced by environmental stimuli such as light, water, and salinity stress [3].

The plant hormone ABA has been identified as a key regulator of multiple stresses. In general, the adaptive responses of plants to various stress conditions can be either ABA-dependent or ABA-independent. However, there is no clear boundary between these two pathways and there is lot of crosstalk between the pathways and the components involved. This review will focus on recent advancements in ABA-mediated stress signaling and the role of ABA in abiotic stress tolerance in the model plant *Arabidopsis thaliana*.

2. The phytohormone Abscisic Acid (ABA)

ABA, a sesquiterpenoid ($C_{15}H_{20}O_4$) with a 15-carbon ring (Figure 1), has a variety of biological functions and is found ubiquitously across several kingdoms, including cyanobacteria, sponges, algae, lichens, mosses, and mammals [4-7]. Discovered in the 1960s and initially named dormin or abscissin, ABA is now established as a widely occurring and important plant growth regulator. Although it was initially identified as an abscission-promoting hormone, later scientists discovered that this was partly due to an indirect effect of inducing ethylene biosynthesis [8]. ABA is an important regulator of plant growth, including embryo and seed development, seedling establishment, vegetative and reproductive growth as well as promoting seed dormancy [9,10]. Seed maturation and promotion of dormancy are important in preventing preharvest sprouting. In addition, ABA has the ability to antagonize the germination promoting effects of gibberellin, regulate guard cells, and regulate stress-responsive gene expression under water-deprived conditions. ABA also has a role in plant pathogen responses in a pathosystem-dependent manner [4,5].

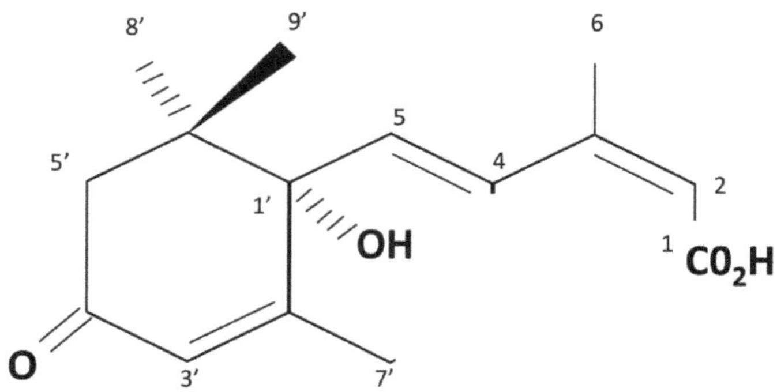

Figure 1. Structure of phytohormone abscisic acid S-(+)-ABA

The molecular structure of ABA has several important features that facilitate its biological functions. The side chain with the two double bonds (Figure 1) and ABA's stereocenter are two such important features. Exposure to UV light changes the conformation from active to inactive form [5].

2.1. ABA signaling in plants

Although ABA has a broad range of functions in plant growth and development, its main function is to regulate plant water balance and osmotic stress tolerance [11]. Thus, understanding ABA signaling is essential to improving plant performance. Genetic screens in *Arabidopsis thaliana* identified many downstream ABA signaling components. Recent findings in the field of ABA signaling reveal a unique hormone perception mechanism (Figure 2) where ABA binds to the ABA receptors Regulatory Components of ABA Receptor/Pyrabactin Resistance Protein1/PYR-like Proteins (RCAR/PYR1/PYLs). RCAR/PYR/PYL proteins belong to the START-domain superfamily and have soluble ligand-binding properties. RCAR/PYR/PYL receptors are found in the cytoplasm as well as in the nucleus. ABA binding to RCAR/PYR/PYLs leads to inactivation of type 2C protein phosphatases (PP2Cs) such as ABSCISIC ACID INSENSITIVE 1 (ABI1) and its close homolog ABI2 [12]. All 14 members of the RCAR family of proteins bind to ABA and interact with PP2Cs. Except for RCAR7/PYL13, all the other RCAR members are positive regulators of ABA signaling. Among the 80 PP2Cs identified in Arabidopsis, six out of nine clade A PP2Cs act as negative regulators of ABA signaling [13]. These phosphatases and RCAR/PYR1/PYLs function as co-receptors and form a high-affinity ABA-binding site. Inactivation of PP2Cs causes suppression of PP2C-mediated dephosphorylation of Sucrose nonfermenting Kinase-1-Related protein kinase 2s (SnRK2s), which are important positive regulators of ABA signaling. As a result, activated SnRK2s target ABA-dependent gene expression and ion channels [5,11]. Table 1 summarizes the major positive and negative regulatory elements in the ABA signaling pathway. Phosphorylated SnRK2s subsequently phosphorylate ABA-responsive element Binding Factors (ABFs), which are basic leucine zipper transcription factors that bind to ABA-Responsive Elements (ABRE) (PyACGTGG/TC), the major *cis*-element in the promoter region of downstream genes that are induced by ABA [19,20].

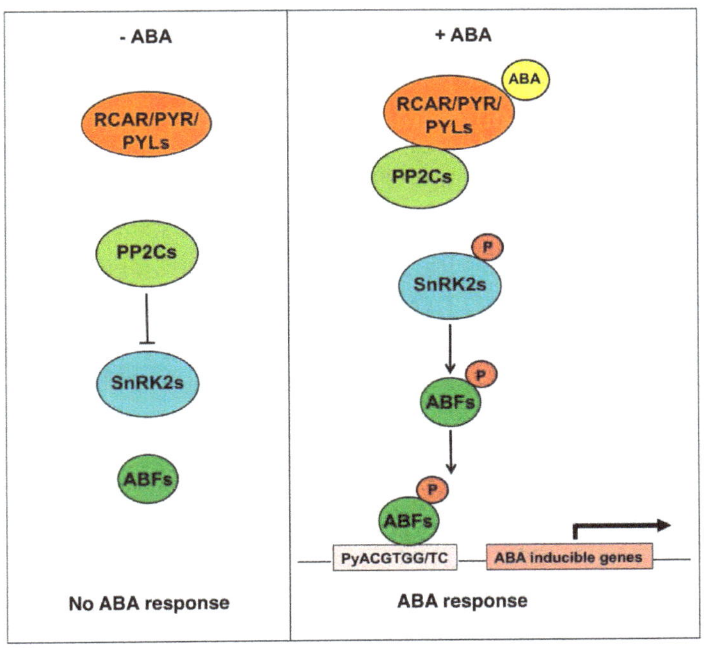

Figure 2. Main components in the core ABA signal transduction pathway

Signaling component	Regulation	Expressed	References
Group A PP2Cs	**Negative regulators**		
ABA INSENSITIVE 1/2 (ABI1/2)		Various tissues and developmental stages	[13-16]
ABA HYPERSENSITIVE GERMINATION 1 (AHG1)			
HYPERSENSITIVE TO ABA 1/2 (HAB1/2)			[17]
SnRK2 subgroup III	**Positive regulators**		
SRK2D/SnRK2.2		Seeds and vegetative tissues	[18]
SRK2I/SnRK2.3		Seeds and vegetative tissues	[18]
SRK2E/OST1/SnRK2.6		Expressed in guard cells and involved in stomatal closure	[15]

Table 1. Major positive and negative regulators of ABA signaling

Therefore, the ABA signaling complex/ABA signalosome is comprised of three major components: (a) RCAR/PYR/PYLs; (b) PP2Cs; and (c) SnRK2s assembled as a double negative regulatory system [7]. In the absence of ABA, PP2Cs dephosphorylate SnRK2s inhibiting kinase activity and thereby preventing downstream gene expression (Figure 2). Several studies showed that these core components are essential for ABA signaling. For instance, Fujita *et al.* [21] showed ABA signaling is completely blocked and *ABF* genes showed reduced expression in the *snrk2.2/2.3/2.6* triple null mutant but not in single or double mutants. In addition, reduced phosphorylation of other bZip transcription factors such as ABSCISIC ACID INSENSITIVE 5 (ABI5), which is a dormancy promoting transcription factor, was also observed [18,22].

In guard cells (Figure 3), ABA binds to the PYR/PYL/RCAR receptor-PP2C complex and blocks its phosphatase activity. Consequently, activated protein kinase SnRK2.6/OPEN STOMATA 1 (OST1) phosphorylates and regulates the key target ion channels, SLOW ANION CHANNEL ASSOCIATED 1 (SLAC1) and K^+ CHANNEL IN ARABIDOPSIS THALIANA 1 (KAT1). SnRK2.6/OST1 acts as a positive regulator of stomatal closure where it activates anion channel SLAC1 and inhibits cation channel KAT1 [23-25].

2.2. ABA-binding proteins and alternate ABA receptors

Identification of putative ABA receptors using forward genetic approaches was not successful for a long time due to genetic redundancy of the genes encoding ABA receptor proteins. However, biochemical approaches leading to purification and analysis of high-affinity ABA-binding proteins have been successful in identification of potential ABA receptor classes [5,26]. Some of these potential ABA receptors are cytosolic while others are on the cell surface,

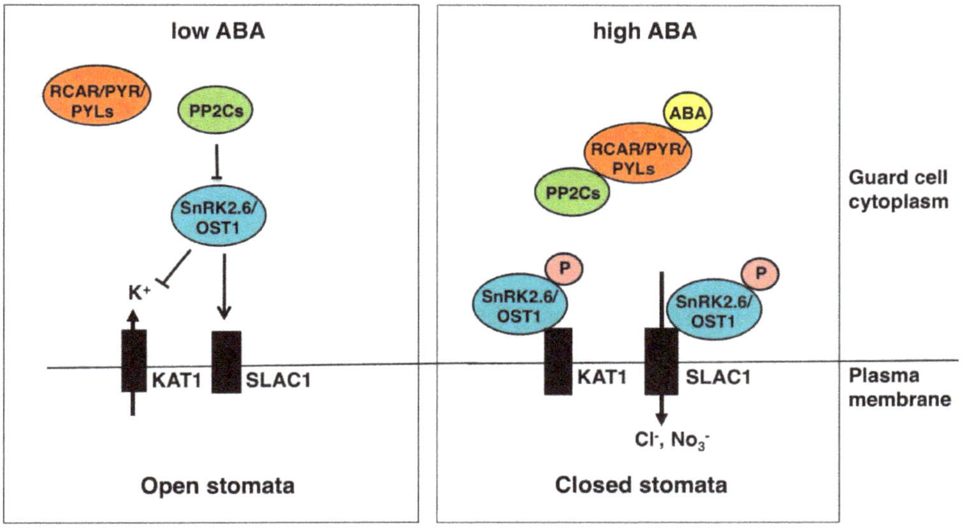

Figure 3. ABA signaling in guard cells

indicating there are extracellular as well as intracellular sites of ABA perception. Studies indicate that there can be multiple ABA receptors at different locations of the cell [27].

2.2.1. *ChlH/ABAR (H subunit of the chloroplast magnesium chelatase/ ABA receptor)*

The primary function of ChlH is chlorophyll synthesis. ChIH was initially identified as an ABA-binding protein in broad bean (*Vicia faba*) and the Arabidopsis protein was named as ABAR. Later it was found that binding of ABA to ChIH/ABAR depends on the stereochemistry and it specifically binds to only (+) ABA to mediate ABA responses. Although ChIH/ABAR is localized in the chloroplast envelope, it functions as a negative regulator of ABA signaling in the nucleus. The cytosolic C-terminus of the ABAR interacts with WRKY transcription factors (WRKY 18, 40, 60) which act as transcriptional repressors, repressing ABA-responsive gene expression in the nucleus. Binding of ABA with ChlH/ABAR promotes interaction with WRKYs, preventing them from repressing downstream genes such as *ABI5* and *DREB2* [28-31]. Thus, it has been proposed that ChlH mediates nuclear-chloroplast signaling. However, another research group has not been able to reproduce these results using *wrky* loss of function mutants. In addition, barley ChlH does not bind to ABA and ChlH loss of function mutants do not show any impaired ABA responses. Despite its ABA-binding properties in Arabidopsis, it is not confirmed whether ChIH functions as an ABA receptor (reviewed in [5,27]). However, ChlH/ABAR mediates ABA-induced stomatal closure and ABA inhibition of blue-light-mediated stomatal opening. In addition, ChIH/ABAR has a role in ABA-mediated fruit ripening in peach and strawberry (reviewed in [31]).

2.2.2. *GTG1/GTG2 (G protein coupled receptor type G protein 1 and 2)*

G protein coupled ABA receptors are plasma membrane localized cell surface receptors that are widely expressed in plants. Both GTG1 and GTG2 showed specific and saturable ABA-binding activity in direct ABA-binding assays. GTGs have GTPase activity and GDP bound GTGs have enhanced ABA-binding ability, which in turn initiates ABA signaling. GTG1/2 bind with GPA1 (G-PROTEIN α SUBUNIT 1), which abolishes its GTPase activity and represses ABA binding. GTP bound GPA1 represses ABA signalling. However, the downstream components of this pathway are not characterized yet [31,32].

2.3. Recent studies on ABA perception and signaling mechanisms

Recent findings that several ABA receptors exist in different parts of the cell provide evidence that ABA is active in a variety of subcellular compartments. ABA synthesis enzymes are present in different compartments, suggesting that ABA synthesis occurs in different parts of the cell and that these ABA levels contribute to overall ABA homeostasis. For example, the ABA biosynthesis enzyme AtABA1 is localized in the chloroplast, whereas AtABA2 is in the cytosol [33,34]. It has also been proposed that ABA produced in cytoplasm, plastids, vacuole, and other subcellular organelles may have different physiological roles initiated by signaling networks via different ABA receptors in each specific compartment [35].

Takeuchi *et al.* [36] identified a potential ABA analog AS6 that can inhibit the activity of PYLs. X-ray crystallography studies showed the structure of ABA facilitates the binding of ABA to PYR/PYL/RCAR receptors and thereby inhibits interaction with PP2Cs. The AS6 ABA analog was able to block PYL-PP2C interaction, indicating that binding of ABA to PYL receptors initiates ABA responses by repressing PP2Cs.

Inhibition of PP2Cs results in autoactivation of SnRK2 kinases and thereby positive regulation of ABA signaling. Recently, the crystal structures of SnRK2.3 and SnRK2.6 were elucidated, providing evidence that kinase activation is a two-step mechanism as well as details of how the ABA signal is transmitted to downstream components [37]. This study also showed that autophosphorylation of SnRK2.6 is more efficient than that of SnRK2.3.

Lumba *et al.* [20] did a comprehensive transcriptomic data analysis in order to generate a mesoscale ABA signaling network. They showed that there are 3 main kinase hubs, MAP3K∂4, SnRK3.15, and SnRK3.22, that interact with PP2Cs and these kinases act as negative regulators of ABA response, in contrast to the SnRK2s involved in ABA signaling. SnRK3.15 and SnRK3.22 also interact with a large number of transcription factors and may have a role in overall ABA responses in the plant [20].

3. ABA in stress signaling

In plants and other organisms, such as algae, cyanobacteria, and fungi, ABA levels tend to increase with exposure to stress, suggesting a potential role of ABA in stress signal transduction [20]. Exogenous ABA application mimics stress conditions in plants and provides a useful means to study the effect of ABA on stress signaling and tolerance [38]. ABA distributes throughout the plant as an inactive glucose sugar conjugate and is converted to the active form by β-glucosidase [4]. ABA acts as an endogenous messenger and salt and drought stress signal transmission to initiate downstream gene expression occurs mainly via ABA signaling. However, cold stress signal transduction occurs in an ABA-independent manner via the C-REPEAT/DEHYDRATION RESPONSIVE ELEMENT BINDING FACTORS (CBFs/DREBs) signaling pathway [39].

A large number of ABA-responsive genes have a common *cis*-element called the ABRE element in their promoter regions. An ABRE together with a Coupling Element (CE) results in ABA induction of gene expression [19]. The **A**BA **R**ESPONSIVE **E**LEMENT **B**INDING FACTOR (AREB/ABF) family of transcription factors are the major transcription factors that regulate ABA-induced gene expression. AREB/ABFs are bZIP transcription factors and their expression is induced by ABA and other potential stress conditions [40]. Different ABFs are induced by ABA at different rates. For instance, *ABF2, ABF3,* and *ABF4* are induced faster than *ABF1.* Moreover, *ABF1* is induced only by cold stress, whereas *ABF2* and *ABF3* are induced by salt stress. *ABF4* levels are induced by salt, drought, and cold stress, suggesting that distinct ABFs have roles in various ABA-dependent stress-responsive pathways [40].

There are nine Group A bZIP transcription factors implicated in ABA signaling and they are subdivided into two groups based on where they are mainly expressed. The *ABI5/AtDPBF* family of genes includes *ABSCISIC ACID INSENSITIVE 5 (ABI5), ENHANCED EM LEVEL (EEL),* and *AREB3* and are expressed in the seed during seed maturation [41]. Other AREB/ABF transcription factors are mainly expressed in vegetative tissues [40].

SnRK2 III is the major subfamily of SnRKs involved in abiotic stress responses. SnRK2 protein kinases phosphorylate AREB/ABFs and regulate their function in ABA-regulated gene expression under stress conditions [10]. SnRK2.6/OST1 is an important regulator of stomatal closure in drought stress. The role of SnRK2.2 and 2.3 is mainly to transmit the ABA signal to inhibit seed germination and seedling growth in response to stress. A decuple *snrk2* mutant in Arabidopsis, which carries mutations for all 10 SnRK2 members, was hypersensitive to osmotic stress and also defective in ABA accumulation and ABA-induced gene expression under osmotic stress, indicating the critical role of SnRK2 kinases in osmotic stress signaling and tolerance. Moreover, the *snrk2.2/3/6* triple mutant had impaired accumulation of proline, which is a compatible osmolite [42].

3.1. ABA and abiotic stress tolerance

In *Arabidopsis thaliana,* about 10% of the genome consists of ABA-regulated genes. Approximately half of these genes are ABA-induced genes and the rest are ABA-repressed. ABA-induced genes code for proteins that confer stress tolerance such as dehydrins, detoxifying enzymes of reactive oxygen species, regulatory proteins (transcription factors, protein kinases, phosphatases) and enzymes required for phospholipid signaling. Genes that are repressed by ABA are mostly related to growth [5]. ABA biosynthesis mutants identified in Arabidopsis [43] and other crop plants [44] wilt and die under prolonged salt and drought stress, suggesting ABA plays an important role in osmotic stress tolerance.

Drought and high salinity generate osmotic stress in plant cells. Endogenous ABA levels are elevated in response to osmotic stress, which in turn coordinates the plant's response to reduced water availability. In addition, seed maturation and post-germinative growth creates cellular dehydration stress, which again results in accumulation of ABA in cells [45]. The role of ABA in drought and salt stress is twofold: water balance and cellular dehydration tolerance. Water balance is achieved through guard cell regulation and the latter role by induction of genes that encode dehydration tolerance proteins in nearly all cells. ABA accumulation is induced by osmotic stress and this is as a result of activation of ABA biosynthesis as well as inhibition of ABA degradation [46]. Thus, ABA-mediated adaptive stress responses of plants to environmental stimuli occur via ABA-responsive gene expression and regulation of stomatal pore size. ABA-responsive gene expression involves various transcription factors, ABA receptors, secondary messengers, protein kinase/phosphatase cascades, and chromatin remodeling factors [45].

Both drought stress and salinity stress upregulate osmotic stress responsive genes that are ABA-inducible. Most of the high-salinity-induced genes are also induced by drought, suggesting there is overlap between salt and drought stress tolerance mechanisms [6]. A large number of transcription factors are induced by multiple stress conditions. AREB1/ABF2, AREB2/ABF4, ABF3, and MYB41 are some of the main transcription factors that are induced by both salt and drought in vegetative tissues [45].

Drought and salt stress results in osmotic imbalance; thus, salt and drought stress tolerance mechanisms aim at restoring cellular homeostasis. These mechanisms are adaptive responses that create either stress tolerance or avoidance of stress conditions. Modifications in metabolic pathways, synthesis of new proteins, changes in ion uptake, and free radical scavenging are some of the stress responses at the cellular level, immediately followed by stress signal transduction [47]. High ABA levels in cells result in synthesis of storage proteins, desiccation tolerance, and dormancy via inhibition of seed germination [22]. In the plant as a whole, key adaptive responses include induction of stomatal closure as well as control of seedling growth and lateral root formation. While the balance between ABA and auxin levels slightly affects primary root growth, ABA represses lateral root formation while auxin promotes it [4,48].

Inhibition of seed germination under abiotic stress is another function of ABA. Seed germination occurs when there is a balance between germination-promoting gibberellin and dormancy-promoting ABA. During late stages of maturation, seeds accumulate ABI5 which in turn activates transcription of LATE EMBRYOGENESIS ABUNDANT (LEA) proteins. LEA proteins confer osmotolerance to the embryo. ABA is necessary for activation of ABI5 via SnRK2.2 and SnRK2.3 phosphorylation of ABI5 [49,50]. When seeds are in unfavorable environmental conditions, elevated endogenous ABA levels results in ABI5 accumulation, preventing seeds from germinating.

3.1.1. ABA and salt tolerance

Salt stress severely impacts plant growth by affecting metabolic processes and photosynthetic efficiency. NaCl initially induces osmotic stress and eventually accumulation of both Na^+ and Cl^- ions generates ionic stress [51]. However, some responses are salt-specific and distinct from responses to osmotic stress (reviewed in [52]). High salinity in the soil is first sensed by the plant roots. Salt and drought stress induce a rapid increase in cytosolic Ca^{2+} levels in the root cells. Ca^{2+} acts a second messenger, inducing salt- and drought-responsive genes [53,54]. Hyperosmotic stress is coupled with Ca^{2+} signaling and Reactive Oxygen Species (ROS) signaling, thereby inducing a cascade of signaling events, which results in downstream gene expression [52].

Biochemical and molecular mechanisms of salt tolerance in plants include exclusion of salt ions, production of suitable osmolytes, changing the structure of the membranes to control ion uptake, and induction of enzymes that produce antioxidants and phytohormones. To manage salt or drought stress, cellular ABA levels increase dramatically. The plant cuticle has been shown to mediate stress signaling as well as ABA biosynthesis and signaling. In addition to its primary function, providing mechanical support to the cell wall and plasma membrane, the cuticle has been implicated in osmotic stress regulation. CED1 (9-CIS EPOXYCAROTENOID DIOXYGENASE DEFECTIVE 1) is an essential protein in cuticle biogenesis. *ced1* mutants are sensitive to osmotic stress, as they are unable to induce ABA biosynthesis in response to osmotic stress [55].

ABA regulates root growth and architecture in plants under stress. Duan *et al.* [56] showed that salt has a strong inhibitory effect on lateral root growth, while primary roots are less sensitive to salt stress. They also showed that endogenous ABA signaling affects root system architecture under stress conditions using ABA biosynthesis mutants (*aba1, aba2*) as well as signal transduction mutants such as *abi1*. Salt stress results in elevated levels of ABA exclusively in lateral root cells and induces a quiescent period in postemergence lateral roots. Lateral roots in a quiescent stage form a thick, well-developed Casparian strip, which acts as a barrier to reduce diffusion of Na^+ ions through the endodermis. In the presence of Na^+ ions, endodermal cells activate ABA signaling and arrest growth so that lateral roots do not elongate into high saline environments. Therefore, ABA is an important signaling molecule in suppressing lateral root growth during salt stress [56].

ABA regulates expression of many salt-stress-responsive genes via transcription factors that are elevated in response to salt. For instance, ABF2/AREB1, ABF3, ABF4/AREB2, ABRE BINDING PROTEIN 9 (ABP9), and MYC/MYB, WRKY, and APETALA2/ETHYLENE RESPONSE FACTOR (AP2/ERF) are some of the salt-stress-responsive transcription factors that enhance stress tolerance. A recent study showed that the PYL8/RCAR3 ABA receptor has a role in ABA-mediated inhibition of primary root growth and also recovery of lateral root growth on exposure to ABA. PYL8/RCAR3 combines the action of ABA and auxin through direct interaction with MYB transcription factors during growth recovery of postemergence lateral roots [48,57].

There are proteins in the cell that are produced in an ABA-dependent manner that have a role in osmotic tolerance. For example, ABI5 activates transcription of LEA proteins. LEA proteins are highly hydrophilic small proteins shown to have an osmoprotectant role against cellular dehydration during late embryogenesis. LEA proteins also have a role in salt stress tolerance [58]. Due to their hydrophilic nature, LEA proteins can sequester ions accumulating in the cell, as well as act as chaperones and retain water molecules to prevent protein aggregation and inactivation of cellular enzymes [59]. In Arabidopsis, 51 LEA proteins have been identified that belong to nine different groups [60]. Jia *et al.* [61] showed overexpression of AtLEA14, which belongs to the LEA group 2 proteins, overactivates salt-stress-inducible genes such as *RD29B*, which encode dehydration protective proteins, and subsequently confers salt tolerance in Arabidopsis.

In addition, ABA has been implicated in histone H3 acetylation and methylation, thereby regulating stress-inducible gene expression at the epigenetic level. Chen *et al.* [62] showed that histone modifications by HISTONE DEACETYLASE 6 (HDA6) are involved in inhibition of seed germination, salt stress responses, and ABA- and salt-mediated gene expression in Arabidopsis.

3.1.2. ABA and drought tolerance

Drought is lack of water in the soil. Drought stress in plants arises due to water deficit conditions and results in removal of water from the cell membranes, disrupting the lipid bilayer structure. In addition, protein denaturation and accumulation of cellular electrolytes results in disruption of cellular metabolism [63]. Therefore, drought causes osmotic stress, and osmotic stress causes dehydration and inhibition of water uptake in plants. ABA accumulates under osmotic stress conditions and plays an important role in the stress response and tolerance of plants. In addition to autoactivation of SnRK2s by inhibition of PP2Cs in the ABA signaling cascade, hyperosmotic stress activates SnRK2s [64]. SnRK2 kinases are a major component of the osmotic stress signaling pathway. The Arabidopsis triple mutant *snrk2.2, snrk2.3, snrk2.6* shows severe drought intolerance and ABA-insensitivity [42]. Also ABF2, ABF3, and ABF4 act as transcriptional activators in mediating ABRE-dependent ABA signaling, which confers drought tolerance in vegetative tissues [40].

ABA induces expression of many transcription factors as well as genes that encode enzymes in the synthesis of osmoprotectants [65]. Osmolytes are compatible solutes such as amino acids (proline), sugar alcohols (mannitol, pinitol), and other sugars that accumulate without disrupting the function of proteins. Osmolytes make an osmotic adjustment facilitating a favorable water potential gradient and promote stress tolerance [66].

Dehydrins and LEA-like proteins act as cellular chaperones that protect cellular membranes and macromolecules in the cell [2]. During seed maturation seeds undergo dehydration stress. LEA proteins accumulate in the embryo as a result of osmotic stress and their functions include protection of enzymes, lipids, and mRNAs from dehydration. LEA proteins have been found to protect mitochondrial membranes from damage. LEA proteins are produced in an ABA-dependent and ABA-independent manner under osmotic stress [47,58].

Under moderate water stress conditions plant root growth has to be maintained in order to keep the plants alive. ABA accumulates under moderate water stress and mediates auxin transport in the root tip, which enhances the proton pumps in the plasma membrane. Proton secretions in the root tip play an important role in primary root growth and root hair development under moderate drought stress [67].

Based on the critical water level, drought tolerance is considered to be mechanisms that confer tolerance to moderate dehydration. Further dehydration requires desiccation tolerance mechanisms in order to restore the ability of cells to rehydrate successfully [68].

3.1.3. ABA and desiccation tolerance

Water loss results in a change in turgor pressure that affects the cell walls. Desiccation tolerance is defined as evolution of cell walls that can withstand extensive water loss without damaging its structure or polymer organization. Desiccation tolerance mechanisms aim to restructure the cells walls and maintain normal growth under water stress conditions [69].

Regulation of the stomatal pore is crucial in adapting plants to abiotic stress by reducing extensive water loss. Stomatal opening and closing occurs as a result of turgor pressure differences in the surrounding guard cells [47]. In response to water stress, ABA concentration is increased in the guard cell cytoplasm and apoplast, which results in a decrease in the turgor pressure due to activation of the K^+ outward rectifying channel and inhibition of the K^+ inward rectifying channel (KAT1 and 2). ABA also induces the anion channel SLAC1 resulting in release of anionic organic acids from the vacuole to the cytoplasm [70]. Reduced turgor pressure initiates closure of stomata as a mechanism of minimizing water loss from the plant. ABA levels rise in leaves immediately following water stress. CHLH/ABAR has been proposed as the chloroplast ABA receptor that links ABA signaling within the chloroplast with ABA signaling in the nucleus. Overexpression of CHLH promotes stomatal closure and thereby dessication tolerance [71].

SnRK2 OPEN STOMATA 1 (OST1) is a key SnRK2 protein kinase involved in regulation of the stomatal aperture by movement of guard cells during ABA signaling [72]. OST1 is activated by ABA, low humidity, and osmotic stress and is an important kinase found in guard cells preventing rapid water loss. Loss of function mutants of SnRK2 do not exhibit ABA-mediated stomatal closure activity and showed a wilty phenotype under dehydration stress conditions [72,73]. Also SnRK2.6/OST1 physically interacts with ABI1 and ABI2. ABI1 is required for ABA-dependent activation of OST1 and both ABI and ABI2 are required for osmotic-stress-induced activation of OST1 [15]. Thus, SnRK2.6/OST1 acts as a positive regulator in ABA-induced stomatal closure. Moreover, Yoshida *et al.* [73] showed that OST1 also positively regulates stress-responsive genes such as *RD29B* and *RD22.*

Reactive Oxygen Species (ROS) have also been identified as secondary messengers in ABA signaling in guard cells. In Arabidopsis, two partially redundant guard cell expressed NADPH oxidase catalytic subunit genes, *AtRbohD* and *AtRbohF,* were found to be involved in ABA signaling in guard cells, ABA-induced stomatal closure and ROS production, ABA activation of Ca^{2+} permeable channels in the plasma membrane of guard cells, and increasing cytosolic Ca^{2+} levels in response to ABA. Thus, these two genes act as positive regulators of ABA signal transduction [74]. Sirichandra *et al.* [72] provided biochemical evidence that OST1 protein kinase physically interacts with AtRbohF NADPH oxidase and phosphorylates it.

4. Conclusions

ABA has a wide range of functions from plant development to biotic and abiotic stress signaling and tolerance. The primary functions of ABA in salt, drought, and desiccation tolerance act via inhibiting seed germination, altering root architecture, and inducing stress-responsive genes as well as gene products that act as osmoprotectants. ABA signaling cascades and stress tolerance mechanisms studied in Arabidopsis provide insight into application of stress tolerance strategies to commercial crops. While ABA is not the only plant hormone involved in stress responses, many of these responses occur in an ABA-dependent manner, indicating the importance of ABA in plant stress response and tolerance.

Author details

V. C. Dilukshi Fernando and Dana F. Schroeder*

*Address all correspondence to: Dana.Schroeder@umanitoba.ca

University of Manitoba, Winnipeg, MB, Canada

References

[1] Wang W, Vinocur B, Altman A. Plant responses to drought, salinity and extreme temperatures: towards genetic engineering for stress tolerance. *Planta*. 2003;218:1–14.

[2] Shinozaki K, Yamaguchi-Shinozaki K. Gene networks involved in drought stress response and tolerance. *J Exp Bot*. 2007;58:221–227.

[3] Cutler AJ, Krochko JE. Formation and breakdown of ABA. *Trends Plant Sci*. 1999;4:472–478.

[4] Wasilewska A, Vlad F, Sirichandra C, Redko Y, Jammes F, Valon C, Frei dit Frey N, Leung J. An update on abscisic acid signaling in plants and more.... *Mol Plant*. 2008;1:198–217. DOI: 10.1093/mp/ssm022

[5] Cutler SR, Rodriguez PL, Finkelstein RR, Abrams SR. Abscisic acid: emergence of a core signaling network. *Annu Rev Plant Biol*. 2010;61:651-679. DOI: 10.1146/annurev-arplant-042809-112122

[6] Roychoudhury A, Paul S, Basu S. Cross-talk between abscisic acid-dependent and abscisic acid-independent pathways during abiotic stress. *Plant Cell Rep*. 2013;32:985–1006. DOI: 10.1007/s00299-013-1414-5

[7] Mehrotra R, Bhalothia P, Bansal P, Basantani MK, Bharti V, Mehrotra S. Abscisic acid and abiotic stress tolerance – different tiers of regulation. *J Plant Physiol*. 2014;171:486–496. DOI: 10.1016/j.jplph.2013.12.007

[8] Cracker LE, Abeles FB. Abscission: role of abscisic acid. *Plant Physiol*. 1969;44:1144–1149.

[9] Barrero JM, Piqueras P, Gonzalez-Guzman M, Serrano R, Rodriguez PL, Ponce MR, Micol JL. A mutational analysis of the *ABA1* gene of *Arabidopsis thaliana* highlights the involvement of ABA in vegetative development. *J Exp Bot*. 2005;56:2071–2083.

[10] Fujii H, Zhu JK. Arabidopsis mutant deficient in 3 abscisic acid-activated protein kinases reveals critical roles in growth, reproduction, and stress. *Proc Natl Acad Sci U S A*. 2009;106:8380–8385. DOI: 10.1073/pnas.0903144106

[11] Raghavendra AS, Gonugunta VK, Christmann A, Grill E. ABA perception and signalling. *Trends Plant Sci*. 2010;15:395–401. DOI: 10.1016/j.tplants.2010.04.006

[12] Nishimura N, Sarkeshik A, Nito K, Park SY, Wang A, Carvalho PC, Lee S, Caddell DF, Cutler SR, Chory J, Yates JR, Schroeder JI. PYR/PYL/RCAR family members are major in-vivo ABI1 protein phosphatase 2C-interacting proteins in Arabidopsis. *Plant J*. 2010;61:290–299. DOI: 10.1111/j.1365-313X.2009.04054.x

[13] Nishimura N, Yoshida T, Kitahata N, Asami T, Shinozaki K, Hirayama T. ABA-Hypersensitive Germination1 encodes a protein phosphatase 2C, an essential component of abscisic acid signaling in Arabidopsis seed. *Plant J*. 2007;50:935–949.

[14] Gosti F, Beaudoin N, Serizet C, Webb AA, Vartanian N, Giraudat J. ABI1 protein phosphatase 2C is a negative regulator of abscisic acid signaling. *Plant Cell*. 1999;11:1897–1910.

[15] Yoshida R, Umezawa T, Mizoguchi T, Takahashi S, Takahashi F, Shinozaki K. The regulatory domain of SRK2E/OST1/SnRK2. 6 interacts with ABI1 and integrates abscisic acid (ABA) and osmotic stress signals controlling stomatal closure in Arabidopsis. *J Biol Chem*. 2006;281:5310–5318.

[16] Umezawa T, Nakashima K, Miyakawa T, Kuromori T, Tanokura M, Shinozaki K, Yamaguchi-Shinozaki K. Molecular basis of the core regulatory network in ABA responses: sensing, signaling and transport. *Plant Cell Physiol*. 2010;51:1821–1830. DOI: 10.1093/pcp/pcq156

[17] Saez A, Apostolova N, Gonzalez-Guzman M, Gonzalez-Garcia MP, Nicolas C, Lorenzo O, Rodriguez PL. Gain of function and loss of function phenotypes of the protein phosphatase 2C HAB1 reveal its role as a negative regulator of abscisic acid signaling. *Plant J*. 2004;37:354–360.

[18] Nakashima K, Fujita Y, Kanamori N, Katagiri T, Umezawa T, Kidokoro S, Maruyama K, Yoshida T, Ishiyama K, Kobayashi M, Shinozaki K, Yamaguchi-Shinozaki K. Three Arabidopsis SnRK2 protein kinases, SRK2D/SnRK2.2, SRK2E/SnRK2. 6/OST1 and SRK2I/SnRK2.3, involved in ABA signaling are essential for the control of seed development and dormancy. *Plant Cell Physiol*. 2009;50:1345–1350. DOI: 10.1093/pcp/pcp083

[19] Busk PK, Pages M. Regulation of abscisic acid-induced transcription. *Plant J*. 1997;11:1285–1295.

[20] Lumba S, Toh S, Handfield LF, Swan M, Liu R, Youn JY, Cutler SR, Subramaniam R, Provart N, Moses A, Desveaux D, McCourt P. A mesoscale abscisic acid hormone interactome reveals a dynamic signaling landscape in Arabidopsis. *Dev Cell.* 2014;29:360–372. DOI: 10.1016/j.devcel.2014.04.004

[21] Fujita Y, Nakashima K, Yoshida T, Katagiri T, Kidokoro S, Kanamori N, Umezawa T, Fujita M, Maruyama K, Ishiyama K, Kobayashi M, Nakasone S, Yamada K, Ito T, Shinozaki K, Yamaguchi-Shinozaki K. Three SnRK2 protein kinases are the main positive regulators of abscisic acid signaling in response to water stress in Arabidopsis. *Plant Cell Physiol.* 2009;50:2123–2132. DOI: 10.1093/pcp/pcp147

[22] Finkelstein RR, Gampala SSL, Rock CD. Abscisic acid signaling in seeds and seedlings. *Plant Cell.* 2002;14 Suppl:S15–45.

[23] Vahisalu T, Kollist H, Wang YF, Nishimura N, Chan WY, Valerio G, Lamminmäki A, Brosché M, Moldau H, Desikan R, Schroeder JI, Kangasjärvi J. SLAC1 is required for plant guard cell S-type anion channel function in stomatal signalling. *Nature.* 2008;452:487–491. DOI: 10.1038/nature06608

[24] Geiger D, Scherzer S, Mumm P, Stange A, Marten I, Bauer H, Ache P, Matschi S, Liese A, Al-Rasheid KA, Romeis T, Hedrich R. Activity of guard cell anion channel SLAC1 is controlled by drought-stress signaling kinase-phosphatase pair. *Proc Natl Acad Sci U S A.* 2009;106:21425–21430. DOI: 10.1073/pnas.0912021106

[25] Sato A, Sato Y, Fukao Y, Fujiwara M, Umezawa T, Shinozaki K, Hibi T, Taniguchi M, Miyake H, Goto DB, Uozumi N. Threonine at position 306 of the KAT1 potassium channel is essential for channel activity and is a target site for ABA-activated SnRK2/OST1/SnRK2.6 protein kinase. *Biochem J.* 2009;424:439–448. DOI: 10.1042/BJ20091221

[26] McCourt P, Creelman R. The ABA receptors–we report you decide. *Curr Opin Plant Biol.* 2008;11:474–478. DOI: 10.1016/j.pbi.2008.06.014

[27] Guo J, Yang X, Weston DJ, Chen JG. Abscisic acid receptors: past, present and future. *J Integr Plant Biol.* 201;53:469–479. DOI: 10.1111/j.1744-7909.2011.01044.x

[28] Shen YY, Wang XF, Wu FQ, Du SY, Cao Z, Shang Y, Wang XL, Peng CC, Yu XC, Zhu SY, Fan RC, Xu YH, Zhang DP. The Mg-chelatase H subunit is an abscisic acid receptor. *Nature.* 2006;443:823–826.

[29] Shang Y, Yan L, Liu ZQ, Cao Z, Mei C, Xin Q, Wu FQ, Wang XF, Du SY, Jiang T, Zhang XF, Zhao R, Sun HL, Liu R, Yu YT, Zhang DP. The Mg-chelatase H subunit of Arabidopsis antagonizes a group of WRKY transcription repressors to relieve ABA-responsive genes of inhibition. *Plant Cell.* 2010;22:1909–1935. DOI: 10.1105/tpc.110.073874

[30] Yan L, Liu ZQ, Xu YH, Lu K, Wang XF, Zhang DP. Auto-and cross-repression of three Arabidopsis WRKY transcription factors WRKY18, WRKY40, and WRKY60 negatively involved in ABA signaling. *J Plant Growth Regul.* 2013;32:399–416. DOI: 10.1007/s00344-012-9310-8

[31] Wang XF, Zhang DP. ABA Signal Perception and ABA Receptors. In: Zhang DP, editor. *Abscisic Acid: Metabolism, Transport and Signaling*. Springer Netherlands; 2014. p. 89–116. DOI: 10.1007/978-94-017-9424-4_6

[32] Pandey S, Nelson DC, Assmann SM. Two novel GPCR-type G proteins are abscisic acid receptors in Arabidopsis. *Cell*. 2009;136:136–148. DOI: 10.1016/j.cell.2008.12.026

[33] Rock CD, Zeevaart JA. The *aba* mutant of *Arabidopsis thaliana* is impaired in epoxy-carotenoid biosynthesis. *Proc Natl Acad Sci U S A*. 1991;88:7496–7499.

[34] Cheng WH, Endo A, Zhou L, Penney J, Chen HC, Arroyo A, Leon P, Nambara E, Asami T, Seo M, Koshiba T, Sheen J. A unique short-chain dehydrogenase/reductase in Arabidopsis glucose signaling and abscisic acid biosynthesis and functions. *Plant Cell*. 2002;14:2723–2743.

[35] Xu ZY, Kim DH, Hwang I. ABA homeostasis and signaling involving multiple subcellular compartments and multiple receptors. *Plant Cell Rep*. 2013;32:807–813. DOI: 10.1007/s00299-013-1396-3

[36] Takeuchi J, Okamoto M, Akiyama T, Muto T, Yajima S, Sue M, Seo M, Kanno Y, Kamo T, Endo A, Nambara E, Hirai N, Ohnishi T, Cutler SR, Todoroki Y. Designed abscisic acid analogs as antagonists of PYL-PP2C receptor interactions. *Nat Chem Biol*. 2014;10:477–482. DOI: 10.1038/nchembio.1524

[37] Ng LM, Soon FF, Zhou XE, West GM, Kovach A, Suino-Powell KM, Chalmers MJ, Li J, Yong EL, Zhu JK, Griffin PR, Melcher K, Xu HE. Structural basis for basal activity and autoactivation of abscisic acid (ABA) signaling SnRK2 kinases. *Proc Natl Acad Sci U S A*. 2011;108:21259–21264. DOI: 10.1073/pnas.1118651109

[38] Bartels D, Souer E. Molecular responses of higher plants to dehydration. In: Hirt H, Shinozaki K, editors. *Plant Responses to Abiotic Stress*. Springer Berlin Heidelberg; 2004. p. 9–38.

[39] Shinozaki K, Yamaguchi-Shinozaki K, Seki M. Regulatory network of gene expression in the drought and cold stress responses. *Curr Opin Plant Biol*. 2003;6:410–417.

[40] Choi H, Hong J, Ha J, Kang J, Kim SY. ABFs, a family of ABA-responsive element binding factors. *J Biol Chem*. 2000;275:1723–1730.

[41] Bensmihen S, Rippa S, Lambert G, Jublot D, Pautot V, Granier F, Giraudat J, Parcy F. The homologous ABI5 and EEL transcription factors function antagonistically to fine-tune gene expression during late embryogenesis. *Plant Cell*. 2002;14:1391–1403.

[42] Fujii H, Verslues PE, Zhu JK. Arabidopsis decuple mutant reveals the importance of SnRK2 kinases in osmotic stress responses in vivo. *Proc Natl Acad Sci U S A*. 2011;108:1717–1722. DOI: 10.1073/pnas.1018367108

[43] Koornneef M, Leon-Kloosterziel KM, Schwartz SH, Zeevaart JAD. The genetic and molecular dissection of abscisic acid biosynthesis and signal transduction in Arabidopsis. *Plant Physiol Biochem*. 1998;36:83–89.

[44] Liotenberg S, North H, Marion-Poll A. Molecular biology and regulation of abscisic acid biosynthesis in plants. *Plant Physiol Biochem*. 1999;37:341–350.

[45] Fujita Y, Fujita M, Shinozaki K, Yamaguchi-Shinozaki K. ABA-mediated transcriptional regulation in response to osmotic stress in plants. *J Plant Res*. 2011;124:509–525. DOI: 10.1007/s10265-011-0412-3

[46] Zhu JK. Salt and drought stress signal transduction in plants. *Annu Rev Plant Biol*. 2002;53:247-273.

[47] Bhattacharjee S, Saha AK. Plant water-stress response mechanisms. In: Gaur RK, Sharma P, editors. *Approaches to Plant Stress and their Management*. Springer India. 2014. p. 149–172. DOI: 10.1007/978-81-322-1620-9_8

[48] Zhao Y, Xing L, Wang X, Hou YJ, Gao J, Wang P, Duan CG, Zhu X, Zhu JK. The ABA receptor PYL8 promotes lateral root growth by enhancing MYB77-dependent transcription of auxin-responsive genes. *Sci Signal*. 2014;7:ra53. DOI: 10.1126/scisignal.2005051

[49] Finkelstein RR, Lynch TJ. The Arabidopsis abscisic acid response gene *ABI5* encodes a basic leucine zipper transcription factor. *Plant Cell*. 2000;12:599–609.

[50] Lopez-Molina L, Mongrand S, Chua NH. A postgermination developmental arrest checkpoint is mediated by abscisic acid and requires the ABI5 transcription factor in Arabidopsis. *Proc Natl Acad Sci U S A*. 2001;98:4782–4787.

[51] Tester M, Davenport R. Na+ tolerance and Na+ transport in higher plants. *Ann Bot*. 2003;91:503–527.

[52] Deinlein U, Stephan AB, Horie T, Luo W, Xu G, Schroeder JI. Plant salt-tolerance mechanisms. *Trends Plant Sci*. 2014;19:371–379. DOI: 10.1016/j.tplants.2014.02.001

[53] Knight H, Trewavas AJ, Knight MR. Calcium signalling in *Arabidopsis thaliana* responding to drought and salinity. *Plant J*. 1997;12:1067–1078.

[54] Tracy FE, Gilliham M, Dodd AN, Webb AA, Tester M. NaCl-induced changes in cytosolic free Ca2+ in *Arabidopsis thaliana* are heterogeneous and modified by external ionic composition. *Plant Cell Environ*. 2008;31:1063–1073. DOI: 10.1111/j.1365-3040.2008.01817.x

[55] Wang ZY, Xiong L, Li W, Zhu JK, Zhu J. The plant cuticle is required for osmotic stress regulation of abscisic acid biosynthesis and osmotic stress tolerance in Arabidopsis. *Plant Cell*. 2011;23:1971–1984. DOI: 10.1105/tpc.110.081943

[56] Duan L, Dietrich D, Ng CH, Chan PM, Bhalerao R, Bennett MJ, Dinneny JR. Endodermal ABA signaling promotes lateral root quiescence during salt stress in Arabidopsis seedlings. *Plant Cell*. 2013;25:324–341. DOI: 10.1105/tpc.112.107227

[57] Antoni R, Gonzalez-Guzman M, Rodriguez L, Peirats-Llobet M, Pizzio GA, Fernandez MA, De Winne N, De Jaeger G, Dietrich D, Bennett MJ, Rodriguez PL. PYRABACTIN RESISTANCE1-LIKE8 plays an important role for the regulation of abscisic acid signaling in root. *Plant Physiol*. 2013;161:931–941. DOI: 10.1104/pp.112.208678

[58] Bhardwaj R, Sharma I, Kanwar M, Sharma R, Handa N, Kaur H, Kapoor D. LEA proteins in salt stress tolerance. In: Ahmad P *et al.*, editors. *Salt Stress in Plants*. Springer New York; 2013. p. 79–112. DOI: 10.1007/978-1-4614-6108-1

[59] Marco F, Bitrián M, Carrasco P, Rajam MV, Alcázar R, Tiburcio AF. Genetic engineering strategies for abiotic stress tolerance in plants. In: Bahadur B *et al.*, editors. *Plant Biology and Biotechnology: Volume II: Plant Genomics and Biotechnology*. Springer India; 2015. p. 579–609. DOI 10.1007/978-81-322-2283-5_29

[60] Hundertmark M, Hincha DK. LEA (late embryogenesis abundant) proteins and their encoding genes in *Arabidopsis thaliana*. *BMC Genomics*. 2008;9:118. DOI: 10.1186/1471-2164-9-118

[61] Jia F, Qi S, Li H, Liu P, Li P, Wu C, Zheng C, Huang J. Overexpression of *Late Embryogenesis Abundant 14* enhances Arabidopsis salt stress tolerance. *Biochem Biophys Res Commun*. 2014;454:505–511. DOI: 10.1016/j.bbrc.2014.10.136

[62] Chen LT, Luo M, Wang YY, Wu K. Involvement of Arabidopsis histone deacetylase HDA6 in ABA and salt stress response. *J Exp Bot*. 2010;61:3345–3353. DOI: 10.1093/jxb/erq154

[63] Mahajan S, Tuteja N. Cold, salinity and drought stresses: an overview. *Arch Biochem Biophys*. 2005;444:139–158.

[64] Monks DE, Aghoram K, Courtney PD, DeWald DB, Dewey RE. Hyperosmotic stress induces the rapid phosphorylation of a soybean phosphatidyl inositol transfer protein homolog through activation of the protein kinases SPK1 and SPK2. *Plant Cell*. 2001;13:1205–1219.

[65] Daszkowska-Golec A, Szarejko I. The molecular basis of ABA-mediated plant response to drought. In: Vahdati K, Leslie C, editors. *Abiotic Stress – Plant Responses and Applications in Agriculture*. Intech; 2013. p. 103–133. DOI: 10.5772/53128

[66] Bray EA. Plant responses to water deficit. *Trends Plant Sci*.1997;2:48–54.

[67] Xu W, Jia L, Shi W, Liang J, Zhou F, Li Q, Zhang J. Abscisic acid accumulation modulates auxin transport in the root tip to enhance proton secretion for maintaining root growth under moderate water stress. *New Phytol*. 2013;197:139–150. DOI: 10.1111/nph.12004

[68] Hoekstra FA, Golovina EA, Buitink J. Mechanisms of plant desiccation tolerance. *Trends Plant Sci*. 2001;6:431–438.

[69] Moore JP, Vicré-Gibouin M, Farrant JM, Driouich A. Adaptations of higher plant cell walls to water loss: drought vs desiccation. *Physiol Plant*. 2008;134:237–245. DOI: 10.1111/j.1399-3054.2008.01134.x

[70] Sirichandra C, Wasilewska A, Vlad F, Valon C, Leung J. The guard cell as a single-cell model towards understanding drought tolerance and abscisic acid action. *J Exp Bot*. 2009;60:1439–1463. DOI: 10.1093/jxb/ern340

[71] Tsuzuki T, Takahashi K, Tomiyama M, Inoue S, Kinoshita T. Overexpression of the Mg-chelatase H subunit in guard cells confers drought tolerance via promotion of stomatal closure in *Arabidopsis thaliana*. *Front Plant Sci*. 2013;4:440. DOI: 10.3389/fpls.2013.00440

[72] Sirichandra C, Gu D, Hu HC, Davanture M, Lee S, Djaoui M, Valot B, Zivy M, Leung J, Merlot S, Kwak JM. Phosphorylation of the Arabidopsis AtrbohF NADPH oxidase by OST1 protein kinase. *FEBS Lett*. 2009;583:2982–2986. DOI: 10.1016/j.febslet.2009.08.033

[73] Yoshida R, Hobo T, Ichimura K, Mizoguchi T, Takahashi F, Aronso J, Ecker JR, Shinozaki K. ABA-activated SnRK2 protein kinase is required for dehydration stress signaling in Arabidopsis. *Plant Cell Physiol*. 2002;43:1473–1483.

[74] Kwak JM, Mori IC, Pei ZM, Leonhardt N, Torres MA, Dangl JL, Bloom RE, Bodde S, Jones JD, Schroeder JI. NADPH oxidase *AtrbohD* and *AtrbohF* genes function in ROS-dependent ABA signaling in Arabidopsis. *EMBO J*. 2003;22:2623–2633. DOI: 10.1093/emboj/cdg277

2

Consequences of Water Deficit on Metabolism of Legumes

Allan Klynger da Silva Lobato, Gélia Dinah Monteiro Viana, Gleberson Guillen Piccinin, Milton Hélio Lima da Silva, Gabriel Mascarenhas Maciel and Douglas José Marques

Abstract

The aim of this chapter is (i) to define water deficit and its consequences on growth and development of higher plants; (ii) to present the interferences induced on metabolism, including gas exchange, biochemical compounds, and osmotic substances; and (iii) to explain how these alterations will affect the growth, development, and production of leguminous plants. This chapter reports that the performance in leguminous is affected by water deficiency, which can cause lower growth and development. For gas exchange, leaf relative water content, stomatal conductance, and transpiration rate suffered decrease when cultivated under water deficit. Biochemical compounds, such as soluble amino acids, soluble proteins, proline, and abscisic acid (ABA), are intensively modified after a period under water restriction. The results prove that ABA mediates actively and significantly the proline accumulation and consequent osmotic adjustment in *Vigna unguiculata* leaves that were induced to water deficit and rehydration.

Keywords: Abiotic stress, leguminous crops, gas exchange, osmotic adjustment, water deficiency

1. Introduction

The water supplement inadequate in soil is considered one of the limiting factors to the productive potential in several species [1–2]. Water deficit is an abiotic factor that affects the agricultural production with high frequency and intensity, influencing morphological, physiological, and biochemical aspects [3–4].

As in other crops, performance in leguminous is affected by water deficiency, which can cause lower growth and development, with progressive reduction in leaf dry matter [5], moreover, to promote the abortion of flowers during drought periods and to affect the yield significantly [6], with consequent repercussion on production parameters, such as number of grains and pods per plant.

The deficit water is characterized by water losses that exceed the absorption rate and by this way acts directly in the plant–water relations [7–8], depending on intense and exposure period, in addition to promote changes in the cell and molecular pathways [9], whereby accumulation of organic solutes with the carbohydrates and proline [10], differential gene expression of DNA [11], and quantity variation in the photosynthetic pigments, mainly chlorophylls and carotenoids [12], in which the stomata enclosed interfere in photosynthetic rates occur [13].

The osmotic adjustment is considered one of the important mechanisms developed by the plants to tolerate the water deficiency [14], which promotes the protection of the plant cell structures with membranes and chloroplasts [15], as well as avoid the cell toxicity provoked by the free radicals and maximize the water retention in cell inside [16]; besides it has the advantage of using carbohydrates as energy source under severe stress [6].

Drought is directly related to the overproduction of reactive oxygen species (ROS) [17], such as hydrogen peroxide (H_2O_2) and superoxide (O_2^-) [18], which are highly toxic compounds. ROS promote the oxidation of membranes and damage essential organelles such as chloroplasts [19] and mitochondria [20], which result in cell damage or death [21,22].

Ascorbate (ASC) and glutathione (GSH) have essential functions in antioxidant metabolism [23,24] because ASC is used as a substrate [25–27]. In addition, GSH produces ASC and glutathione disulfide (GSSG), which is used to regenerate GSH via glutathione reductase (GR) [28,29].

The soybean is considered a species sensitive to several abiotic stresses [30], when compared with other tropical legumes, such as *Vigna unguiculata* and *Phaseolus vulgaris* [31,32], as well as other species such as *Gossypium hirsutum* and *Sorghum bicolor* [33,34], in which the sensitivity at water deficit can be emphasized, mainly during the growth and development period, which might cause strong reduction in the yield [35]. However, *V. unguiculata* (L.) Walp. is a species tolerant to drought due to rusticity, and it presents large protein content in grain. This crop is frequently found in agricultural areas in Brazil that are under the influence of abiotic stresses. These areas present small rain index and high temperature. In addition, the soil is susceptible to salinity or to fertility loss [36].

2. Objectives

The aim of this chapter is (i) to define what is water deficit and the consequences on growth and development of higher plants; (ii) to present the interferences induced on metabolism, including gas exchange, biochemical compounds, and osmotic substances; and (iii) to explain how these alterations will affect the growth, development, and production of leguminous plants.

3. Interference of water deficit on growth and development

Lizana et al. [37] while working with two varieties of *P. vulgaris* under water deficit observed paraheliotropic leaf movement, which was previously described by Pastenes et al. [38]. Leaf movements in Arroz and Orfeo subjected to drought were shown and compared. Figure 1B presents the evolution of the movement of leaves after increasing periods of drought, being determined that the variety Arroz is more sensitive than Orfeo [37].

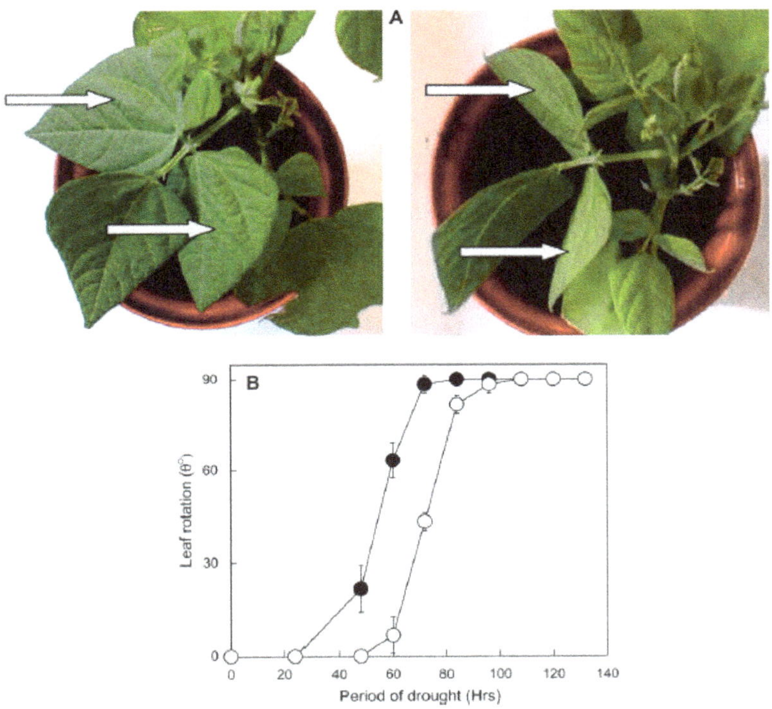

Figure 1. Leaf movement in different drought times. (A) Plants before (left) and after (right) drought-induced leaf movement. Leaf rotation was measured on flanking leaves (arrows) of the first mature trifoliate leaves. (B) Relationship between period of drought and leaf rotation angle (h) in Arroz (closed symbols) and Orfeo (open symbols) [37].

Lobato et al. [39] while studying morphological alterations in *Glycine max* under progressive water stress found variations in the following parameters: (A) evaluated height of plants, (B) shoot dry matter, (C) number of leaves, and (D) root dry matter.

The lower height and shoot dry matter in the plants under water deficiency occurred, probably due to the abscisic acid (ABA) action, in which case it is produced in the cells under abnormal conditions and this way inhibited the cell division and/or DNA synthesis [39].

The smaller number of leaves showed in the plants under water stress occurred with consequence a lower or void extension rate of the leaf area existent in the plant, moreover probably increase in the ABA levels in roots, in which it will be transported from roots to shoot and act in the apical region of the plant with antagonist of the auxin and cytokinin, responsible for growth and cell division, respectively [40]; through these hormonal mechanisms, the buds remain dormant and develop not the leaf news. In the period between 0 and 2 days of water stress (Figure 2D), the weight higher of the root dry matter. According to Kerbauy [41], studies with gene-modified plants describe a decrease in the ethylene levels and increase in the ABA in plant roots under water stress, when compared with plants normally irrigated; hence, it proves the different behavior of these hormones, besides it are attributes at ABA the capacity of the remain ethylene normal levels produced in root of plants under normal conditions.

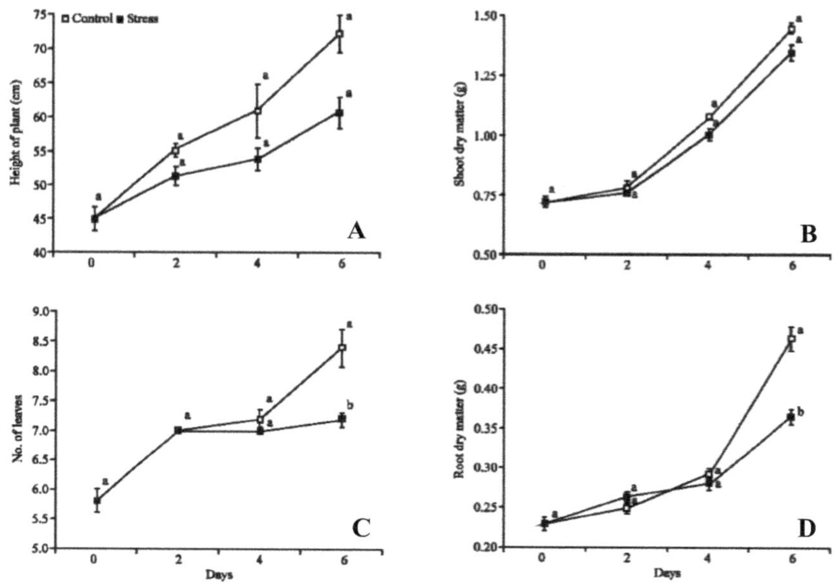

Figure 2. (A) Height of plants, (B) shoot dry matter, (C) number of leaves, and (D) root dry matter in plants of *Glycine max* cultivar sambaiba under 0, 2, 4, and 6 days of water stress. Averages followed by the same letter do not differ among themselves by the Tukey's test at 5% of probability, and the bars represent the mean standard error [39].

4. Modifications on gas exchange

Barbosa et al. [42] evaluated the root contribution to water relations and shoot in two contrasting *V. unguiculata* cultivars and showed that water deficit promoted significant decrease in leaf relative water content (Figure 3A) in tolerant and sensitive cultivars. Inoculated plants of control treatment presented higher values of leaf relative water content, when compared with same treatments of non-inoculated plants.

The tolerant cultivar showed better performance in this parameter, when compared with that of same treatments the cultivar that is sensitive to water stress. In both tolerant and sensitive cultivars, stomatal conductance had a significant reduction in plants exposed to water deficiency (Figure 3B). Plants that were inoculated presented non-significant difference, when compared with that of non-inoculated plants.

Water restriction produced a significant decrease in transpiration rates in both cultivars (Figure 3C). The inoculation provoked non-significant changes in tolerant and sensitive plants. When the tolerant cultivar was submitted to water deficit, the values were higher than those found in the sensitive cultivar, this behavior being similar in inoculated and non-inoculated plants.

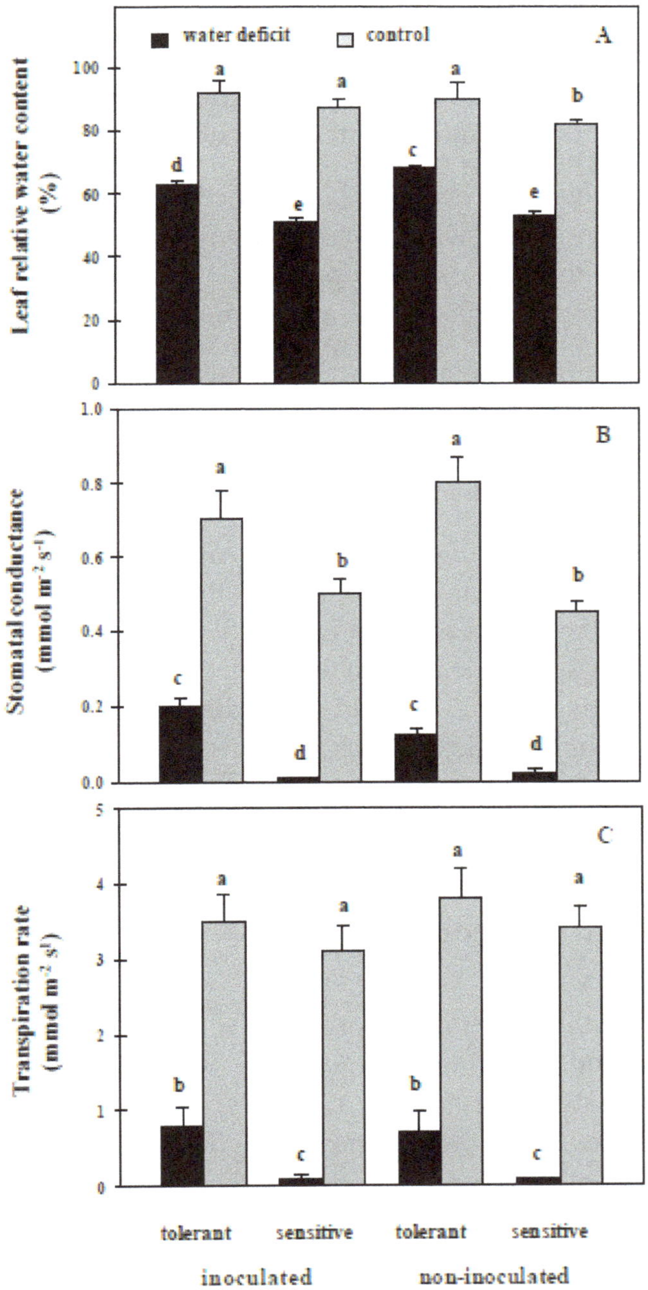

Figure 3. (A) LRWC, (B) g_s, and (C) *E* in two contrasting *Vigna unguiculata* cultivars under water deficit and subjected to inoculation. Means followed by the same letter are not significantly different by the Scott–Knott test at 5% of probability. The bars represent the mean standard error [42].

The reduction in relative water content in leaf is because of lower absorption rate of water by plant via roots and water loss occasioned by gas exchanges through stomata [43]. Similar results were reported by Maia et al. [44] when working with *Zea mays*.

Water deficit promoted a significant fall in stomatal conductance of the two cultivars, but tolerant plants presented higher values of this variable, probably by maintaining better plant water condition. This study revealed that root dry matter exercises influence on stomatal conductance in *V. unguiculata* plants submitted to 5 days of water deficiency, and this fact is based on the indirect effect produced by root on stomatal mechanisms. In other words, an insufficient root system developed during water deficiency will supply lower amount of water to shoot and, consequently, will promote reduction in stomatal conductance.

Decrease in stomatal conductance is explained by reduction in water availability in substrate, and it produces a reduction in leaf water potential, with consequent stomatal closing. The results described by Santos and Carlesso [4] reported that on conditions of water deficit, there is an increase in ABA concentration in xylem sap, promoting stomata closing.

Gholz et al. [45] reported that stomatal closing reduces the CO_2 influx to leaf, affecting production, transport, and utilization of photo-assimilates, and hence the yield. Results similar to those found in this study were found by Santos et al. [46], who studied five *P. vulgaris* genotypes subjected to water deficiency.

Decrease in transpiration rate of *V. unguiculata* plants can be attributed to stomatal behavior, because under water deficit, stomata are kept partially closed, contributing to change in transpiration behavior of plant [47]. Leite and Filho [48] reported that reduction of transpiration is an important mechanism of tolerance to drought. Values of transpiration demonstrated direct relation with stomatal conductance and also with leaf relative water content. Similar results were shown by Nogueira et al. [49] in a study oftwo *Arachis hypogeae* cultivars exposed to water deficiency.

5. Water deficit on nitrogen compounds

Lobato et al. [50] evaluated the effects of the progressive water deficit, as well as investigated the physiological and biochemical behavior in *G. max* cv. Sambaiba that was submitted to water restriction during the vegetative phase (Table 1). The increase in the levels of free amino acids is due to high synthesis of amino acids from protein hydrolyses, in which case the free amino acids are utilized by the plant to reduce the effects of the water deficit through organic solute accumulation, thereby increasing the water retention capacity [51].

Under water stress, the free amino acids such as proline and glycinebetaine are strongly influenced and consequently quickly accumulated [52], as well as of secondary form occur the increase of aspartate, glutamate and alanine [53]. The result on increase in the free amino acids found by Asha and Rao [54] while working with *Arachis hypogaea* under water deficit corroborates the results of this study.

Free amino acids (μmol g^{-1} DM)		
Days	Control	Stress
0	10.1 ± 3.5 a	10.1 ± 3.5 a
2	10.2 ± 2.8 a	41.6 ± 3.0 b
4	9.9 ± 1.9 a	41.2 ± 3.1 b
6	10.1 ± 1.5 a	49.3 ± 2.4 b

Table 1. Free amino acids and proline in *Glycine max* plants (cv. Sambaiba) under 0, 2, 4, and 6 days of water deficit. Averages followed by the same letter do not differ among themselves by Tukey's test at 5% of probability [50].

The reduction in the total soluble proteins showed in the plants under water stress is due to probable increase in the proteases enzyme activity (Table 2), in which case this proteolytic enzyme promotes the breakdown of the proteins and, consequently, decreases the protein amount presents in the plant under abiotic stress conditions [55]. In inadequate conditions to the plant is active the pathway of proteins breakdown, because the plant use the proteins to the synthesis of nitrogen compounds as amino acids that might auxiliary the plant osmotic adjustment [56]. Similar results on reduction in the proteins were found by Ramos et al. [57], investigating the effects of the water stress in *P. vulgaris*.

Soluble proteins (mg g^{-1} DM)		
Days	Control	Stress
0	9.74 ± 0.11 a	9.74 ± 0.11 a
2	10.05 ± 0.37 a	7.69 ± 0.09 b
4	9.87 ± 0.26 a	7.73 ± 0.19 b
6	9.71 ± 0.22 a	7.79 ± 0.21 b

Table 2. Total soluble proteins in *Glycine max* plants (cv. Sambaiba) under 0, 2, 4, and 6 days of water deficit. Averages followed by the same letter do not differ among themselves by Tukey's test at 5% of probability [50].

6. Relation between abscisic acid and proline

The ABA hormone is synthesized in the plastids and is linked to the stomatal mechanism [58] and quickly responds to water deficiency [59]. The ABA can be produced in the roots and/or shoots, but this hormone is usually synthesized under water deficiency in the roots and translocated to leaves in order to improve stomatal control. The ABA signalization pathway depends on the Ca^{+} influx into the cytosol [60], activating the K^{+}, Cl^{-}, and malate^{-2} efflux channels to external medium, through plasmatic membranes and concomitantly blocking the K^{+} entrance to cytosol. Therefore, the cytosol solute flux in the direction of the cell wall results in a decrease of turgescence pressure in the guard cells, and, consequently, the stomata are closed [61].

The progressive increase in ABA concentration in the plants of the stress treatment is related to the stomatal mechanism because this hormone, under these conditions, provokes stomatal closing [62], consequently reducing the water losses during the gas exchanges in essential physiological processes such as transpiration and photosynthesis [59].

Based in study carried out by Costa et al. [5] on impact of water deficit and rehydration on nitrogen compounds and ABA in *V. unguiculata* leaves (Figure 4), the research detected that leaf relative water content influences ABA concentration present in the leaf.

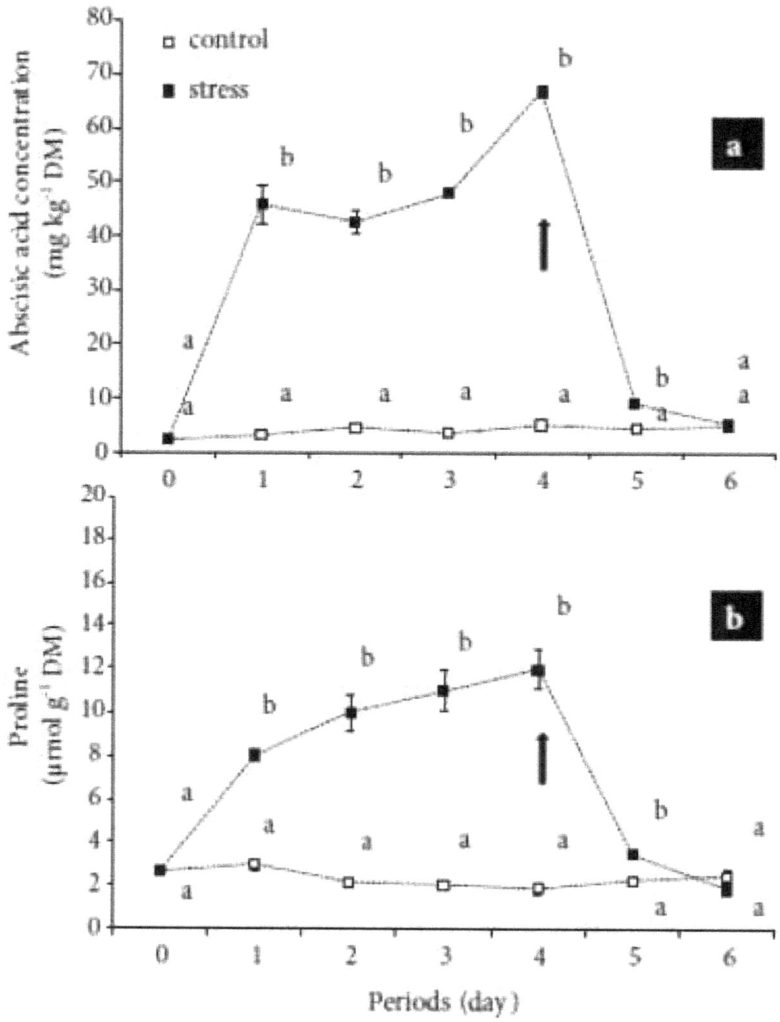

Figure 4. (a) Abscisic acid concentration and (b) proline in *Vigna unguiculata* plants cv. Vita 7 subjected to 4 days of water restriction and 2 days of rehydration. Means followed by the same letter are not significantly different by Tukey's test at 5% of probability. The bars represent the mean standard error, and the arrow indicates the rehydration point [5].

Therefore, the relative water content acts as a signal, and the ABA works during the signal transduction due to the easy and fast movement of this compound into plant tissue, and as a response, the stomatal closing occurs in *V. unguiculata* plants subjected to water deficit.

The fast decrease in the ABA concentration after rehydration indicates the efficiency of the signalization pathway, transduction, and consequent response of this compound. The results reported in this study on ABA are corroborated by Hsu et al. [63] evaluating the consequences of water stress in *Oryza sativa* L. and the effects of heavy metal stress in *Cicer arietinum* L.[64].

The results obtained by Costa et al. [5] prove that ABA mediates actively and significantly the proline accumulation and consequent osmotic adjustment in *V. unguiculata* leaves induced to water deficit and rehydration (Figure 5). A recent study indicated that *V. unguiculata* plants considered resistant to water deficit presented proline accumulation [65] and, consequently, are more adapted to environments with low water supplement, when compared to sensitive plants. The rehydration reduced the proline levels, suggesting that this nitrogen compound participates actively in the osmotic adjustment in this species. The proline accumulation during water deficit presented in this study is similar with results reported by Sarker et al. [66], investigating *Triticum aestivum*, and Smita and Nayyar [67], evaluating *C. arietinum*.

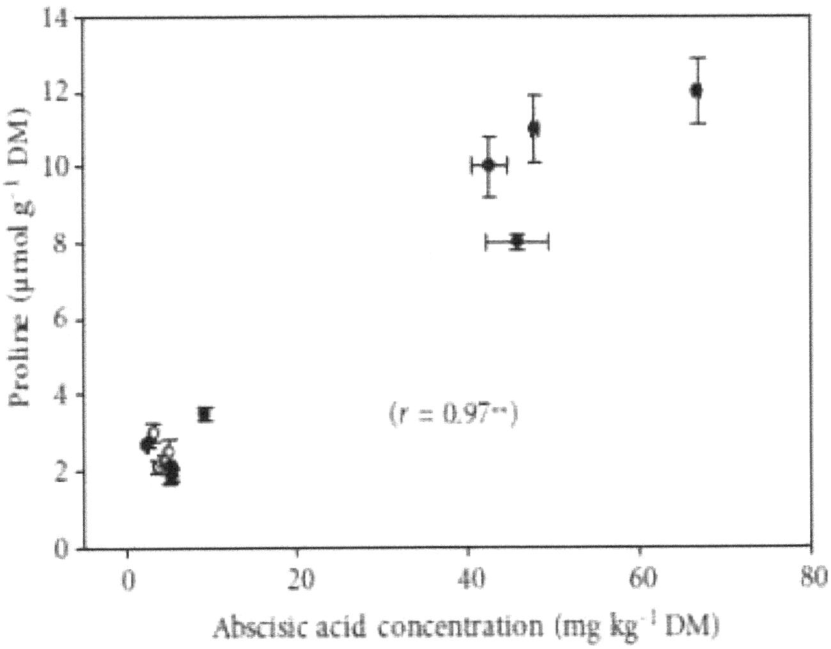

Figure 5. Relationship between abscisic acid concentration and proline in *Vigna unguiculata* plants cv. Vita 7 subjected to 4 days of water restriction and 2 days of rehydration. The bars represent the mean standard error, and the asterisks (**) indicate the significance at 0.01 probability level [5].

7. Final considerations

This chapter reports that the performance in leguminous is affected by water deficiency, which can cause lower growth and development, with progressive reduction in leaf dry matter, moreover to promote abortion of flowers during drought periods, and to affect the yield significantly, with consequent repercussion on production parameters, such as number of grains and pods per plant. In relation to morphological parameters, negative alterations related to height of plants, shoot dry matter, number of leaves, root dry matter, and paraheliotropic leaf movement were described. For gas exchange, leaf relative water content, stomatal conductance, and transpiration rate suffered decrease when cultivated under water deficit. Biochemical compounds, such as soluble amino acids, soluble proteins, proline, and ABA, are intensively modified after a period under water restriction. The results prove that ABA mediates actively and significantly the proline accumulation and consequent osmotic adjustment in *V. unguiculata* leaves that were induced to water deficit and rehydration.

Acknowledgements

This chapter had financial support from Fundação Amazônia Paraense de Amparo à Pesquisa (FAPESPA/Brazil), Universidade Federal Rural da Amazônia (UFRA/Brazil), and Conselho Nacional de Pesquisa (CNPq/Brazil) to Lobato AKS.

Author details

Allan Klynger da Silva Lobato[1*], Gélia Dinah Monteiro Viana[1], Gleberson Guillen Piccinin[2], Milton Hélio Lima da Silva[3], Gabriel Mascarenhas Maciel[4] and Douglas José Marques[5]

*Address all correspondence to: allanllobato@yahoo.com.br

1 Núcleo de Pesquisa Vegetal Básica e Aplicada, Universidade Federal Rural da Amazônia, Paragominas, Brazil

2 Campus de Parauapebas, Universidade Federal Rural da Amazônia, Belém, Brazil

3 Coordenação de Botânica, Museu Paraense Emílio Goeldi, Belém, Brazil

4 Instituto de Ciências Agrárias, Universidade Federal de Uberlândia, Monte Carmelo, Brazil

5 Setor de Olericultura e Experimentação, Universidade José do Rosário Vellano (UNIFENAS), Alfenas, Brazil

References

[1] Boyer, J.S. Plant production. Environmental Science. 1982; 218(4671):443–448.

[2] Lechinoski, A., Freitas, J.M.N., Castro, D.S., Lobato, A.K.S., Oliveira Neto, C.F. and Cunha, R.L.M. Influência do estresse hídrico nos teores de proteínas e aminoácidos solúveis totais em folhas de teca (*Tectona grandis* L. f.). Revista Brasileira de Biotecnologia. 2007;5(2):927–929.

[3] Fontana, D.C., Berlato, M.A. and Bergamaschi, H. Micrometeorological alterations in soybeans grown under different water regimes. Pesquisa Agropecuária Brasileira. 1992;27(5):661–669.

[4] Santos, R.F. and Carlesso, R. Water deficit and morphologic and physiologic behaviour of plants. Revista Brasileira de Engenharia Agricola e Ambiental. 1992;2(3):287–294.

[5] Costa, R.C.L., Lobato, A.K.S., Silveira, J.A.G. and Laughinghouse IV, H.D. ABA-mediated proline synthesis in cowpea leaves exposed to water deficiency and rehydration. Turkish Journal of Agriculture and Forestry. 2011;35:309–317.

[6] Pimentel, C. The water and plant relation. EDUR, Seropédica. 2004.

[7] Costa, R.C.L., Lobato, A.K.S., Oliveira Neto, C.F., Maia, P.S.P., Alves, G.A.R. and Laughinghouse IV, H.D. Biochemical and physiological responses in two *Vigna Unguiculata* (L.) Walp. cultivars under water stress. Journal of Agronomy. 2008;7(1):98–101.

[8] Fernández, C.J., Mcinnes, K.J. and Cothren, J.T. Water status and leaf area production in water-and nitrogen-stressed cotton. Crop Science. 1996;36:1224–1233.

[9] Zhu, J.K. Salt and drought stress signal transduction in plants. Plant Biology Revistas. 2002;53(1):247–273.

[10] Araújo, S.A.C. and Deminicis, B.B. Fotoinibição da fotossíntese. Revista Brasileira de Biociência. 2009; 7(4):463–472.

[11] Casagrande, E.C., Farias, J.R.B., Neumaier, N., Oya, T., Pedroso, J., Martins, P.K., Breton, M.C. and Nepamuceno, A.L. Differential gene expression in soybean during water deficit. Revista Brasileira de Fisiologia Vegetal. 2001;13(2):168–184.

[12] Chandrasekar, V., Sairam, R.K. and Srivastava, G.C. Physiological and biochemical responses of hexaploid and tetraploid wheat to drough stress. Journal of Agronomy and Crop Science. 2000;185(4):168–184.

[13] Ribas, C.M., Taylor, N.L., Giles, L., Busquets, S., Finnegan, P.M., Day, D.A., Lambers, H., Medrano, H., Berry, J.A. and Flexas, J. Effects of water stress on respiration in soybean leaves. Journal of Plant Physiology. 2005;139(1):466–473.

[14] Nitrogen assimilation and osmotic adjustment in nodulated plants of bean [*Vigna unguiculata* (L.) Walp] under water stress, Ph.D Thesis. Universidade Federal do Ceará, Brasil.

[15] Martínez, J.P., Lutts, S., Schanck, A., Bajj, I.M. and Kinet, J.M. Is osmotic adjustment required for water stress resistance in the Mediterranean shrub *Atriplex halimus* L. Journal of Plant Physiology. 2004;161(9):1041–1051.

[16] Here, P.D., Cress, W.A. and Van Standen, J. Dissecting the roles of osmolyte accumulation during stress. Plant Cell Environment. 1998;21(6):535–553.

[17] Asada, K. Production and scavenging of reactive oxygen species in chloroplasts and their functions. Plant Physiology. 2006;141(2):391–396.

[18] Queiroz, C.G.S., Garcia, Q.S. and Lemos Filho, J.P. Stress-induced morphogenic responses: Growing out of trouble. Trends in Plant Science. 2007;12(3):98–105.

[19] Carvalho, M.H.C. Drought stress and reactive oxygen species. Plant Signaling & Behavior. 2008;3:156–165.

[20] Moller, I.M. Plant mitochondria and oxidative stress: Electron transport, NADPH turnover, and metabolism of reactive oxygen species. Review of Plant Physiology and Plant Molecular Biology. 2001;52:561–591.

[21] Mittler, R. Oxidative stress, antioxidants and stress tolerance. Trends in Plant Science. 2002;7(9):405–410.

[22] Pereira, J.W.L., Melo Filho, P.A., Albuquerque, M.B., Nogueira, R.J.M.C. and Santos, R.C. Mudanças bioquímicas em genótipos de amendoim submetidos a déficit hídrico moderado. Revista Ciência Agronômica. 2012;43(4):766–773.

[23] Wang, C.Q., Liu, T.L. and Xu, H.J. Ascorbate – glutathione metabolism during PEG-induced water deficit in *Trifolium repens.* Journal of Plant Physiology. 2011;58(4):597–602.

[24] Barbosa, M.R., Ramos, H.M.M., Andrade Júnior, A.S., Do Nascimento, F.N.E. and Cardoso, M.J. Geração e desintoxicação enzimática de espécies reativas de oxigênio em plantas. Ciência Rural. 2014;44(3):453–460.

[25] Mehlhorn, H., Lelandais, M., Korth, H.G. and Foyer, C.H. Ascorbate is the natural substrate for plant peroxidase. FEBS Letters. 1996;378(3):203–206.

[26] Shamsi, T.N. and Fatima, S. Metalotioneína: Classification, clinical and biochemical applications. Jornal de Proteínas e Proteômica. 2014;5(1):26–33.

[27] Slesak, I., Libik, M., Karpinska, B., Karpinski, S. and Miszalski, Z. O papel de peróxido de hidrogênio na regulação do metabolismo da planta e sinalização celular em resposta a estresses ambientais. Acta Biochimica Polonica. 2007;54:39–50.

[28] Abreu, V.M., Von Pinho, E.V.R., Von Pinho, R.G., Naves, G.M.F., Silva Neta, I.C., Guimarães, R.M. and Carvalho, M.R. Physiological performance and expression of isozymes in maize seeds subjected to water stress. Journal of Seed Science. 2014;36(1):40–47.

[29] Creissen, G., Firmin, J., Fryer, M., Kular, B., Leyland, N., Reynolds, H., Pastori, G., Wellburn, F., Baker, N., Wellburn, A., et al. Elevated glutathione biosynthetic capacity in the chloroplasts of transgenic tobacco plants. Plant Cell. 199;11(7):1277–1291.

[30] Van Heerden, P.D.R. and Kruger, G.H.J. Photosynthetic limitation in soybean during cold stress. South African Journal of Science. 2000;96(2):201–206.

[31] Roy, H.M., Zuily, Y.F., Kidric, M., Pham Thi, A.T., Silva, J.V. Effects of drought stress on proteolytic activies in *Phaseolus* and *Vigna* leaves from sensitive and resistant plants. Physiologia Plantarum. 1992;85(1):847–855.

[32] Silveira, J.A.G., Costa, R.C.L., Viegas, R.A., Oliveira, J.T.A. and Figueredo, M.V.B. *N*-compound accumulation and carhydrate shortage on N_2 fixation in drought-stressed and rewatered cowpea plants. Journal of Agricultural Research. 2003;1(3):65–75.

[33] Inamullah, I.A. Adaptive responses of soybean and cotton two waters stress. Plant Production Science. 2005;8(1):16–26.

[34] Younis, M.E., El-Shahaby, O.A., Albo-Hamed, S.A. and Ibrahim, H. Effects of water stress on growth, pigments and CO_2 assimilation in three sorghum cultivars. Journal of Agronomy and Crop Science. 2000;185(2):73–82.

[35] Van Heerden, P.D.R. and Kruge, G.H.J. Separately and simultaneously induced dark chilling and drought stress effects on photosynthesis, proline accumulation and antioxidant metabolism in soybean. Journal of Plant Physiology. 2002;159:1077–1086.

[36] Lobato, A.K.S., Oliveira Neto, C.F., Costa, R.C.L., Santos Filho, B.G., Cruz, F.J.R. and Laughinghouse IV, H.D. Biochemical and physiological behavior of *Vigna unguiculata* (L.) Walp. Under water stress during the vegetative phase. Asian Journal of Plant Sciences. 2008;7(1):44–49.

[37] Lizana, C., Wentworth, M., Martinez, J.P., Villegas, D., Meneses, R., Murchie, E.H., Pastenes, C., Lercari, B., Vernieri, P., Horton, P. and Pinto, M. Differential adaptation of two varieties of common bean to abiotic stress. Journal of Experimental Botany. 2006;57(3):685–697.

[38] Pastenes, C., Pimentel, P. and Lillo, J. Leaf movements and photoinhibition in relation to water stress in field-grown beans. Journal of Experimental Botany. 2005;56(411):425–433.

[39] Lobato, A.K.S., Costa, R.C.L., Oliveira Neto, C.F., Santos Filho, B.G., Cruz, F.J.R., Freitas, J.M.N. and Cordeiro, F.C. Morphological changes in soybean under progressive water stress. International Journal of Botany. 2008;4(2):231–235.

[40] Taiz, L. and Zeiger, E. Plant physiology. Sinauer Associates, Massachusetts.

[41] Kerbauy, G.B. Plant Physiology. Guanabara. Koogan S.A., Rio de Janeiro.

[42] Barbosa, M.A.M., Lobato, A.K.S., Viana, G.D.M., Coelho, K.N.N., Barbosa, J.R.S., Costa, R.C.L., Santos Filho, B.G. and Neto, C.F.O. Root contribution to water relations

and shoot in two contrasting *Vigna unguiculata* cultivars subjected to water deficit and inoculation. Romanian Agricultural Research. 2013;30:155–161.

[43] Lobato, A.K.S., Costa, R.C.L., Neto, C.F.O., Filho, B.G.S., Alves, G.A.R., Freitas, J.M.N., Cruz, F.J.R., Marochio, C.A. and Coimbra, G.K. Responses of the pigments and carbono metabolismo in *Vigna unguiculata* cultivars submitted to water deficit. Research Journal of Biological Sciences. 2009;4:593–598.

[44] Maia, P.S.P., Neto, C.F.O., Castro, D.S., Freitas, J.M.N., Lobato, A.K.S. and Costa, R.C.L. Leaf relative water contente, proline level, and total soluble carbohydrates in leaves of two maize cultivars submitted to water stress. Revista Brasileira de Biociências. 2007;5:918–920.

[45] Hoagland, D.R., Amon, D.I. The water-culture method for growing plants without soil. California Agricultural Experiment Station, San Francisco. 1950. 34 pp.

[46] Santos, M.G., Ribeiro, R.V., Machado, E.C. and Pimentel, C. Photosynthetic parameters and leaf water potential of five common bean genotypes under mild water deficit. Biologia Plantarum. 2009;53:229–236.

[47] Oliveira, A.D., Fernandes, E.J. and Rodrigues, T.J. Stomatal conductance as indicator of water stress in bean. Engenharia Agrícola. 2005;25:86–97.

[48] Leite, M.L. and Filho, J.S.V. Dry matter production of cowpea (*Vigna unguiculata* L. Walp.) plants submitted to water deficit. Ciencias Exata da Terra, UEPG. 2004;10:43–51.

[49] Nogueira, R.J.M.C., Santos, R.C., Neto, E.B. and Santos, V.F. Physiological behavior of two peanut cultivars submitted to suppression. Pesquisa Agropecuária Brasileira. 1998;33:1963–1969.

[50] Lobato, A.K.S., Santos Filho, B.G., Costa, R.C.L., Oliveira Neto, C.E., Meirelles, A.C.S., Cruz, F.J.R., Alves, G.A.R., Neves, H.K.B., Pita, J.D., Lopes, M.J.S., Freitas, J.M.N., Monteiro, B.S. and Ferreira Ramos, R. Physiological and biochemical changes in soybean (*Glycine max*) plants under progressive water deficit during the vegetative phase. Agricultural Journal. 2008;3(5):327–333.

[51] Sircelj, H., Tausz, M., Grill, D. and Batic, F. Biochemical responses in leaves of two apple tree cultivars subjected to progressing drought. Journal of Plant Physiology. 2005;162(12):1308–1318.

[52] Carceller, M., Prystupa, P. and Lemcoff, J.H. Remobilization of proline and other nitrogen compounds from senescing leaves of maize under water stress. Journal of Agronomy and Crop Science. 1999;183(1):61–66.

[53] Ramos, M.L.G., Parson, R. and Sprent, J.I. Differences in ureide and amino acid content of water stressed soybean inoculated with *Bradyrhizobium japonicum* and *B. elkanii*. Pesquisa Agropecuaria Brasileira. 2005;40(5):453–458.

[54] Asha, S. and Rao, K.N. Effect of simulated water logging on the levels of amino acids in groundnut at the time of sowing. Journal of Plant Physiology. 2002;7(3):288–291.

[55] Debouba, M., Gouia, H., Suzuki, A. and Ghorbel, M.H. NaCl stress effects on enzymes involved in nitrogen assimilation pathway in tomato *Lycopersicon esculentum*. Journal of Plant Physiology. 2006;163(12):1247–1258.

[56] Sankar, B., Jaleel, C.A., Manivannan, P., Kishorekumar, A., Somasundaram, R. and Panneerselvan, R. Drought-induced biochemical modifications and proline metabolism in *Abelmoschus esculentus* (L.) Moench. Acta Botanica Croatica. 2007;66(1):43–56.

[57] Ramos, M.L.G., Gordon, A.J., Minchin, F.R., Sprent, J.I. and Parsons, R. Effect of water stress on nodule physiology and biochemistry of a drought tolerant cultivar of common bean (*Phaseolus vulgaris* L.). Annals of Botany. 1999;83(1):57–63.

[58] Kirda, C., Topaloglu, F., Topçu, S. and Kaman, H. Mandarin yield response to partial root drying and conventional deficit irrigation. Turkish Journal of Agriculture and Forestry. 2007;31:1–10.

[59] Liu, F., Jensen, C.R., Shahanzari, A., Anderson, M.N. and Sven-Erik, J. ABA regulated stomatal control and photosynthetic water use efficiency of potato (*Solanum tuberosum* L.) during progressive soil drying. Journal of Plant Science. 2005;168(3):831–836.

[60] Nayyar, H. and Kaushal, S.K. Alleviation of negative effects of water stress in two contrasting wheat genotypes by calcium and abscisic acid. Biology Plantarum. 2002;45:65–70.

[61] Assmann, S.M. Open stomata opens the door to ABA signaling in *Arabidopsis* guard cells. Trends in Plant Science. 2003;8:151–153.

[62] Stikic, R. and Davies, W.J. Stomatal reactions of two different maize lines to osmotically induced drought stress. Biology Plant. 2000; 43:399–405.

[63] Hsu, S.Y., Hsu, Y.T. and Kao, C.H. Ammonium ion, ethylene, and abscisic acid in polyethylene glycol-treated rice leaves. Biology Plantarum. 2003;46(2):239–242.

[64] Atici, O., Agar, G. and Battal P. Changes in phytohormone contents in chickpea seeds germinating under lead or zinc stress. Biology Plantarum. 2005;49(2):215–222.

[65] Costa, R.C.L. Nitrogen assimilation and osmotic adjustment in nodulated plants of stringed beans (*Vigna unguiculata* (L.) Walp.) under water stress. Journal of Agronomy. 2008;7(1):98–101.

[66] Sarker, A.M., Rahman, M.S. and Paul, N.K. Effect of soil moisture on relative leaf water content, chorophyll, proline and sugar accumulation in wheat. Journal of Agronomy and Crop Science. 1999;183(4):225–229.

[67] Smita, N.H. Carbendazim alleviates effects of water stress on chickpea seedlings. Biology Plantarum. 2005;49:289–291.

3

Evaluation of TMV Lesion Formation and Timing of Signal Transduction during Induction of Systemic Acquired Resistance (SAR) in Tobacco with a Computer-Assisted Method

Zoltán Á. Nagy, György Kátay, Gábor Gullner and Attila L. Ádám

Abstract

Nicotiana tabacum L. cv. Xanthi nc plants were inoculated with tobacco mosaic virus (TMV) in order to develop a method for evaluation of lesion size and its distribution characteristics during the induction of systemic acquired resistance (SAR). All necrotic lesions were scored with an image analysis software and subjected to statistical analysis. The diminished lesion size and its right-skewed, non-normal distribution seem to be an important feature of SAR response. The results showed that the degree of induced resistance differs according to the position of the leaf on the plant's shoot. In order to detect the timing of signal transduction from TMV infected leaves to distant ones, the infected leaves were removed from the tobacco plants at different time intervals. When the infected leaves were removed after 4 days, the SAR was always induced on the distant leaves indicating complete signal transduction within 4 days.

Keywords: Lesion size and distribution, signal transduction, SAR, TMV, tobacco

1. Introduction

Systemic acquired resistance (SAR) is a defence response that initiates immunity to a wide range of pathogens in distant uninfected leaves after a former localised necrotic infection of plants. It was described after tobacco mosaic virus (TMV) infection in tobacco by Ross [1]. In agricultural practice, SAR has been recognised as a strategy to control plant pathogens because of its evolutionary stability [2], long-lasting effectiveness [3] and putative transgenerational effect [4]. Not only pathogen infections but also a number of chemicals and biotic factors including hormonal compounds (salicylic acid and methyl jasmonate), benzothiadiazole, Ningnanmycin (an antiviral agent against TMV) and insect eggs have been shown to induce SAR [5-9]. The signal transduction process and its timing are essential components of SAR induction that moves from induced to distant leaves. Recent studies indicated that the movement of SAR signal(s) is connected to plasmodesmata [10].

Induction of SAR is often validated by the increased expression of marker genes in distant tissues and/or by limited symptom expression/multiplication of the pathogen. However, the exact evaluation of symptom expression often has serious limitations. Visual assessment of leaf spots may lead to false analysis [11]. The number of developing TMV lesions is influenced by many factors (especially by the inoculation method and physiological state of plants) and therefore not well suitable as a single factor for characterization of the development of SAR [1,12].

In the present contribution we analysed local TMV lesion size formation and its distribution after SAR induction by a computer-assisted method. The method can detect nearly all lesions on tobacco leaves including the smallest ones (ca. 0.2 mm) and consequently is suitable for the exact determination and comparison of lesion size distribution of differentially treated plants. It is particularly important if a certain component is not normally distributed and mean of data is masking the fine differences. Sherwood [13] rejected normal distribution of fungal lesions (orchardgrass—*Stagonospora arenaria* interaction) in resistant genotypes, but not in susceptible ones. However, to the best of our knowledge, there is no systematic research on lesion size distribution of viral local necrotic infections.

Therefore, our tasks were as follows:

a. To develop a reliable and well-adaptable method for semi-automated measurement of lesion size and its distribution using an appropriate statistical analysis (see Materials and methods)

b. To evaluate and compare TMV lesion size distribution in control versus SAR plants

c. To determine the timing of signal transduction process(es) by comparing lesion size distributions of distant leaves after removal of TMV-infected signal-inducing leaves at different time intervals

2. Materials and methods

Nicotiana tabacum L. cv. Xanthi nc plants were grown in greenhouse and inoculated with TMV U1 strain at 6 leaf stage (fully expanded leaves) as we have described earlier [12]. The four bottom-most leaves of plants were inoculated for induction of systemic acquired resistance without any further treatments (treatment: SAR) [1]. A subset of these plants was further treated: the inoculated leaves were removed from the plants 2 and 4 days after inoculation (treatments: leaf removal, LR2+TMV and LR4+TMV, respectively). As a control, leaves were removed similarly from uninoculated plants (treatments LR2 and LR4). Seven days after inoculation, the two fully expanded distant leaves (the 5th and 6th leaf levels) were used for challenge inoculation and data analysis. Abrasive (carborundum, 50 mg/100ml) was added to the inoculum. Mock inoculation was performed only with abrasive in 10 mM K-phosphate buffer (pH = 7.0).

Four days after challenge inoculation two or three detached leaves (of the same leaf level) per treatment, were directly scanned to obtain high resolution (300 dpi) digital images with a scanner (HP Scanjet G2k710). Data were collected from these images after threefold magnification on the computer screen. All dark, brownish black, round-shaped TMV lesions typical of this infection on tobacco plants were selected visually for analysis. The ImageJ 1.48v image analysis software [14] was used for lesion selection and size calculations. However, due to the low contrast of small TMV spots, lesion selection was done manually using a drawing tablet that outperforms the precision of a standard computer mouse. Lesion size was expressed as the mean of the major and minor axes of the best fitting ellipse having equal area to the lesion. Considering all lesions per leaf resulted in more accurate estimation of lesion size compared to former approaches [7,12,13]. Figure 1 demonstrates this process on TMV-infected tobacco leaves.

All calculations (lesion size distribution, mean and variance of lesion size) were carried out with R [15]. Shapiro-Wilk w test for normal distribution and density estimation of data were calculated with functions 'shapiro.test' and 'density', respectively with their default settings.

For comparison of sample means a multiple comparison procedure was used with the R package multcomp [16]. The method allows simultaneous comparisons while the family-wise error rate, used as the standard measure for false positive results in multiple testing, remains well controlled [17]. The method uses the HC3 covariance estimation [18,19]. Furthermore, this method tolerates unequal variances, non-normal distribution of data and unbalanced group sizes, which often occur in biological datasets.

Figure 1. Leaf necrosis on *Nicotiana tabacum* cv. Xanthi nc leaves after challenge inoculation. (A): control, (B): SAR, (C): the plants were not inoculated with TMV, but their leaves were removed after 4 days (LR4), (D): the four inoculated leaves were removed from the plants 4 days after inoculation (LR4+TMV). Bar: 3 cm. Red circles on (A) mark the processed lesions during image analysis.

3. Results and discussion

3.1. Effect of SAR induction on TMV lesion size and its distribution

Data from Figs. 2, 3 and 4 are based on one representative experiment (Table 1) and serve as an example of three separate experiments. Mock inoculation with abrasive had no effect on lesion diameter and its distribution compared to control plants (Figs. 2, 3A, B and 4A, B). Induction of SAR was clearly manifested in differences of lesion development (Fig. 1) and in lesion size (Fig. 2) 4 days after challenge inoculation in all experiments. Although in control plants no significant differences were found between 5th and 6th leaf levels (Fig. 2), induction of SAR caused significant differences in degree of diminishing TMV lesion size between the 5th and 6th leaf levels. The effect of SAR induction often was more pronounced on the 5th leaf level than the 6th one (Fig. 2). Therefore, data for 5th and 6th leaves are presented separately in Figs. 2, 3 and 4. Generally, mean lesion diameter of leaves with SAR (0.528 and 0.659 mm for 5th and 6th leaf, respectively) was about half of the leaves from control plants (1.099 – 1.110 mm) (Fig. 2, Table 1).

Treatments	Leaf No.	n[1]	Lesion diameter (mm)			Normality test	
			Min	Max	Mean	Shapiro-Wilk's w	p
Control	5	343	0.25	2.01	1.099	0.9906	0.027
LR2	5	656	0.20	2.30	0.793	0.9396	< 0.001
LR2 + TMV	5	471	0.28	2.35	1.019	0.9747	< 0.001
LR4	5	428	0.29	2.27	1.024	0.9897	0.004
LR4 + TMV	5	187	0.22	1.32	0.525	0.8847	< 0.001
Mock	5	311	0.32	2.08	1.169	0.9929	0.150
SAR	5	226	0.23	1.31	0.528	0.8570	< 0.001
Control	6	132	0.43	1.80	1.110	0.9781	0.031
LR2	6	250	0.38	1.66	1.000	0.9835	0.005
LR2 + TMV	6	408	0.41	2.07	1.087	0.9851	< 0.001
LR4	6	145	0.46	1.96	1.182	0.9891	0.320
LR4 + TMV	6	145	0.29	1.35	0.666	0.9159	< 0.001
Mock	6	154	0.32	1.89	1.170	0.9889	0.266
SAR	6	103	0.25	1.55	0.659	0.9195	< 0.001

[1]For abbreviations and explanations of treatments see Fig. 2.

[2]Number of lesions.

Table 1. Test for normal distribution and lesion size data of TMV inoculated *Nicotiana tabacum* cv. Xanthi nc plants.

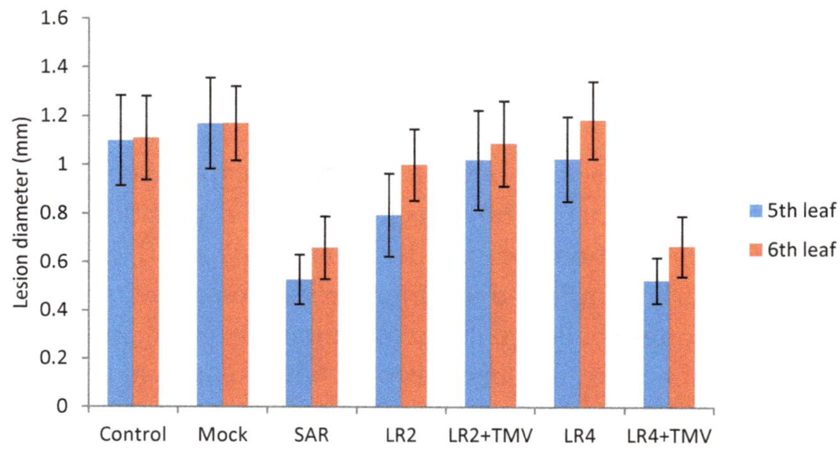

Figure 2. Mean lesion size (and standard deviation) after challenge inoculation with TMV on *Nicotiana tabacum* cv. Xanthi nc plants at leaf level 5 and 6. The four bottom-most leaves of plants were first inoculated with TMV and challenged on the 7th day. The four inoculated leaves were removed from the plants 2 and 4 days after inoculation (LR2+TMV and LR4+TMV, respectively). LR2 and LR4: the plants were not inoculated with TMV, but their leaves were removed at similar time intervals. Control: untreated plants; mock inoculated plants were treated with abrasive only; SAR: inoculated with TMV and challenged 7 days later without further treatments [1,12].

The distribution of TMV lesion size in most cases did not follow a normal distribution neither in control nor in leaves with SAR and other treatments as indicated by the results of Shapiro-Wilk w test (Table 1). Therefore, we used a statistical method suitable for comparison of non-normally distributed data. Comparison of multiple sample means under heteroscedasticity also showed highly significant differences ($P < 0.001$) between control leaves and leaves with SAR both in the 5th and 6th leaf levels (Fig. 4A, B).

Supporting the above data, the distribution of lesion sizes in control leaves and in leaves with SAR was massively different both in 5th and 6th leaves (Fig. 3A, B). The lesion sizes in control plants showed a plateau-like distribution and covered a wide range from 0.25 to 2.01 mm and 0.43 to 1.80 mm on leaves 5 and 6, respectively (Table 1). The distribution of lesions in resistant leaves with SAR was completely different showing a peak at 0.3—0.8 mm range (about 70-80% of total number of lesions) and above 1.5 mm size, practically no lesions were detected (Fig. 3A, B, Table 1).

TMV causes local hypersensitive necrotic lesions in tobacco plants carrying *N* gene from *Nicotiana glutinosa* L. [20], accompanied by programmed cell death and development of symptoms within 2 days after inoculation. This resistant response is further strengthened during SAR induction as indicated by limited lesion size and a different type of lesion size distribution as compared to control plants. Similar to our results, a non-normal distribution of necrotic spots was reported in a resistant plant—fungus interaction [13]. This shift in lesion size distribution is probably due to biochemical responses that are manifested in more effective restriction of lesion development, multiplication/growth and/or movement of the pathogen in resistant genotypes. It has been reported recently that an antiviral agent against TMV can induce SAR in tobacco [9].

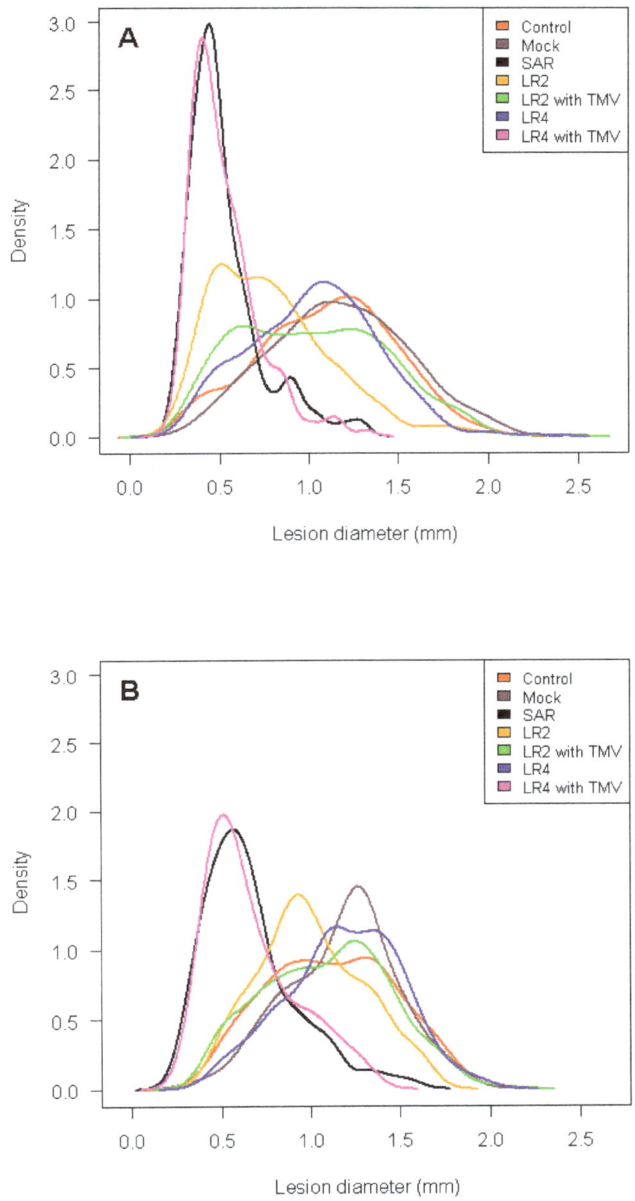

Figure 3. Kernel density estimation of TMV lesion size distribution on *Nicotiana tabacum* cv. Xanthi nc leaves at leaf level 5 (upper panel, A) and 6 (lower panel, B) after induction of systemic acquired resistance (SAR). For abbreviations and explanations of treatments see Fig. 2.

3.2. Effect of sequential removal of inducing leaves on SAR development

In order to detect the timing of signal transduction process from infected leaves to distant leaves, four bottom-most infected leaves were removed from tobacco plants at different time intervals, 2 (leaf removal, LR2) or 4 (LR4) days after TMV infection (Figs. 2, 3 and 4). The leaf removal without TMV infection after 4 days did not result in a significant shift in lesion development (LR4, Figs. 1 and 2, Table 1). Moreover, distribution of lesion sizes showed a plateau-like picture, comparable to control leaves (Fig. 3). Simultaneous comparison of treatments also showed that LR4 treated plants did not significantly differ from control plants at least on the 6th leaf level ($p = 0.3500$ with confidence interval: [-0.0353; 0.1805], Fig. 4B). On leaf level 5 the effect was almost the same as on level 6 indicating a limited effect ($p = 0.0346$ with confidence interval: [-0.146; -0.003], Fig.4A). On the contrary, removal of TMV-infected leaves after 4 days (LR4+TMV) mimicked the development of SAR in all characteristics in all three experiments. Lesion development was considerably, about 50% inhibited (Figs. 1, 2 and Table 1). The statistical analysis of data clearly showed highly significant differences between LR4+TMV and control plants ($p < 0.001$ for both leaf levels) but no significant differences between LR4+TMV and SAR treatments at both leaf levels (Fig. 4A, B). Not surprisingly, the distribution of lesion development of LR4+TMV plants was nearly the same as in leaves with SAR showing a characteristic peak at about 0.5 mm of lesion diameter. (Fig. 3A, B). These results clearly indicate that a 4-day period after the inducing infection of lower 4 leaves is enough for complete signal transduction of SAR in distant leaves. Consequently, the movement of signal molecule(s) should be detectable before this time point. These results also indicate that lesion size distribution as a resistance marker is a suitable tool for prediction of signalling events. Similar experiments with removal of leaves after a 2-day interval (LR2 and LR2+TMV) showed less clear evidences. Leaf removal without TMV infection (LR2) considerably influenced lesion development in all experiments (Fig. 2, Table 1) and somewhat shifted distribution of lesion size at both leaf levels (Fig. 3A, B). LR2 plants showed significant differences as compared either to control or SAR treated plants (Fig. 4A, B). This fact could be related to a different mechanism as compared to SAR induction, for example differences in hormone balance of distant leaves during longer incubation period after leaf removal. In LR2+TMV plants, the development of SAR was not detected on the basis of lesion size and its distribution characteristics (Figs. 2, 3A, B). Family-wise comparison of data rather suggests that LR2+TMV plants did not significantly differ from control ones but differed from the SAR treatment ($p = 0.0316$, confidence interval: [-0.1543; -0.0044]), Fig. 4A, B).

Altogether, these results indicate that the signal transduction starts probably only after visual appearance of local TMV symptoms (40—48 h post inoculation) and it is completed within the next 2 days. The identification of the exact timing of signal transduction from induced leaves is necessary for the further characterization of signal molecule(s) in phloem sap-enriched petiolar exudates.

In conclusion, we developed an easily applicable semi-automated method for the detection of the size of necrotic lesions and its distribution in tobacco leaves after TMV inoculation using appropriate statistical analysis. Decreased lesion size diameter and its characteristic non-normal, right-skewed distribution seem to be an accurate and important feature of the resistant response in distant leaves with SAR. Application of this method during SAR induction indicated that signal transduction is completed in distant leaves by the 4th day after inducing TMV inoculation. Further experiments are in progress to characterize the chemical nature of this signal.

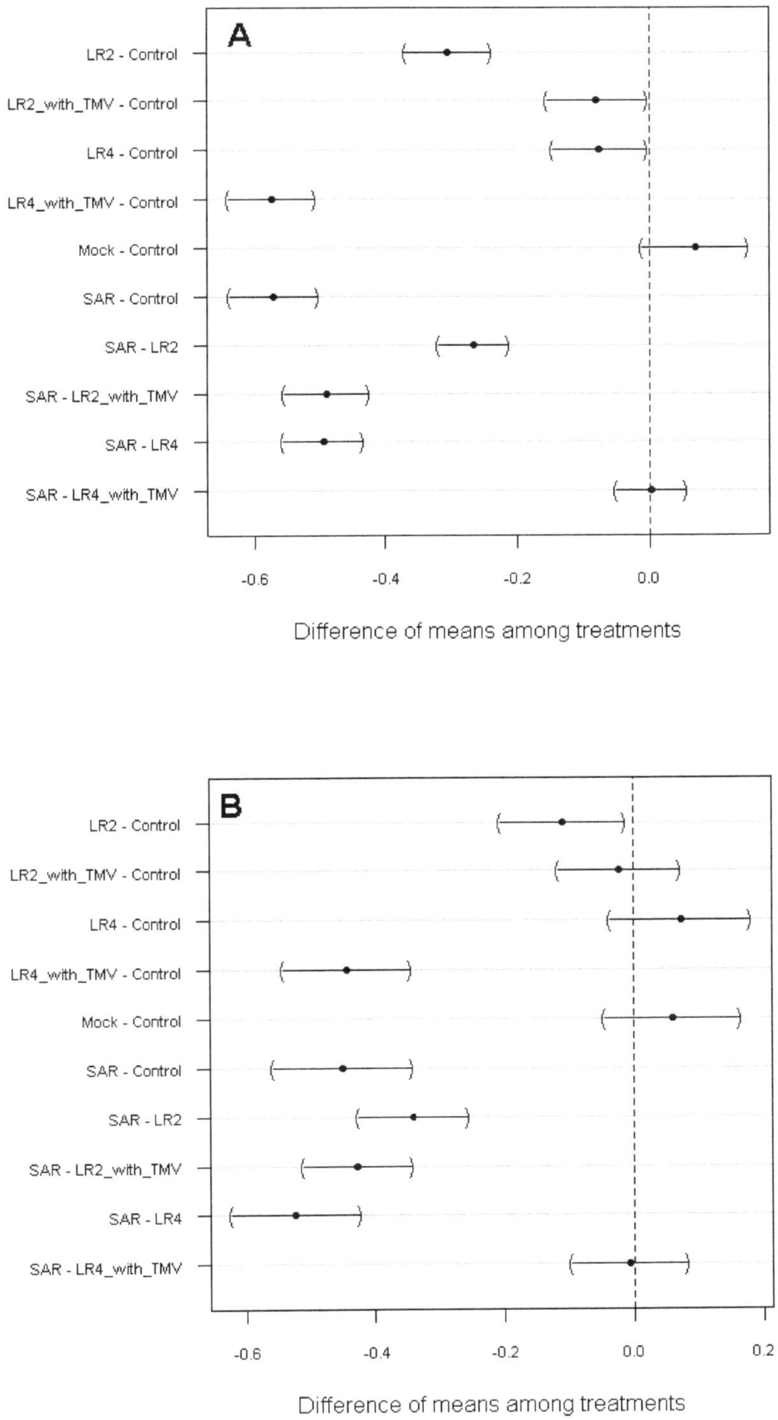

Figure 4. Multiple comparisons of group means of selected treatments on leaf level 5 (A) and 6 (B). Dots represent the difference of the estimated means between treatments. Brackets flank the 95% confidence intervals. The difference is considered significant if the confidence interval does not contain the 0, represented by a vertical dashed line. Abbreviations and explanations are the same as on Fig. 2.

Acknowledgements

The authors wish to thank the National Research, Development and Innovation Office (ALÁ, OTKA-K112146) for funding.

Author details

Zoltán Á. Nagy*, György Kátay, Gábor Gullner and Attila L. Ádám

*Address all correspondence to: adam.attila@agrar.mta.hu

Plant Protection Institute, Centre for Agricultural Research, Hungarian Academy of Sciences, Budapest, Hungary

References

[1] Ross AF.: Systemic acquired resistance induced by localized virus infections in plants. Virology. 1961;14:340–358.

[2] Winter PS, Bowman CE, Villani PJ, Dolan TE, Hauck NR.: Systemic acquired resistance in moss: further evidence for conserved defense mechanisms in plants. PLoS ONE, 2014;7:e101880. DOI: 10.1371/journal.pone.0101880.

[3] Hammerschmidt R, Kuć J.: Induced resistance to disease in plants. Kluwer Academic Publishers; Dordrecht, The Netherlands: 1995.

[4] Luna E, Bruce TJ, Roberts MR, Flors V, Ton J.: Next generation systemic acquired resistance. Plant Physiol. 2012;158:844–853.

[5] Chen Z, Silva H, Klessig DF.: Active oxygen species in the induction of plant systemic acquired resistance by salicylic acid. Science. 1993;262:1883–1886.

[6] Dietrich R, Ploss K, Heil M.: Growth responses and fitness costs after induction of pathogen resistance depend on environmental conditions. Plant Cell Environ. 2005;28:211–222.

[7] Shulaev V, Leon J, Raskin I.: Is salicylic acid a translocated signal of systemic acquired resistance in tobacco? The Plant Cell. 1995;7:1691–1701. doi:10.1105/tpc.7.10.1691.

[8] Hilfiker O, Groux R, Bruessow F, Kiefer K, Zeier J, Reymond P.: Insect eggs induce a systemic acquired resistance in *Arabidopsis*. Plant J. 2014;80:1085–94.

[9] Han Y, Luo Y, Qin S, Xi L, Wan B, Du L.: Induction of systemic acquired resistance against tobacco mosaic virus by Ningnanmycin in tobacco. Pestic Biochem Physiol. 2014;111:14–18.

[10] Carella P, Isaacs M, Cameron RK.: Plasmodesmata-located protein overexpression negatively impacts the manifestation of systemic acquired resistance and the long-distance movement of Defective in Induced Resistance1 in *Arabidopsis*. Plant Biology. 2015;17:395–401.

[11] Sherwood RT, Berg CC, Hoover MR, Zeiders KE.: Illusions in visual assessment of *Stagonospora* leaf spot of orchardgrass. Phytopathology. 1983;73:173–177.

[12] Ádám A, Barna B, Farkas T, Király Z.: Effect of TMV-induced systemic acquired resistance and removal of the terminal bud on membrane lipids of tobacco leaves. Plant Science. 1990;66:173–179.

[13] Sherwood RT.: Weibull distribution of lesion size in the *Stagonospora* leaf spot of orchardgrass. Phytopathology. 1987;77:715–717.

[14] Schneider CA, Rasband WS, Eliceiri KW: NIH Image to ImageJ: 25 years of image analysis, Nature Methods. 2012;9:671–675.

[15] R Development Core Team: R: A language and environment for statistical computing. R Foundation for Statistical Computing, Vienna, Austria. 2012. ISBN 3-900051-07-0, URL http://www.R-project.org/.

[16] Hothorn T, Bretz F, Westfall P.: Simultaneous inference in general parametric models. Biometrical J. 2008;50:346–363.

[17] Herberich E, Sikorski J, Hothorn T.: A robust procedure for comparing multiple means under heteroscedasticity in unbalanced designs. PLoS ONE. 2010;5:e9788. doi: 10.1371/journal.pone.0009788.

[18] Long JS, Ervin LH.: Using heteroscedasticity consistent standard errors in the linear regression model. American Statistician. 2000;54:217–224

[19] MacKinnon JG, White H.: Some heteroscedasticity consistent covariance with improved finite sample properties. Journal of Econometrics. 1985;29:53–57.

[20] Holmes FO.: Inheritance of resistance to tobacco-mosaic disease in tobacco. Phytopathology. 1938;28:553–561.

4

Abiotic and Biotic Elicitors–Role in Secondary Metabolites Production through In Vitro Culture of Medicinal Plants

Poornananda M. Naik and Jameel M. Al-Khayri

Abstract

Plant secondary metabolites are having the great application in human health and nutritional aspect. Plant cell and organ culture systems are feasible option for the production of secondary metabolites that are of commercial importance in pharmaceuticals, food additives, flavors, and other industrial materials. The stress, including various elicitors or signal molecules, often induces the secondary metabolite production in the plant tissue culture system. The recent developments in elicitation of plant tissue culture have opened a new avenue for the production of secondary metabolite compounds. Secondary metabolite synthesis and accumulation in cell and organ cultures can be triggered by the application of elicitors to the culture medium. Elicitors are the chemical compounds from abiotic and biotic sources that can stimulate stress responses in plants, leading to the enhanced synthesis and accumulation of secondary metabolites or the induction of novel secondary metabolites. Elicitor type, dose, and treatment schedule are major factors determining the effects on the secondary metabolite production. The number of parameters, such as elicitor concentrations, duration of exposure, cell line, nutrient composition, and age or stage of the culture, is also important factors influencing the successful production of biomass and secondary metabolite accumulation. This chapter reviews the various abiotic and biotic elicitors applied to cultural system and their stimulating effects on the accumulation of secondary metabolites.

Keywords: Cell culture, elicitor, organ culture, secondary metabolites, stress

1. Introduction

The total mankind is dependent on plants as a source of carbohydrates, proteins, vitamins, food, and shelter. Plants are studied for their important constituents and the nutritional factors for over decades. Along with the essential primary metabolites, higher plants are also capable to produce a number of low molecular weight compounds. A diverse group of organic compounds that are produced by plants to facilitate interaction with the biotic environment and the establishment of a defense mechanism are called as plant secondary metabolites [1–3]. The production of these metabolites is very low (less than 1% dry weight) and depends greatly on the physiological and developmental stage of the plant [4,5]. Plant natural products have been an important part of medicine throughout human history. In recent years, the use of herbal medicines has steadily increased worldwide [6]. With this increasing demand comes growing concerns about the safety and efficacy of herbal medicines. Although the potential for medicinal plants seems almost limitless, there are a few major obstacles that hinder large-scale utilization by the western medical system. Among them is the lack of reproducibility common in testing many plant extracts (up to 40%), which has limited the enthusiasm for developing plant-based pharmaceuticals [7]. Unlike standardized single-entity pharmaceutical drugs, herbal medicines consist of complex mixtures with multiple compounds responsible for therapeutic activity, making standardization difficult [8]. Further complicating the issue is the fact that plants, unlike synthetic medicines, are living organisms, with inherent biological variation [9]. Just because plant material originates from the same species, it does not necessarily mean that the chemical content will be identical. This lack of reproducibility may be due to two main factors, genetic variability and differences in growing conditions.

In addition, plants are a valuable source of a wide range of secondary metabolites, which are used as pharmaceuticals, agrochemicals, flavors, fragrances, colors, biopesticides, and food additives. Plants are producing new compounds and in future new chemical models are drawing for new drugs because the most of the plants chemistry is yet to be explored [10]. The characterization of molecular structures and chemical analysis helped us to pinpoint the activities of plants under controlled conditions. Although all these advancements, we still depend on the secondary metabolites of biological sources including pharmaceuticals [11].

Due to various agro alimentary, perfumes, flavors, colors, and pharmacological effects, the secondary metabolites are having extensive demand and various commercial preparations are available in the market. Besides, the appeal of using natural products for medicinal purposes is increasing, and metabolic engineering can alter the production of pharmaceuticals and help to design new therapies. The evolving commercial importance of secondary metabolites has in recent years resulted in a greater interest in secondary metabolism, particularly in the possibility of altering the production of bioactive plant metabolites [12]. Secondary metabolites are separated into nitrogen compounds (alkaloids, nonprotein amino acids, amines, alcamides, cyanogenic glycosides, and glucosinolates) and nonnitrogen compounds (monoterpenes, diterpenes, triterpenes, tetraterpenes, sesquiterpenes, saponins, flavonoids, steroids, and coumarins).

The plant tissue culture plays an important role in the rapid clonal propagation, regeneration of genetically manipulated superior clones, conservation of germplasm, production of secondary metabolites, and ex vitro conservation of valuable phytodiversity [13,14]. The plant, cell, tissue and organ culture techniques have come up with an escapable tool with the possibilities of acclaiming and supplementing the conventional method in plant breeding, plant improvement, and biosynthetic pathways. This technique has several potential applications in crop improvement, and efficient regeneration is a prerequisite in such improvement programs. The biotechnological production of secondary metabolites in plant cell and organ cultures is an attractive alternative to the extraction of the whole plant material [15]. In particular, plant–specific important compounds are obtained by using the plant cell and organ cultures [2]. The faster proliferation rates and shorter biosynthetic cycle of cell and organ cultures leads to have a higher rate of metabolism when compared to field grown plants [16]. Further, plant cell/organ cultures are under controlled conditions proliferates at their optimum growth rates when compared to the cultivated plants, which are facing environmental, ecological, and climatic variations. In recent years, various strategies have been developed for use in biomass accumulation and the synthesis of secondary compounds, such as strain improvement, optimization of medium, and culture environments, elicitation, precursor feeding, metabolic engineering, permeabilization, immobilization, and biotransformation methods, bioreactor cultures, and micropropagation [17]. The focus of the present chapter is the influence of abiotic and biotic elicitors on the secondary metabolite production in the in vitro cultured medicinal plants.

2. Classification of elicitors and secondary metabolite production via in vitro culture of medicinal plants

Stress is an important factor in determining the chemical composition and therapeutic activity of medicinal plants. Actively stimulating, or eliciting, the plant stress response to induce the desired chemical response is called elicitation, harnessing the connection between plant stress and phytochemistry. "Elicitor may be defined as a substance for stress factors which, when applied in small quantity to a living system, it induces or improves the biosynthesis of specific compound which do have an important role in the adaptations of plants to a stressful conditions" [18]. Elicitation is the induced or enhanced biosynthesis of metabolites due to addition of trace amounts of elicitors [18]. Several biotechnological strategies have been hypothesized and applied for the productivity enhancement, and elicitation is recognized as the most practically feasible strategy for increasing the production of desirable secondary compounds from cell, organ, and plant systems [19–21].

On the basis of nature, elicitors can be divided into two types abiotic and biotic (Figure 1). Abiotic elicitors comprise of substances that are of nonbiological origin and are grouped in physical, chemical, and hormonal factors. Biotic elicitors are the substances of biological origin that include polysaccharides originated from plant cell walls (e.g. chitin, pectin, and cellulose) and micro–organisms.

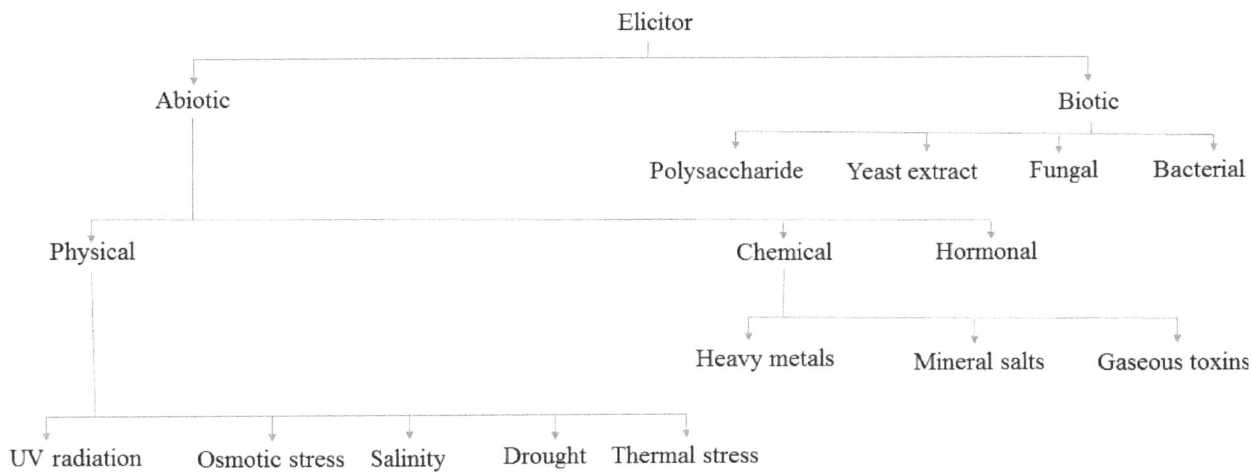

Figure 1. Elicitors Classification Based on their Nature

3. Abiotic elicitors

As mentioned above, the abiotic elicitors are categorized into physical, chemical, and hormonal elicitors. Abiotic elicitors have wide range of effects on the plants and in the production of secondary metabolites (Table 1).

Elicitor	Plant species	Nature of culture	Compounds	References
Ozone (O_3)	*Melissa officinalis*	Shoot	Rosmarinic acid	[163]
	Hypericum perforatum	Cell suspension	Hypericin	[164]
	Pueraria thomsnii	Cell suspension	Puerarin	[165]
pH	*Bacopa monnieri*	Shoot	Bacoside A	[166]
	Withania somnifera	Hairy root	Withanolide A	[167]
	Withania somnifera	Cell suspension	Withanolide A	[168]
Sucrose	*Hypericum adenotrichum*	Seedling	Hypericin and pseudohypericin	[41]
	Corylus avellana	Cell suspension	Paclitaxel	[169]
	Bacopa monnieri	Shoot	Bacoside A	[166]
	Withania somnifera	Cell suspension	Withanolide A	[168]
Ultraviolet C	*Vitis vinifera*	Cell suspension	Stilbene	[34]
Proline	*Stevia rebaudiana*	Callus and suspension	Steviol glycoside	[40]
Polyethylene glycol	*Stevia rebaudiana*	Callus and suspension	Steviol glycoside	[40]

Elicitor	Plant species	Nature of culture	Compounds	References
	Hypericum adenotrichum	Seedling	Hypericin and pseudohypericin	[41]
Jasmonic acid	*Bacopa monnieri*	Shoot	Bacoside A	[170]
	Plumbago indica	Hairy root	Plumbagin	[88]
	Plumbago rosea	Cell suspension	Plumbagin	[86]
Methyl jasmonate	*Salvia miltiorrhiza*	Hairy root	Tanshinone	[94]
	Perovskia abrotanoides	Adventitious roots	Cryptotanshinone and tanshinone IIA	[78]
	Vitis vinifera	Cell suspension	Stilbene	[34]
	Bacopa monnieri	Shoot	Bacoside	[96]
	Salvia officinalis	Shoot	Diterpenoid	[171]
	Silybum marianum	Cell suspension	Silymarin	[172]
	Salvia castanea	Hairy root	Tanshinone	[173]
	Gymnema sylvestre	Cell suspension	Gymnemic acid	[148]
	Withania somnifera	Hairy root	Withanolide A, withanone, and withaferin A	[95]
	Andrographis paniculata	Cell suspension	Andrographolide	[97]
	Vitis vinifera	Cell suspension	trans-Resveratrol	[91]
	Taverniera cuneifolia	Root	Glycyrrhizic acid	[135]
Gibberelic acid	*Salvia miltiorrhiza*	Hairy root	Tanshinones	[174]
	Echinacea pupurea	Hairy root	Caffeic acid derivatives	[175]
Salicylic acid	*Salvia miltiorrhiza*	Hairy root	Tanshinone	[94]
	Vitis vinifera	Cell suspension	Stilbene	[34]
	Digitalis purpurea	Shoot	Digitoxin	[176]
	Hypericum hirsutum	Shoot	Hypericin and pseudohypericin	[177]
	Gymnema sylvestre	Cell suspension	Gymnemic acid	[148]
	Withania somnifera	Hairy root	Withanolide A, withanone, and withaferin A	[95]
	Datura metel	Root	Hyoscyamine and scopolamine	[178]
	Glycyrrhiza uralensis	Adventitious root	Glycyrrhizic acid	[179]

Elicitor	Plant species	Nature of culture	Compounds	References
Sodium salicylate	*Salvia officinalis*	Shoot	Carnosol	[180]
Sodium chloride	*Catharanthus roseus*	Embryogenic tissues	Vinblastine and vincristine	[55]
Sorbitol	*Perovskia abrotanoides*	Adventitious roots	Cryptotanshinone and tanshinone IIA	[78]
Silver (Ag)	*Perovskia abrotanoides*	Adventitious roots	Cryptotanshinone and tanshinone IIA	[78]
	Vitis vinifera	Cell suspension	Resveratrol	[67]
	Salvia castanea	Hairy root	Tanshinone	[173]
	Datura metel	Hairy root	Atropine	[150]
Cadmium (Cd)	*Vitis vinifera*	Cell suspension	Resveratrol	[67]
	Datura stramonium	Root	Sesquiterpenoid	[79]
Cobalt (Co)	*Vitis vinifera*	Cell suspension	Resveratrol	[67]
Copper (Cu)	*Ammi majus*	Shoot	Xanthotoxin	[181]
	Bacopa monnieri	Shoot	Bacoside	[170]
	Datura stramonium	Root	Sesquiterpenoid	[79]

Table 1. Effect of Different Abiotic Elicitors on the Production of Various Secondary Metabolites in Plants

3.1. Physical elicitors

Physical elicitors include light, osmotic stress, salinity, drought, and thermal stress.

3.1.1. Light

The light is a physical factor that can affect the metabolite production. Light can stimulate such secondary metabolites include gingerol and zingiberene production in *Zingiber officinale* callus culture [22]. The effect of light irradiation on anthocyanin production in cell suspension cultures of *Perilla frutescens* was reported [23]. The effect of light and hormones on the digitoxin accumulation in *Digitalis purpurea* L. was reported by Hagimori et al. [24]. Moreover, in hairy root cultures of *Artemisia annua,* the effect of light irradiation influenced the artemisinin biosynthesis [25]. The effect of white light on taxol and baccatin III accumulation in cell cultures of *Taxus cuspidate* was reported by Fett–Neto et al. [26]. Ultraviolet (UV) radiation stimulates secondary metabolite production. Increasing UV–B exposure in field–grown plants not only increased the total essential oil and phenolic content but also decreased the amount of the possibly toxic beta–asarone [27]. These findings are to be expected as phenolics are known UV protectants [28]. *Catharanthus roseus* plants, exposed to UV–B light, show significant increases in the production of vinblastine and vincristine, which have proven effective in the treatment of leukemia and lymphoma [29]. UV–C irradiation promotes the phenylpropanoid pathway and stimulates flavonoid synthesis [30]. UV–C irradiation is an effective method to enhance stilbene production in *Vitis vinifera* berries [31], *V. vinifera* leaves [32], and *V. vinifera* callus of different genotypes [33]. UV–C together with methyl jasmonate (MeJA) or salicylic acid (SA) also used to enhance stilbene production in *V. vinifera* cell cultures [34].

3.1.2. Osmotic stress

Osmotic stress (water stress) is an abiotic physical elicitor [35] and is one of the important environmental stresses that can alter the physiological and biochemical properties of plants and increase the concentration of secondary metabolites in plant tissues [36]. Proline acts as an osmolyte, as protective agent for cytoplasmic enzymes, as a reservoir of nitrogen and carbon sources for post–stress growth, or even as a stabilizer of the machinery for protein synthesis, regulation of cytosolic acidity and scavenging of free radicals [37]. However, the various roles of proline have been proposed, but the main role could be the osmotic adjustment in osmotically stressed plant tissues and the protection of plasma membrane integrity [38]. Polyethylene glycol (PEG) is an osmotic agent (nonpenetrating osmoticum) that has been used for induction of water stress in many plants [39]. The proline and PEG enhanced the production of steviol glycosides content in both callus as well as suspension culture of *Stevia rebaudiana* [40]. PEG elicited the pharmacologically active compounds, such as hypericin and pseudohypericin, in *Hypericum adenotrichum* [41]. Sucrose is a typical osmotic stress agent used for the induction of water stress in plants that also serves as a vital carbon and energy source [42]. It has been shown that water and osmotic imbalance can strongly influence the synthesis of hypericin and hyperforin in *Hypericum perforatum* plants [43]. In addition, it has been reported that both hypericin and pseudohypericin concentrations decreased, while hyperforin concentration increased significantly in the plants grown under water stress conditions [36].

3.1.3. Salinity

Salinity reduces plant growth and development and alters a wide array of physiological and metabolic processes [44,45]. Plants have developed complex mechanisms for adaptation to the osmotic, ionic, and oxidative stresses that are induced by the salt stress. Exposure to salinity is known to induce or stimulate the production of secondary plant products, such as phenols, terpenes, and alkaloids [46–48]. *C. roseus* grown under salt stress showed increased levels of the alkaloid vincristine [49]. In *Grevillea*, a significant increase in anthocyanin concentration was reported under salinity exposure in both the salt–tolerant *Grevillea ilicifolia* and the salt–sensitive *Grevillea arenaria* [50]. In contrast to this, salt stress decreased the anthocyanin level in the salt–sensitive species [51]. In *Datura innoxia*, salt treatment increased the total alkaloid content in young leaves, and the results indicated that at the organ level, tropane alkaloid accumulation was related to plant growth [52]. Glycine betaine was increased under salinity in numerous species including *Triticum aestivum* [53] and *Trifolium repens* [54]. Salinity also increased the diamine and polyamine content in *Oryza sativa* [53]. An improved synthesis of vinblastine and vincristine was observed in *C. roseus* embryogenic tissue culture by using NaCl as an elicitor [55].

3.1.4. Drought stress

One of the most important abiotic stress is drought, which affect plant growth and their developmental process [56]. The available water in the soil is reduced to such critical levels, and atmospheric conditions add to the continuous loss of water; the situation is called drought stress. The high temperature in the environment and solar radiations add up the water deficit in the soil, which leads to drought stress. Drought stress tolerance is observed in all types of plants, but its extent varies from species to species [56]. Drought stress, which can also greatly reduce plant growth, can increase secondary metabolite content. Mild water stress significantly increased the content of the anti–inflammatory saikosaponins in *Bupleurum chinense* [57]. Moderate water stress increased the content of salvianolic acid in roots of *Salvia miltiorrhiza*, although the content of other bioactives, including tanshinone, was lowered [58]. Moderate drought stress also increased the production of rosmarinic, ursolic, and oleanolic acid in *Prunella vulgaris* [59]. A weak water deficit greatly increased the glycyrrhizic acid content in roots of *Glycyrrhiza uralensis* [60]. In *Hypericum brasiliense*, the amounts of various phenols and betulinic acid were drastically increased under drought stress [61].

3.1.5. Thermal stress

Although thermal stress can greatly reduce plant growth and induce senescence, elevated temperatures (heat stress) or low temperatures (cold stress) have also been shown to increase secondary metabolite production. Temperature strongly influences metabolic activity and plant ontology, and high temperatures can induce premature leaf senescence [62]. Elevated temperatures increase leaf senescence and root secondary metabolite concentrations in the herb *Panax quinquefolius* [63]. A 5°C increase in temperature significantly increased the ginsenoside content in roots of *P. quinquefolius* [63]. A temperature variation has multiple effects on the metabolic regulation, permeability, and rate of intracellular reactions in plant cell cultures [62]. Temperature range of 17–25°C is normally used for the induction of callus tissues and growth of cultured cells [16]. The temperature and light quality influences on the production of ginsenoside in hairy root culture of *Panax ginseng* [64]. The *Melastoma malabathricum* cell cultures incubated at a lower temperature range (20 ± 2°C) grew better and had higher anthocyanin production than those grown at 26 ± 2°C and 29 ± 2°C [65]. Fifteen days at 35°C significantly increased the hypericin and hyperforin content in shoots of *Hypericum perforatum* [66].

4. Chemical elicitors

Heavy metals have become one of the main abiotic stress agents for living organisms because of their increasing use in the developing fields of industry and agrotechnics and high bioaccumulation and toxicity [67]. Although a lot of information is available concerning the effects of heavy metals on plant growth and physiology, much less is known regarding their effects on the production of secondary metabolites. Heavy metal–induced changes in metabolic activity of plants can affect the production of photosynthetic pigments, sugars, proteins, and nonprotein thiols. These effects can result from the inhibition of enzymes involved in the production of these natural products, likely through impaired substrate utilization [68]. Metals may alter the production of bioactive compounds by changing aspects of secondary metabolism [2]. Metals including Ni, Ag, Fe, and Co have been shown to elicit the production of secondary metabolites in a variety of plants [69].

An increased oil content up to 35% in *Brassica juncea* was seen due to the effective accumulation of metals (Cr, Fe, Zn, and Mn) [70]. The highest accumulations of secondary metabolites such as shikonin [71] and also the production of digitalin [72] were observed by treating Cu^{2+} and Cd^{2+}. The production of betalains in *Beta vulgaris* also stimulated by Cu^{2+} [73]. Co^{2+} and Cu^{2+} have a stimulatory effect on the production of secondary metabolites in *Beta vulgaris* [73]. The betalaines production was enhanced by exposing the hairy root culture to metal ions [74]. The stimulatory effects of Cu^{2+} on the accumulation of betacyanins in callus cultures of *Amaranthus caudatus* were reported by Obrenovic [75]. The addition of Zn^{2+} (900 μM) improved the yield of lepidine in cultures of *Lepidium sativum* [76]. However, Cu proved more effective than Zn in enhancing the yield product [76]. In hairy root cultures of *Brugmansia candida*, silver nitrate ($AgNO_3$) or cadmium chloride ($CdCl_2$) elicited the overproduction of two tropane alkaloids, scopolamine, and hyoscyamine [20]. The production of taxol in cell culture of *Taxus* sp. was enhanced by the rare–earth metal (lanthanum) [77]. $AgNO_3$ stimulated the production of tanshinone in the root culture of *Perovskia abrotanoides* [78]. The treatment of root cultures of *Datura stramonium* with cadmium salts at external concentrations of approximately 1 mM has been found to induce the rapid accumulation of high levels of sesquiterpenoid–defensive compounds, notably lubimin and 3–hydroxylubimin, but not alkaloid [79].

5. Hormonal elicitors

Various plant hormones have been extensively used in elicitation studies. The most studied, because of their key roles in the plant defense response, are jasmonic acid (JA) and SA and its derivatives.

5.1. Jasmonates

Jasmonates, including JA and MeJA, are a family of cyclopentanone compounds that modulate a wide range of plant responses [80,81] and act as effective elicitors to enhance secondary metabolites in in vitro cultures. They constitute an important class of elicitors for many plant secondary metabolic pathways, which are typically manifested by the elicitation of secondary metabolite biosynthesis when plants face particular environmental stresses [82]. JA is an important signal molecule of plant in response to wound and pathogen attack [83]. JA and its more active derivative MeJA can induce the production of a wide range of plant secondary metabolites such as rosmarinic acid, terpenoid indole alkaloid, and plumbagin in various cell cultures [84–86]. JA elicitation are reported to induce the production of rosmarinic acid in *Mentha piperita* [84], anthocyanin in *V. vinifera* [87], and plumbagin in hairy roots of *Plumbago indica* [88]. JA and MeJA have been used as elicitors for stilbene biosynthesis in *V. vinifera* foliar cuttings [89], *V. vinifera* cell cultures [90,91], and *Vitis rotundifolia* hairy root cultures [92]. The addition of MeJA to *V. vinifera* cell cultures also promoted anthocyanin accumulation [93]. MeJA with transgenic technology highly enhanced the production of tanshinones in *Salvia miltiorrhiza* hairy roots [94]. In the hairy root culture of *Withania somnifera*, MeJA elicited the production of withanolide A, withanone, and withaferin A [95]. The MeJA enhanced the production of bacoside A, a valuable triterpenoid saponin having nootropic therapeutic activity in in vitro shoot cultures of *Bacopa monnieri* [96]. In *Andrographis paniculata* cell culture, the MeJA induced the production of andrographolide content [97]. In *Glycyrrhiza glabra*, methyl jasmonate induced the production of the saponin soyasaponin [98]. It enhanced the production of paclitaxel in *Tacus canadensis* and *T. cuspidate* [99], and in *Rubus idaeus*, it stimulated the production of the raspberry ketone benzalacetone [100].

5.2. Salicylic acid

Salicylic acid, well known for the systemic acquired resistance it induces in the plant response to many pathogens, can also elicit the production of secondary metabolites in plants [101,102]. SA with transgenic technology highly enhanced the production of tanshinones in *S. miltiorrhiza* hairy roots [94]. The higher production of withanolide A, withanone, and withaferin A was reported in the elicited–hairy roots of *W. somnifera* [95]. SA induced the stilbene production in the cell suspension of *V. vinifera* [34]. It stimulated the production of alkaloids such as vincristine and vinblastine in periwinkle [103], the tropane alkaloid scopolamine in hairy root cultures of *Brugmansia candida* [77], and pilocarpine in jaborandi leaves [104]. Anthraquinone production was greatly increased in *Rubia cordifolia* after a SA treatment [105]. SA also affects terpenoid secondary metabolism in plants. It induced accumulation of the triterpenoids ginsenosides in ginseng and glycyrrhizin in licorice [106,107]. Recent evidence demonstrated that suitable concentrations of SA can also promote monoterpene production [108].

5.3. Gibberellic acid

Gibberellin (GA), a phytohormone, is also well known as an effective elicitor for the production of secondary metabolites [109].

6. Biotic elicitors

In the production of secondary metabolites from plants, the use of biotic elicitors had an important role (Table 2).

Elicitor	Plant species	Nature of culture	Compounds	References
Chitin	*Hypericum perforatum*	Shoot	Hypericin and pseudohypericin	[182]
	Hypericum perforatum	Cell suspension	Phenylpropanoid and naphtodianthrone	[183]
	Vitis vinifera	Cell suspension	trans-Resveratrol and viniferins	[91]
Pectin	*Hypericum perforatum*	Shoot	Hypericin and pseudohypericin	[182]
Dextran	*Hypericum perforatum*	Shoot	Hypericin and pseudohypericin	[182]
Yeast extract	*Perovskia abrotanoides*	Adventitious roots	Cryptotanshinone and tanshinone IIA	[78]
	Plumbago rosea	Cell suspension	Plumbagin	[86]
	Silybum marianum	Cell suspension	Silymarin	[172]
Trichoderma atroviride	*Salvia miltiorrhiza*	Hairy root	Tanshinone	[184]
Protomyces gravidus	*Ambrosia artemisiifolia*	Hairy root	Thiarubrine A	[134]
Claviceps purpurea	*Azadirachta indica*	Hairy root	Azadirachtin	[136]
Mucor hiemalis	*Taverniera cuneifolia*	Root	Glycyrrhizic acid	[135]
Fusarium oxysporum	*Hypericum perforatum*	Cell suspension	Phenylpropanoid and naphtodianthrone	[183]
Phoma exigua	*Hypericum perforatum*	Cell suspension	Phenylpropanoid and naphtodianthrone	[183]
Botrytis cinerea	*Hypericum perforatum*	Cell suspension	Phenylpropanoid and naphtodianthrone	[183]
Aspergillus niger	*Gymnema sylvestre*	Cell suspension	Gymnemic acid	[149]
Saccharomyces cerevisiae	*Gymnema sylvestre*	Cell suspension	Gymnemic acid	[149]
Agrobacterium rhizogenes	*Gymnema sylvestre*	Cell suspension	Gymnemic acid	[149]
Bacillus subtilis	*Gymnema sylvestre*	Cell suspension	Gymnemic acid	[149]
Escherichia coli	*Gymnema sylvestre*	Cell suspension	Gymnemic acid	[149]
	Datura metel	Hairy root	Atropine	[150]
Bacillus cereus	*Datura metel*	Hairy root	Atropine	[150]
Staphylococcus aureus	*Datura metel*	Hairy root	Atropine	[150]
Rhizobium leguminosarum	*Taverniera cuneifolia*	Root	Glycyrrhizic acid	[135]

Table 2. Effect of Different Biotic Elicitors on the Production of Various Secondary Metabolites in Plants

6.1. Polysaccharide

The biotic elicitors have been utilized to increase secondary metabolite production in medicinal plants. In a *Panax ginseng* cell suspension, the cell wall–derived elicitor oligogalacturonic acid significantly increased the ginseng saponin content [110]. The treatment of cultured cells of *Lithospermum erythrorhizon* with the polysaccharide agropectin induced the production of the naphthoquinone shikonin [111]. The chitosan treatment of cultures of *Plumbago rosea* increased the plumbagin content [112]. The application of chitin or chitosan induced the production of coumarins and fluoroquinolone alkaloids in shoot cultures of *Ruta graveolens* [113]. Chitosan enhanced the production of trans–resveratrol and viniferins in the cell system of *V. vinifera* [91]. Chitin induced the phenylpropanoid and naphtodianthrone production in cell suspension cultures of *H. perforatum* [114].

6.2. Yeast origin

For decades, scientists are using yeast extract as one of the biotic elicitors. Yeast extracts stimulated ethylene biosynthesis in tomato [115] and bacterial resistance in bean (*Phaseolus vulgaris*) [116]. Yeast extract elicited the production of tanshinone in the root culture of *Perovskia abrotanoides* [78].

6.3. Fungal origin

Biotic elicitors produced by pathogens have mainly been used to induce the plant defense response. In the past, biological mixtures were prepared from pathogens without identification of the active compounds. The use of pathogenic and nonpathogenic fungal preparations as elicitors has become one of the most effective strategies to induce phenylpropanoid/flavonoid biosynthetic pathways in plant cells [117,118]. Necrotrophic pathogens such as *Botrytis* sp. usually kill the host cells often through secretion of toxins before deriving nutrients from them [119]. On the other hand, biotrophic pathogens *Fusarium* sp. or *Phoma* sp. try to avoid killing the host cells, and derive their nutritional benefits from extensive contact with them and by altering the host metabolism and secretion systems [120,121]. An early defense reaction of the plant cell attacked by fungal pathogen includes the rapid and transient production of reactive oxygen species (ROS). Plant cells are usually protected against the detrimental effects of ROS by a complex of nonenzymatic and enzymatic antioxidant systems [122]. It has been demonstrated that the phenylalanine ammonia lyase (PAL) enzyme that catalyses the entry of L–phenylalanine into the phenylpropanoid pathway has reputedly a crucial role in the synthesis of antioxidant/defense–related compounds [117]. The mycelia extracts from the above mentioned fungi induced partitioning of the phenylpropanoid pathway and a rapid stimulation of the monolignol pathway in *Linum usitatissimum* cultured cells [123]. Cultures of *Phytophthora* elicited microbial resistance in soybean [124] and potato [125]. Extracts from microbial–enriched composts stimulated systemic resistance to *Phytophthora* in pepper (*Capsicum annuum*) [126]. As the plant defense response and the production of secondary metabolites are closely related, it is not surprising that a number of elicitors have also been shown to increase the production of secondary metabolites in medicinal plant cell culture. Similar to plant defense, initial work on secondary metabolite elicitation was performed using biological mixtures. Fungal cell wall fragments increased the production of the indole alkaloids ajmalicine, serpentine, and catharanthine by up to five times in cell suspensions of *C. roseus* [69,127] and the 12–oxo–phytodienoic acid, raucaffrincine, in *Rauwolfia canescens* [128]. Fungal mycelia increased the diosgenin content in *Dioscorea deltoidea* cells by 72% [129]. In *Papaver somniferum*, fungal spores increased the content of codeine, morphine, and sanguinarine by over eightfold [130,131]. A mixture of fungal polysaccharides increased the amount of the antimicrobial alkaloid acridone epoxide up to 100–fold in cultures of *R. graveolens* [132]. *Taxus chinensis* cells treated with an endophytic fungus found in the bark of the *T. chinensis* tree produced three times as much taxol as nonelicited cells [133].

The content of thiarubrine A was enhanced 3–fold in *Ambrosia artemisiifolia* hairy root cultures through the utilization of autoclaved cell wall filtrates from the fungus *Protomyces gravidus*, a pathogen of *Ambrosia artemisiifolia* [134]. The fungal challenged root cultures of *Taverniera cuneifolia*, increased the glycyrrhizic acid content [135]. Moreover, maximum increase in glycyrrhizic acid was noticed in *Mucor hiemalis* treated cultures. In case of *Fusarium moniliforme* and *Aspergillus niger*, threefold increase in glycyrrhizic acid was observed as compared to control unchallenged root culture. However, marginal increase in glycyrrhizic acid content was noticed, in *Penicillium fellutanum* and *Aspergillus tenuis* challenged cultures [135]. Similarly, biotic elicitors from *Claviceps purpurea* were included in *Azadirachta indica* hairy root cultures, leading to a 5–fold increase in the production of azadirachtin [136]. The transformed cell suspension cultures of *W. somnifera* when treated with the dual elicitation of copper sulfate (100 μM) and the cell extract of *Verticilium dahaliae* (5% v/v) showed highest production of withaferin A content when compared with the individual elicitors [137].

6.4. Bacterial origin

The bacterial elicitors stimulated the biosynthesis of scopolamine in adventitious hairy root cultures of *Scopolia parviflora* via the inhibition of H6H (hyoscyamine 6β–hydoxylase) expression [138]. In bacterial elicitation, the maximum glycyrrhizic acid increase was observed in *Rhizobium leguminosarum* challenged culture as compared to unchallenged control roots of *Taverniera cuneifolia* [135]. Furthermore, in *Bacillus aminovorans*, *Agrobacterium rhizogenes*, and *Bacillus cereus* challenged cultures, significant increase in glycyrrhizic acid content was observed. However, root culture challenged by *Agrobacterium tumefaciens* did not show any significant increase in glycyrrhizic acid content [135]. The gradual increase in hypericin and pseudohypericin was observed in seedlings of *H. perforatum* after challenging with Rhizobacterium [139]. Coronatine, phytotoxin produced by the *Pseudomonas syringae* species significantly induced taxane synthesis in taxane media cell cultures [140], also induced the viniferins production in the cell culture of *V. vinifera* [91].

7. Parameters of elicitors

Elicitation has been widely used to increase the production or to induce de novo synthesis of secondary metabolites in in vitro plant cell cultures [141]. This opened up a new area of research that could have important economic benefits for pharmaceutical industry. Several parameters such as elicitor concentration and selectivity, duration of elicitor exposure, age of culture, cell line, growth regulation, nutrient composition, and quality of cell wall materials are also important factors influencing the successful production of secondary metabolite [142]. Some of these parameters were highlighted on elicitation of some medicinal plants for the production of secondary metabolites.

7.1. Elicitor concentration

Elicitor concentration plays a very important role in elicitation process. High dosage of elicitor has been reported to induce hypersensitive response leading to cell death, whereas an optimum level was required for induction [143–145]. At 0.1% (w/v) sodium chloride, ginseng saponin content and productivity were increased to approximately 1.15 and 1.13 times control values, respectively [146]. In the cell culture of *S. miltiorrhiza*, the effects of different concentrations of SA were affected the accumulation of salvianolic acid B and of caffeic acid. The increased accumulation of salvianolic acid B and of caffeic acid was observed in the applications of 3.125–25.0 mg/L of SA at 8 and 96 h when compared to the 32.0–50.0 mg/L of SA. After 96 h treatments with 3.125–25.0 mg/L of SA, the concentration of the phenolic acids decreased drastically compared to the amount 8 h after the treatments but still accumulated the higher concentrations of compound than that of the control [147]. The various concentrations (50, 100, 150, 200, and 250 μM) of MeJA and SA were used in the cell suspension cultures of *Gymnema sylvestre*. The MeJA at 150 μM and SA at 200 μM enhanced the accumulation of gymnemic acid content [148]. In the hairy root culture of *W. somnifera*, the MeJA (15 μM) and SA (150 μM) enhanced the production of withanolide A, withanone, and withaferin A content [95]. In the cell suspension culture of *V. vinifera*, the cobalt at all three used concentrations (5.0, 25, and 50 μM), Ag, and Cd at low concentration (5.0 μM) were most effective to stimulate the phenolic acid production, increasing the 3–O–glucosyl–resveratrol up to 1.6–fold of the control level (250.5 versus 152.4 μmol/g), 4 h after the treatments [67]. In the *A. paniculata* cell culture, MeJA at 5 μM showed 5.25 higher accumulation of andrographolide content compared with control [97]. The root cultures of *Taverniera cuneifolia* treated with increasing concentrations of MeJA (1.0, 2.5, 5, 10, 100, and 1000 μM) [135]. The glycyrrhizic acid content increased gradually with increase in MeJA (1–100 μM) concentration. Approximately 2.5–fold increase in glycyrrhizic acid production was noticed in MeJA (100 μM) treated roots, as compare to the unchallenged root culture. However, further increase in MeJA (1000 μM) concentration resulted in decrease in glycyrrhizic acid production [135].

7.2. Duration of elicitor exposure

The cell suspension culture of *G. sylvestre* was treated with MeJA and SA for 24 h, 48 h, and 72 h. With the MeJA treatment, the maximum gymnemic acid production was recorded 72 h after treatment with 150 μM (135.41 ± 0.43 mg/g DCW). The gymnemic acid content was 15.4–fold higher than the control cultures that were free of the elicitor. When the MeJA concentration exceeded 150 μM, there was a drastic fall (36.3%) in the gymnemic acid accumulation [148]. A high concentration of 200 μM SA was required to induce substantial quantities of gymnemic aid (43.27 ± 0.80 mg/g DCW) in the suspensions that reached a maximum after 48 h treatment. The SA–induced response toward gymnemic acid accumulation resulted in a 4.9–fold increase in comparison to the control cultures [148]. The different biotic elicitors (*A. rhizogenes, Bacillus subtilis, Escherichia coli, Aspergillus niger*, and *Saccharomyces cerevisiae*) required different duration of time (24, 48, 48, 72, and 72 h, respectively) to elicit the gymnimic acid in the cell suspension culture of *G. sylvestre* [149]. The MeJA and SA for 4 h exposure time enhanced the production of withanolide A, withanone and withaferin A content in the hairy root culture of *W. somnifera* [95]. The yields of atropine content in the *Datura metel* hairy roots were increased by nanosilver as an elicitor after 12, 24, and 48 h, but atropine accumulation in *D. metel* hairy roots was reduced by $AgNO_3$, *Bacillus cereus*, and *Staphylococcus* after 12, 24, and 48 h [150]. In the cell culture of *Andrographis paniculata*, the MeJA induced the highest accumulation of andrographolide at 24 h compared with 48 and 72 h of treatments [97]. In the cell system of *V. vinifera*, a rapid accumulation of trans–resveratrol was recorded with MeJA treatment, starting from 2 h and reaching its maximum value at 96 h and the highest levels of viniferins recorded in cell cultures elicited with chitin (chitosan) for 144 h [91]. The MeJA produced the highest amount of bacoside A, 1.5–fold higher than the control shoots in the *B. monnieri* shoot culture after 48 h [96]. MeJA elicitation can cause an initial rapid increase in amount of various secondary metabolites from 24 to 72 h compared to controls after which a subsequent decrease can be found [151].

7.3. Age of culture

Age of culture plays is an important parameter in the production of bioactive compounds by elicitation. The treatment with MeJA and SA in the hairy root culture of *W. somnifera* showed highest accumulation of withanolide A, withanone, and withaferin A content after 40 days of culture [95]. In a study, 20–day–old cell cultures of *C. roseus* showed higher yields of ajmalicine on elicitation. The optimum level of ajmalicine (166 μg/g DW) was observed in 20–day–old cells elicited with extracts of *Trichoderma viride* followed by 90 and 88 μg/g DW ajmalicine in cells elicited with *A. niger* and *F. moniliforme*, respectively [127,152]. A similar type of observation was noticed from various workers Rijhwani and Shanks [153] and Ganapathi and Kargi [142]. The selenium addition at inoculum time did not significantly affect ginseng saponin accumulation. However, the addition of 0.5 mM selenium as an elicitor, after 21 days of culture, ginseng saponin content and productivity increased to about 1.31 and 1.33 times control levels, respectively [146]. The MeJA, at a concentration of 10 μM and 100 μM when introduced to cell suspension of *C. roseus* on day 6 of cell growth increased ajmalicine and serpentine production, respectively, re–elicitation showed a negative effect on both growth and alkaloid synthesis [154].

7.4. Nutrient composition

The composition of the medium or selection of medium also played a vital role in elicitation process. In the callus culture of *Erythroxylum coca*, the amounts of cocaine, cinnamoylcocaine, chlorogenic acid (CGA), and 4–coumaroyl quinate (CQA) were significantly affected by the culture medium [155]. Cocaine production was nearly an order of magnitude greater on Anderson rhododendron medium (ARM) [156], Gamborg B5 (GB5) [157], and Murashige–Tucker medium (MMT) [158], but the amounts produced on MMT and GB5 were not significantly different from each other. Cinnamoylcocaine was affected in the same way. The major factor controlling tropane alkaloid (TA) accumulation was medium composition, with cocaine levels on ARM being nearly an order of magnitude greater than on the other media. Many nutrients, including cobalt, copper, molybdenum, calcium, magnesium, iron, boron, iodine, manganese, zinc, and myo–inositol, and the growth regulators and the ammonium:nitrate ratio are at equivalent levels in ARM as in one of the other media, and can therefore be excluded as factors promoting TA accumulation. However, a number of factors differed between ARM and the other media, and might be responsible for the elevated TA content. Total ion concentration is lower in ARM, and could be an important factor given the importance of salt content in controlling secondary metabolism [155]. Nitrate concentration was also lower in ARM, and there are numerous reports in the literature of an inverse relationship between nitrate availability and accumulation of secondary metabolites in many plant species, including *Arabidopsis thaliana* [159], *Hordeum vulgare* [160], and *Nicotiana tabacum* [161]. Similarly, the reduction of nitrate concentration in the culture medium of *Atropa belladonna* hairy roots increases alkaloid content [162]. In regards to CGA, the media were all significantly different from each other with the lowest production on ARM and highest on MMT. Less CQA was produced on ARM than on either of the other two media, which did not differ from each other [155]. Apart from these characteristics, the efficiency of elicitation also depends on elicitor specificity, cell line or clones of microbial elicitor used the presence of growth regulators, nutrient composition of the medium, and the environmental conditions.

8. Conclusion

The development of plant tissue cultures for the production of secondary metabolites has been underway for more than three decades. Although there are well–established plant tissue culture techniques, their application to large scale production is still limited to a few processes. Various stimulation and process strategies have been exercised to improve secondary metabolite production in plant tissue cultures. Elicitation has been widely applied for enhancement of secondary metabolite production in plant cell and organ cultures. The effects of various abiotic and biotic elicitors on secondary metabolite production in plant tissue cultures are dependent on the specific secondary metabolites. The exploration of the production of useful secondary metabolites through regulation of biosynthetic pathway of the various plant cell and tissue cultures of medicinal plants has been carried out by a group of plant scientists in several countries during the last decade. Although, elicitation enhances secondary metabolism in plant cells in vitro, but the exact mechanism is not exactly understood. There is a tremendous scope for the large–scale production of secondary metabolites in the plant tissue culture system by using the elicitors as an agent.

Author details

Poornananda M. Naik and Jameel M. Al–Khayri*

*Address all correspondence to: jmkhayri@yahoo.com; jkhayri@kfu.edu.sa

Department of Agricultural Biotechnology, College of Agriculture and Food Sciences, King Faisal University, Al–Hassa, Saudi Arabia

References

[1] Wink M. Plant breeding: importance of plant secondary metabolites for protection against pathogens and herbivores. Theoretical and Applied Genetics. 1988;75:225–233.

[2] Verpoorte R, Contin A, Memelink J. Biotechnology for the production of plant secondary metabolites. Phytochemistry Reviews. 2002;1:13–25.

[3] Murthy HN, Lee EJ, Paek KY. Production of secondary metabolites from cell and organ cultures: strategies and approaches for biomass improvement and metabolite accumulation. Plant Cell Tissue and Organ Culture. 2014;118:1–16.

[4] Dixon RA. Natural products and plant disease resistance. Nature. 2001;411:843–847.

[5] Oksman–Caldentey1 KM, Inze D. Plant cell factories in the post–genomic era: new ways to produce designer secondary metabolites. Trends in Plant Science. 2004;9:9.

[6] Mosihuzzaman M. Herbal medicine in healthcare—an overview. Natural Product Communications. 2012;7:807–812.

[7] Cordell GA. Biodiversity and drug discovery—a symbiotic relationship. Phytochemistry. 2000;55:463–480.

[8] Schmidt BM, Ribnicky DM, Lipsky PE, Raskin I. Revisiting the ancient concept of botanical therapeutics. Nature Chemical Biology. 2007;3:360–366.

[9] Shaw D, Graeme L, Pierre D, Elizabeth W, Kelvin C. Pharmacovigilance of herbal medicine. Journal of Ethnopharmacology. 2012;140:513–518.

[10] Cox PA, Balick MJ. The ethanobotanical approach to drug discovery. Scientific American. 1994;270(6): 82–87.

[11] Pezzuto JM. Natural product cancer chemoprotective agents. In: Arnason JT, Mata R, Romeo JT, editors. Recent advances in phytochemistry. Phytochemistry of Medicinal Plants. New York: Plenum.1995. p. 19–45.

[12] Vanisree M, Lee CY, Lo SF, Nalawade SM, Lin CY, Tsay HS. Studies on the production of some important secondary metabolites from medicinal plants by plant tissue cultures. Botanical Bulletin of Academia Sinica. 2004;45:1–22.

[13] Anis M, Husain MK, Faisal M, Shahzad A, Ahmad N, Siddique I, Khan H, In: Kumar A, Sopory SK. editors. Recent Advances in Plant Biotechnology and its Application, New Delhi: IK International Pvt. Ltd; 2009. p. 397–410.

[14] Anis M, Husain MK, Siddique I, Varshney A, Naz R, Perveen S, Khan MI, Ahmed MR, Husain MK, Khan PR, Aref IM. Biotechnological approaches for the conservation of forestry species. In: Jenkins JA. editor. Forest Decline: Causes and Impacts. USA: Nova Science Publishers Inc. 2011. p. 1–39.

[15] Skrzypczak–Pietraszek E, Słota J, Pietraszek J. The influence of L–phenylalanine, methyl jasmonate and sucrose concentration on the accumulation of phenolic acids in *Exacum affine* Balf. f. ex Regel shoot culture. Acta Biochimica Polonica. 2014;61(1): 47–53.

[16] Rao RS, Ravishankar GA. Plant cell cultures: chemical factories of secondary metabolites, Biotechnology Advances. 2002;20:101–153.

[17] Sarin R. Useful metabolites from plant tissue cultures. Biotechnology. 2005;4:79–93.

[18] Radman R, Saez T, Bucke C, Keshavarz T. Elicitation of plant and microbial systems. Biotechnology and Applied Biochemistry. 2003;37:91–102.

[19] Poulev A, O'Neal JM, Logendra S, Pouleva RB, Timeva V, Garvey AS, Gleba D, Jenkins IS, Halpern BT, Kneer R, Cragg GM, Raskin I. Elicitation, a new window into plant chemodiversity and phytochemical drug discovery. Journal of Medicinal Chemistry. 2003;46: 2542–2547.

[20] Angelova Z, Georgiev S, Roos W. Elicitation of plants. Biotechnology and Biotechnological Equipment. 2006;20:72–83.

[21] Namdeo AG. Plant cell elicitation for production of secondary metabolites: a review. Pharmacognosy Reviews. 2007;1:69–79.

[22] Anasori P, Asghari G. Effects of light and differentiation on gingerol and zingiberene production in callus culture of *Zingiber officinale* Rosc. Research in Pharmaceutical Sciences. 2008;3:59–63.

[23] Zhong JJT, Seki SI, Kinoshita, Yoshida T. Effect of light irradiation on anthocyanin production by suspended culture of *Perilla frutescens*. Biotechnology and Bioengineering. 1993;38:653–658.

[24] Hagimori M, Matsumoto T, Obi Y. Studies on the production of *Digitalis cardenolides* by plant tissue culture III. Effects of nutrients on digitoxin formation by shoot–forming cultures of *Digitalis purpurea* L. grown in liquid media. Plant and Cell Physiology. 1982;23:1205–1211.

[25] Liu CZ, Guo C, Wang Y, Ouyang F. Effect of light irradiation on hairy root growth and artemisinin biosynthesis of *Artemisia annua* L. Process Biochemistry. 2002;38:581–585.

[26] Fett–Neto AG, Pennington JJ, Di Cosmo F. Effect of white light on taxol and baccatin III accumulation in cell cultures of *Taxus cuspidata* Sieb and Zucc. Journal of Plant Physiology. 1995;146:584–590.

[27] Kumari RSB, Agrawal S, Singh NKD. Supplemental ultraviolet–B induced changes in essential oil composition and total phenolics of *Acorus calamus* L. (sweet flag). Ecotoxicology and Environmental Safety. 2009;72:2013–2019.

[28] Rozema J, Bjorn LO, Bornman JF, Gaberscik A, Ha°der DP, Trost T, Germ M, Klisch M, Gro°niger A, Sinha RP, Lebert M, He YY, Buffoni–Hall R, Bakker NVJ, de Staaij J, van de Meijkamp BB. The role of UV–B radiation in aquatic and terrestrial ecosystems—an experimental and functional analysis of the evolution of UV–absorbing compound. Journal of Photochemistry and Photobiology B. 2002;66:2–12.

[29] Bernard YKB, Christie AMP, Jacqueline VS, Ka–Yiu S. The effects of UV–B stress on the production of terpenoid indole alkaloids in *Catharanthus roseus* hairy roots. Biotechnology Progress, 2009;25:8615.

[30] Dixon RA, Paiva NL. Stress–induced phenylpropanoid metabolism. Plant Cell. 1995;7:1085–1097.

[31] Wang LJ, Ma L, Xi HF, Duan W, Wang JF, Li SH. Individual and combined effects of $CaCl_2$ and UV–C on the biosynthesis of resveratrols in grape leaves and berry skins. Journal of Agricultural and Food Chemistry. 2013;61:7135–7141.

[32] Wang W, Tang K, Yang HR, Wen PF, Zhang P, Wang HL, Huang WD. Distribution of resveratrol and stilbene synthase in young grape plants (*Vitis vinifera* L. cv. Cabernet Sauvignon) and the effect of UV–C on its accumulation. Plant Physiology and Biochemistry. 2010;48:142–152.

[33] Liu W, Liu CY, Yang CX, Wang LJ, Li SH. Effect of grape genotype and tissue type on callus growth and production of resveratrols and their piceids after UV–C irradiation. Food Chemistry. 2010;122:475–481.

[34] Xu A, Zhan JC, Huang WD. Effects of ultraviolet C, methyl jasmonate and salicylic acid, alone or in combination, on stilbene biosynthesis in cell suspension cultures of *Vitis vinifera* L. cv. Cabernet Sauvignon plant cell tissue and organ culture. 2015;122:197–211.

[35] Vasconsuelo A, Boland R. Molecular aspects of the early stages of elicitation of secondary metabolites in plants. Plant Science. 2007;172(5):861–875.

[36] Zobayed SMA, Afreen F, Kozai T. Phytochemical and physiological changes in the leaves of St. John's wort plants under a water stress condition. Environmental and Experimental Botany. 2007;59:109–116.

[37] Verbruggen N, Hermans C. Proline accumulation in plants: a review. Amino Acids. 2008;35:753–759.

[38] Xu J, Yin HX, Li X. Protective effects of proline against cadmium toxicity in micropropagated hyperaccumulator *Solanum nigrum* L. Plant Cell Reports. 2009;28:325–333.

[39] Van den Berg L, Zeng YJ. Response of South African indigenous grass species to drought stress induced by polyethylene glycol (PEG) 6000. The South African Journal of Botany. 2006;72:284–286.

[40] Pratibha G, Satyawati S, Sanjay S. Biomass yield and steviol glycoside production in callus and suspension culture of *Stevia rebaudiana* treated with proline and polyethylene glycol. Applied Biochemistry and Biotechnology. 2015; DOI: 10.1007/s12010–015–1616–0.

[41] Omer Y, Bengi E. Effects of sucrose and polyethylene glycol on hypericins content in *Hypericum adenotrichum.* Eurasian Journal of Biosciences. 2013;7:101–110.

[42] Liu CZ, Cheng XY. Enhancement of phenylethanoid glycosides biosynthesis in cell cultures of *Cistanche deserticola* by osmotic stress. Plant Cell Reports. 2008;27: 357–362.

[43] Pavlik M, Vacek J, Klejdus B, Kuban V. Hypericin and hyperforin production in St. John's wort *in vitro* culture: influence of saccharose, polyethylene glycol, methyl jasmonate, and *Agrobacterium tumefaciens*. Journal of Agricultural and Food Chemistry. 2007;55:6147–6153.

[44] Bernstein N. Effects of salinity on root growth. In: Eshel A, Beeckman T. editors. Plant Roots: The Hidden Half, 4th ed. Boca Raton: CRC Press; 2013. p. 784.

[45] Munns R, Tester T. Mechanisms of salinity tolerance. Annual Review of Plant Biology. 2008;59:651–681.

[46] Haghighi Z, Karimi N, Modarresi M, Mollayi, S, Enhancement of compatible solute and secondary metabolites production in *Plantago ovata* Forsk. by salinity stress. Journal of Medicinal Plants Research. 2012;6:3495–3500.

[47] Selmar D. Potential of salt and drought stress to increase pharmaceutical significant secondary compounds in plants. Agriculture and Forestry Research. 2008;58:139–144.

[48] Winkel–Shirley B. Biosynthesis of flavonoids and effects of stress. Current Opinion in Plant Biology. 2002;5:218–223.

[49] Misra N, Gupta AK. Effect of salinity and different nitrogen sources on the activity of antioxidant enzymes and indole alkaloid content in *Catharanthus roseus* seedlings. Journal of Plant Physiology. 2006;163:11–18.

[50] Kennedy BF, De Filippis LF. Physiological and oxidative response to NaCl of the salt tolerant Grevillea ilicifolia and the salt sensitive *Grevillea arenaria*. Journal of Plant Physiology. 1999; 155:746–754.

[51] Daneshmand F, Arvin MJ, Kalantari KM. Physiological responses to NaCl stress in three wild species of potato in vitro. Acta Physiologiae Plantarum. 2010;32:91–101.

[52] Brachet J, Cosson L. Changes in the total alkaloid content of *Datura innoxia* Mill. Subjected to salt stress. Journal of Experimental Botany. 1986;37:650–656.

[53] Krishnamurthy R, Bhagwat KA. Polyamines as modulators of salt tolerance in rice cultivars. Plant Physiology. 1989;91:500–504.

[54] Varshney KA, Gangwar LP. Choline and betaine accumulation in *Trifolium alexandrinum* L. during salt stress. Egyptian Journal of Botany. 1988;31:81–86.

[55] Fatima S, Mujib A, Dipti T. NaCl amendment improves vinblastine and vincristine synthesis in *Catharanthus roseus*: a case of stress signalling as evidenced by antioxidant enzymes activities Plant Cell Tissue and Organ Culture. 2015;121:445–458.

[56] Xu Z, Zhou G, Shimizu H. Plant responses to drought and rewatering. Plant Signaling and Behavior. 2010;5:649–54. PMID:20404516. DOI: 10.4161/psb.5.6.11398.

[57] Zhu Z, Liang Z, Han R, Wang X. Impact on fertilization on drought response in the medicinal herb *Bupleurum chinense* DC: growth and saikosaponin production. Industrial Crops and Products. 2009;29: 629–633.

[58] Liu H, Wang X, Wang D, Zou, Z, Liang Z, Effect of drought stress on growth and accumulation of active constituents in *Salvia miltiorrhiza* Bunge. Industrial Crops and Products. 2011;33:84–88

[59] Chen Y, Guo Q, Liu L, Liao L, Zhu Z. Influence of fertilization and drought stress on the growth and production of secondary metabolites in *Prunella vulgaris* L. Journal of Medicinal Plants Research 2011;5:1749–1755.

[60] Li W, Hou J, Wang W, Tang X, Liu C, Xing D, Effect of water deficit on biomass production and accumulation of secondary metabolites in roots of *Glycyrrhiza uralensis*. Russian Journal of Plant Physiology. 2011;58:538–542.

[61] de Abreu, IN, Mazzafera P. Effect of water and temperature stress on the content of active constituents of *Hypericum brasiliense* Choisy. Plant Physiology and Biochemistry. 2005;43:241–248.

[62] Morison JIL, Lawlor DW. Interactions between increasing CO_2 concentration and temperature on plant growth. Plant Cell and Environment. 1999;22:659–682.

[63] Jochum GM, Mudge KW, Thomas RB. Elevated temperatures increase leaf senescence and root secondary metabolite concentration in the understory herb *Panax quinquefolius* (Araliaceae). American Journal of Botany. 2007;94:819–826. PMID: 21636451.

[64] Yu K, Niranjana MH, Hahn E, Paek K. Ginsenoside production by hairy root cultures of *Panax ginseng*: influence of temperature and light quality. Biochemical Engineering Journal. 2005;23:53–56.

[65] Chan LK, Koay SS, Boey PL, Bhatt A. Effects of abiotic stress on biomass and anthocyanin production in cell cultures of *Melastoma malabathricum*. Biological Research 2010;43:127–135. PMID:21157639.

[66] Zobayed SMA, Afreen F, Kozai T. Temperature stress can alter the photosynthetic efficiency and secondary metabolite concentrations in St. John's wort. Plant Physiology and Biochemistry. 2005;43:977–984.

[67] Cai Z, Kastell A, Speiser C, Smetanska I. Enhanced resveratrol production in *Vitis vinifera* cell suspension cultures by heavy metals without loss of cell viability. Applied Biochemistry and Biotechnology. 2013;171:330–340.

[68] Nasim SA, Dhir B. 2010 Heavy metals alter the potency of medicinal plants. Reviews of Environmental Contamination and Toxicology. 2010;203:139–149.

[69] Zhao J, Zhu W, Hu Q. Selection of fungal elicitors to increase indole alkaloid accumulation in *Catharanthus roseus* suspension cell culture. Enzyme and Microbial Technology. 2001;28:666–672.

[70] Singh S, Sinha S. Accumulation of metals and its effects in *Brassica juncea* (L.) *Czern.* (cv. *Rohini*) grown on various amendments of tannery waste. Ecotoxicology and Environmental safety. 2005;62:118 27. PMID:15978297.

[71] Mizukami H, Konoshima M, Tabata M. Effect of nutritional factors on shikonin derivative formation in *Lithospermum callus* cultures. Phytochemistry. 1977;16:1183–1186.

[72] Ohlsson AB, Berglund T. Effect of high $MnSO_4$ levels on cardenolide accumulation by *Digitalis lanata* tissue cultures in light and darkness. Journa of Plant Physiology. 1989;135:505–507.

[73] Trejo–Tapia G, Jimenez–Aparicio A, Rodriguez–Monroy M, De Jesus–Sanchez A, Gutierrez–Lopez G. Influence of cobalt and other microelements on the production of betalains and the growth of suspension cultures of *Beta vulgaris*. Plant Cell Tissue and Organ Culture. 2001;67:19–23.

[74] Thimmaraju BN, Ravishankar GA. In situ and ex situ adsorption and recovery of betalains from hairy root cultures of *Beta vulgaris*. Biotechnology Progress. 2004;20:777-85. PMID:15176882. DOI: 10.1021/bp0300570.

[75] Obrenovic S. Effect of Cu (11) D–penicillanine on phytochrome mediated betacyanin formation in *Amaranthus caudatus* seedlings. Plant Physiology and Biochemistry. 1990;28:639–646.

[76] Saba PD, Iqbal M, Srivastava PS. Effect of $ZnSO_4$ and $CuSO_4$ on regeneration and lepidine content in *Lepidium sativum*. Biologia Plantarum. 2000;43:253–256.

[77] Pitta–Alvarez SI, Spollansky TC, Giullietti AM. The influence of different biotic and abiotic elicitors on the production and profile of tropane alkaloids in hairy root cul-

tures of *Brugmansia candida*. Enzyme and Microbial Technology. 2000;26:252–258. PMID:10689085.

[78] Arehzoo Z, Christina S, Florian G, Parvaneh A, Javad A, Seyed H M, Christoph W. Effects of some elicitors on tanshinone production in adventitious root cultures of *Perovskia abrotanoides* Karel. Industrial Crops and Products, 2015;67:97–102.

[79] Furze JM, Rhodes MJC, Parr AJ, Robins RJ, Withehead IM., Threlfall DR. Abiotic factors elicit sesquiterpenoid phytoalexin production but not alkaloid production in transformed root cultures of *Datura stramonium*. Plant Cell Reports. 1991;10(3):111–114.

[80] Creelman RA, Mullet JE. Biosynthesis and action of jasmonates in plants. Annual Review of Plant Biology. 1997;48(1):355–381.

[81] Sembdner G, Parthier B. The biochemistry and the physiological and molecular actions of jasmonates. Annual Review of Plant Biology. 1993;44(1):569–589.

[82] Pauwels L, Inze´ D, Goossens A. Jasmonate–inducible gene: what does it mean? Trends in Plant Science. 2009;14:87–91.

[83] Wasternack C, Parthier B. Jasmonate–signalled plant gene expression. Trends in Plant Science. 1997;2:302–307.

[84] Krzyzanowska J, Czubacka A, Pecio L, Przybys M, Doroszewska T, Stochmal A, Oleszek W. The effects of jasmonic acid and methyl jasmonate on rosmarinic acid production in *Mentha piperita* cell suspension cultures. Plant Cell Tissue and Organ Culture. 2012:108;73–81.

[85] Almagro L, Gutierrez J, Pedren˜o MA, Sottomayor M. Synergistic and additive influence of cyclodextrins and methyl jasmonate on the expression of the terpenoid indole alkaloid pathway genes and metabolites in *Catharanthus roseus* cell cultures. Plant Cell Tissue and Organ Culture. 2014;119:543–551.

[86] Silja PK, Gisha GP, Satheeshkumar K. Enhanced plumbagin accumulation in embryogenic cell suspension cultures of *Plumbago rosea* L. following elicitation. Plant Cell Tissue and Organ Culture. 2014;119:469–477.

[87] Curtin C, Zhang W, Franco C. Manipulating anthocyanin composition in *Vitis vinifera* suspension cultures by elicitation with jasmonic acid and light irradiation. Biotechnology Letters. 2003;25:1131–1135.

[88] Gangopadhyay M, Dewanjee S, Bhattacharya S. Enhanced plumbagin production in elicited *Plumbago indica* hairy root cultures. Journal of Bioscience and Bioengineering 2011;111:706–710.

[89] Belhadj A, Saigne C, Telef N, Cluzet S, Bouscaut J, Corio–Costet MF, Me´rillon JM. Methyl jasmonate induces defense responsesin grapevine and triggers protection

against Erysiphe necator. Journal of Agricultural and Food Chemistry. 2006;54:9119–9125.

[90] Tassoni A, Fornale S, Franceschetti M, Musiani F, Michael AJ, Perry B, Bagni N. Jasmonates and Na–orthovanadate promote resveratrol production in *Vitis vinifera* cv. Barbera cell cultures. New Phytologist. 2005;166:895–905.

[91] Taurino M, Ingrosso I, D'amico L, Domenico1 S. D, Nicoletti I, Corradini D, Santino A, Giovinazzo G. Jasmonates elicit different sets of stilbenes in *Vitis vinifera* cv. Negramaro cell cultures. SpringerPlus. 2015;4:49.

[92] Nopo–Olazabal C, Condori J, Nopo–Olazabal L, Medina–Bolivar F. Differential induction of antioxidant stilbenoids in hairy roots of *Vitis rotundifolia* treated with methyl jasmonate and hydrogen peroxide. Plant Physiology and Biochemistry. 2014;74:50–69.

[93] Tassoni A, Durante L, Ferri M. Combined elicitation of methyl–jasmonate and red light on stilbene and anthocyanin biosynthesis. Journal of Plant Physiology. 2012;169:775–781.

[94] Xiaolong H, Min S, Lijie C, Chao X, Yanjie Z, Guoyin K. Effects of methyl jasmonate and salicylic acid on tanshinone production and biosynthetic gene expression in transgenic *Salvia miltiorrhiza* hairy roots. Biotechnology and Applied Biochemistry. 2015;62(1):24–31. DOI: 10.1002/bab.1236.

[95] Sivanandhan G, Dev G, K, Jeyaraj M, Rajesh M, Arjunan A, Muthuselvam M, Manickavasagam M, Selvaraj N, Ganapathi A. Increased production of withanolide A, withanone, and withaferin A in hairy root cultures of *Withania somnifera* (L.) Dunal elicited with methyl jasmonate and salicylic acid. Plant Cell Tissue and Organ Culture. 2013;114:121–129.

[96] Sharma P, Yadav S, Srivastava A, Shrivastava N. Methyl jasmonate mediates upregulation of bacoside A production in shoot cultures of *Bacopa monnieri*. Biotechnology Letters. 2013;35:1121–1125.

[97] Sharma SN, Jhaa Z, Sinhab RK, and Gedac AK. Jasmonate–induced biosynthesis of andrographolide in *Andrographis paniculata*. Physiologia Plantarum. 2015;153: 221–229.

[98] Hayashi H, Huang P, Inoue K, Up–regulation of soyasaponin biosynthesis by methyl jasmonate in cultured cells of *Glycyrrhiza glabra*. Plant and Cell Physiology. 2003;44:404–411.

[99] Ketchum REB, Tandon M, Gibson DM, Begley T, Shuler ML. Isolation of labeled 9–dihydrobaccatin III and related taxoids from cell cultures of taxus cell cultures of *Taxus canadensis* elicited with methyl jasmonate. Journal of Natural Products. 1999;62:1395–1398.

[100] Pedapudi S, Chin CK, Pedersen H. Production and elicitation of benzalacetone and the raspberry ketone in cell suspension cultures of *Rubus idaeus*. Biotechnology Progress. 2000;16:346–349.

[101] Hayat Q, Hayat S, Irfan M, Ahmad A. Effect of exogenous salicylic acid under changing environment: a review. Environmental and Experimental Botany. 2010;68:14–25.

[102] Pieterse CMJ, van Loon LC. Salicylic acid–independent plant defense pathways. Trends in Plant Science. 1999;4:52–58.

[103] Idrees M, Naeem M, Aftab T, Khan MM. Salicylic acid mitigates salinity stress by improving antioxidant defence system and enhances vincristine and vinblastine alkaloids production in periwinkle [*Catharanthus roseus* (L.) G. Don]. Acta Physiologia Plantarum. 2010;33:987–999.

[104] Avancini G, Abreu IN, Saldan˜a MDA, Mohamed RS, Mazzafera P. 2003 Induction of pilocarpine formation in jaborandi leaves by salicylic acid and methyl jasmonate. Phytochemistry. 2003;63:171–175.

[105] Bulgakov VP, Tchernoded GK, Mischenko NP, Khodakovskaya MV, Glazunov VP, Radchenko SV, Zvereva EV, Fedoreyev SA, Zhuravlev YN. Effect of salicylic acid, methyl jasmonate, ethephon and cantharidin on anthraquinone production by *Rubia cordifolia* callus cultures transformed with the rolB and rolC genes. Journal of Biotechnology. 2002;97:213–221.

[106] Ali M, Yu KW, Hahn EJ, Paek KY. Methyl jasmonate and salicylic acid elicitation induces ginsenosides accumulation, enzymatic and non–enzymatic antioxidant in suspension culture *Panax ginseng* roots in bioreactors. Plant Cell Reports. 2006;25:613–620.

[107] Shabani L, Ehsanpour A, Asghari G, Emami J. Glycyrrhizin production by in vitro cultured *Glycyrrhiza glabra* elicited by methyl Jasmonate and salicylic acid. Russian Journal of Plant Physiology. 2009;56:621–626.

[108] Xu YW, Lv SS, Zhao D, Chen JW, Yang WT, Wu W. Effects of salicylic acid on monoterpene production and antioxidant systems in *Houttuynia cordata*. African Journal of Biotechnology. 2012;11:1364–1372.

[109] Liang Z, Ma Y, Xu T, Cui B, Liu Y, Guo Z, Yang D. Effects of abscisic acid, gibberellin, ethylene and their interactions on production of phenolic acids in *Salvia miltiorrhiza* bunge hairy roots. PLoS One. 2013;8(9):e72806.

[110] Hu X, Neill S, Cai W, Tang Z. Hydrogen peroxide and jasmonic acid mediate oligogalacturonic acid–induced saponin accumulation in suspension–cultured cells of *Panax ginseng*. Physiologia Plantarum. 2003;118:414–421.

[111] Fukui H, Yoshikawa N, Tabata M. Induction of shikonin formation by agar in *Lithospermum erythrorhizon* cell suspension cultures. Phytochemistry. 1983;22:2451–2453.

[112] Komaraiah P, Ramakrishna SV, Reddanna P, Kavi Kishor PB. Enhanced production of plumbagin in immobilized cells of *Plumbago rosea* by elicitation and in situ adsorption. Journal of Biotechnology. 2003;101:181–187.

[113] Orlita A, Sidwa–Gorycka M, Paszkiewicz M, Malinski E, Kumirska J, Siedlecka EM, Łojkowska E, Stepnowski P. Application of chitin and chitosan as elicitors of coumarins and fluoroquinolone alkaloids in *Ruta graveolens* L. (common rue). Biotechnology and Applied Biochemistry. 2008;51:91–96.

[114] Sonja GS, Oliver T, Stéphane M, Alain D, Eric L, Claude J, and Daniel H. Polysaccharide elicitors enhance phenylpropanoid and naphtodianthrone production in cell suspension cultures of *Hypericum perforatum*. Plant Cell Tissue and Organ Culture. 2015; DOI: 10.1007/s11240–015–0798–z.

[115] Felix G, Grosskopf DG, Regenass M, Basse CW, Boller T. Elicitor–induced ethylene biosynthesis in tomato cells: characterization and use as a bioassay for elicitor action. Plant Physiology. 1991;97:19–25.

[116] Stangarlin JR, Kuhn OJ, Assi L, Schwan–Estrada KRF. Control of plant diseases using extracts from medicinal plants and fungi. In: Mendez–Vilas A. editor. Science Against Microbial Pathogens: Communicating Current Research and Technological Advances. Formatex Research Center, Badajoz, Spain; 2011. p. 1033–1042.

[117] Dixon RA, Achnine L, Kota P, Liu CJ, Reddy MKS, Wang L. The phenylpropanoid pathway and plant defence—a genomics perspective. Molecular Plant Pathology. 2002;3:371–390.

[118] Lattanzio V, Lattanzio VM, Cardinali A. Role of phenolics in the resistance mechanisms of plants against fungal pathogens and insects. Phytochemistry Advances in Research. 2006;661:23–67.

[119] Glazebrook J. Contrasting mechanisms of defense against biotrophic and necrotrophic pathogens. Annual Review of Phytopathology. 2005;43:205–227.

[120] Leonard KJ, Bushnell WR. Fusarium Head Blight of Wheat and Barley. APS Press, USA; 2004.

[121] Boerema GH, de Gruyter J, Noordeloos ME, Hamers MEC. Phoma Identification Manual. Differentiation of Specific and Intra–pecific Taxa in Culture. CABI Publishing, Wallingford; 2004.

[122] Gill SS, Tuteja N. Reactive oxygen species and antioxidant machinery in abiotic stress tolerance in crop plants. Plant Physiology and Biochemistry. 2010;48:909–930.

[123] Hano C, Addi M, Bensaddek L, Cronier D, Baltora–Rosset S, Doussot J, Maury S, Mesnard F, Chabbert B, Hawkins S, Laine´ E, Lamblin F. Differential accumulation of monolignol–derived compounds in elicited flax (*Linum usitatissimum*) cell suspension cultures. Planta. 2006;223:975–989.

[124] Keen NT. Specific elicitors of plant phytoalex in production: determinants of race specificity in pathogens? Science. 1975;187:74–75.

[125] Bostock RM, Kuc JA, Laine RA. Eicosapentaenoic and arachidonic acids from *Phytophthora infestans* elicit fungitoxic sesquiterpenes in the potato. Science. 1981;212(4490):67–69.

[126] Sang MK, Kim JG, Kim KD. Biocontrol activity and induction of systemic resistance in pepper by compost water extracts against *Phytophthora capsici*. Phytopathology. 2010;100:774–783.

[127] Namdeo A, Patil S, Fulzele DP. Influence of fungal elicitors on production of ajmalicine by cell cultures of *Catharanthus roseus*. Biotechnology Progress 2002;18:159–162.

[128] Parchmann S, Gundlach H, Mueller MJ. Induction of 12–oxo–phytodienoic acid in wounded plants and elicited plant cell cultures. Plant Physiology. 1997;115:1057–1064.

[129] Rokem JS, Schwarzberg J, Goldberg I. Autoclaved fungalmycelia increase diosgenin production in cell suspension cultures of *Dioscorea deltoidea*. Plant Cell Reports. 1984;3:159–160.

[130] Balazova A, Bilka F, Blana′rikova′ V, Psena′k M. Effect of a fungal elicitor on levels of sanguinarine and polyphenoloxidase activity in a suspension culture of *Papaver somniferum* L. Ceská a Slovenská Farmacie. 2002;51:182–185.

[131] Heinstein PF. Future approaches to the formation of secondary natural products in plant cell suspension cultures. Journal of Natural Products. 1985;48:1–9.

[132] Eilert U, Ehmke A, Wolters B. Induced accumulation of acridone alkaloid epoxides in *Ruta graveolens* suspension cultures. Planta Medica. 1984;50:508–512.

[133] Wang C, Wu J, Mei X. Enhancement of taxol production and excretion in *Taxus chinensis* cell culture by fungal elicitation and medium renewal. Applied Microbiology and Biotechnology. 2001;55:404–410.

[134] Bhagwath SG, Hjortsø MA. Statistical analysis of elicitation strategies for thiarubrine A production in hairy root cultures of *Ambrosia artemisiifolia*. Journal of Biotechnology. 2000;80: 159–167.

[135] Awad V, Kuvalekar A, Harsulkar A. Microbial elicitation in root cultures of *Taverniera cuneifolia* (Roth)Arn. for elevated glycyrrhizic acid production. Industrial Crops and Products. 2014;54:13–16.

[136] Satdive RK, Fulzele DP, Eapen S. Enhanced production of azadirachtin by hairy root cultures of *Azadirachta indica* A. Juss by elicitation and media optimization. Journal of Biotechnology. 2007;128:281–289.

[137] Baldi A, Singh D, Dixit VK. Dual elicitation for improved production of withaferin A by cell suspension cultures of *Withania somnifera*. Applied Biochemistry and Biotechnology. 2008;151:556–564.

[138] Jung HY, Kang SM, Kang YM, Kang MJ, Yun DJ, Bahk JD, Yang JK, Choi MS. Enhanced production of scopolamine by bacterial elicitors in adventitious hairy root cultures of *Scopolia parviflora*. Enzyme and Microbial Technology. 2003;33:987–990.

[139] Manero FJ, Algar E, Martin Gomez MS, Saco Sierra MD, Solano BR. Elicitation of secondary metabolism in *Hypericum perforatum* by rhizospherebacteria and derived elicitors in seedlings and shoot cultures. Pharmaceutical Biology. 2012;50:1201–1209.

[140] Onrubia M, Moyano E, Bonfill M, Cusido′ RM, Goossens A, Palazo′n J. Coronatine, a more powerful elicitor for inducing taxane biosynthesis in Taxus media cell cultures than methyl jasmonate. Journal of Plant Physiology. 2013;170:211–219.

[141] Dicosmo F, Misawa M. Eliciting secondary metabolism in plant cell cultures. Trends in Biotechnology. 1985;3:318–322.

[142] Ganapathi G, Kargi F. Recent advances in indole alkaloid production by *Catharanthus roseus* (Periwinkle). Journal of Experimental Botany. 1990;41:259–267.

[143] Collinge DB, Susarenka AJ. Plant gene expression in response to pathogens. Plant Molecular Biology. 1987;9:389–410.

[144] Mukandan U, Hjorosto MA. Effect of fungal elicitor on thiophene production in hairy root cultures of *Tagetes patula*. Applied Microbiology and Biotechnology. 1990;33:145–147.

[145] Roewer IA, Cloutier N, Van der Heijden R. Transient induction of tryptophan decarboxylase (TDC) and strictosidine synthase, (SS) genes in cell suspension cultures of *Catharanthus roseus*. Plant Cell Reports. 1992;11(2): 86–89.

[146] Jeong GA, Park DH. Enhanced secondary metabolite biosynthesis by elicitation in transformed plant root system. Applied Biochemistry and Biotechnology. 2007;130:436–446.

[147] Dong J, Wan G, Liang Z. Accumulation of salicylic acid–induced phenolic compounds and raised activities of secondary metabolic and antioxidative enzymes in *Salvia miltiorrhiza* cell culture. Journal of Biotechnology. 2010;148:99–104.

[148] Chodisetti B, Rao K, Gandi S, Giri A. Gymnemic acid enhancement in the suspension cultures of *Gymnema sylvestre* by using the signaling molecules—methyl jasmonate and salicylic acid. In Vitro Cellular and Developmental Biology—Plant. 2015;51:88–92.

[149] Chodisetti B, Rao K, Gandi S, Giri A. Improved gymnemic acid production in the suspension cultures of *Gymnema sylvestre* through biotic elicitation. Plant Biotechnology Reports. 2013;7:519–525.

[150] Zahra S, Mehrnaz K, Gholamreza A, Mustafa G. Improvement of atropine production by different biotic and abiotic elicitors in hairy root cultures of *Datura metel.* Turkish Journal of Biology. 2015;39:111–118.

[151] Wungsintaweekul J, Choo–malee J, Charoonratana T, Niwat N. Methyl jasmonate and yeast extract stimulate mitragynine production in *Mitragyna speciosa* (Roxb.) Korth shoot culture. Biotechnology Letters. 2012;34:1945–1950.

[152] Namdeo AG. Investigation on pilot scale bioreactor with reference to the synthesis of bioactive compounds from cell suspension cultures of *Catharanthus roseus* Linn. [thesis]. Indore: Devi Ahilya Vishwavidyalaya; 2004.

[153] Rijhwani SK, Shanks JV. Effect of subculture cycle on growth and indole alkaloid production by *Catharanthus roseus* hairy root cultures. Enzyme and Microbial Technology. 1998;22(7): 606–611.

[154] Gautam S, Mishra A, Tiwari A. Catharanthus alkaloids and their enhanced production using elicitors: a review. International Journal of Pharmacy and Technology. 2011;3(1):713–724.

[155] Docimo T, Davis AJ, Luck K, Fellenberg C, Reichelt M, Phillips M, Gershenzon J, D'Auria JC. Influence of medium and elicitors on the production of cocaine, amino acids and phytohormones by *Erythroxylum coca* calli. Plant Cell Tissue and Organ Culture. 2015;120:1061–1075.

[156] Anderson WC. Tissue culture propagation of Rhododendrons. In Vitro Journal of Tissue Culture Association. 1978;14:334.

[157] Gamborg OL, Murashige T, Thorpe TA, Vasil IK. Plant–tissue culture media. In Vitro Journal of Tissue Culture Association. 1976;12:473–478.

[158] Murashige T, Tucker DPH. Growth factor requirements of citrus tissue culture. In: Chapman HD, editor. Proceedings of the 1st international citrus symposium; University of California–Riverside Publication, Riverside. California; 1969. p. 1155–1161.

[159] Scheible WR, Morcuende R, Czechowski T, Fritz C, Osuna D, Palacios–Rojas N, Schindelasch D, Thimm O, Udvardi MK, Stitt M. Genome–wide reprogramming of primary and secondary metabolism, protein synthesis, cellular growth processes, and the regulatory infrastructure of *Arabidopsis* in response to nitrogen. Plant Physiology. 2004;136:2483–2499.

[160] Kova´cˇik J, Klejdus B, Babula P, Jarosˇova´ M. Variation of antioxidants and secondary metabolites in nitrogen–deficient barley plants. Journal of Plant Physiology. 2014;171:260–268.

[161] Fritz C, Palacios–Rojas N, Feil R, Stitt M. Regulation of secondary metabolism by the carbon–nitrogen status in tobacco: nitrate inhibits large sectors of phenylpropanoid metabolism. Plant Journal. 2006;46:533–548.

[162] Bensaddek L, Gillet F, Saucedo JEN, Fliniaux MA. The effect of nitrate and ammonium concentrations on growth and alkaloid accumulation of *Atropa belladonna* hairy roots. Journal of Biotechnology. 2001;85:35–40.

[163] Tonelli M, Pellegrini E, D'Angiolillo F, Nali C, Pistelli L, Lorenzini G. Ozone–elicited secondary metabolites in shoot cultures of *Melissa officinalis* L. Plant Cell Tissue and Organ Culture. 2015;120:617–629.

[164] Xu M, Yang B, Dong J, Lu D, Jin H, Sun L, Zhu Y, Xu X. Enhancing hypericin production of *Hypericum perforatum* cell suspension culture by ozone exposure. Biotechnology Progress 2011;27:1101–1106.

[165] Sun L, Su H, Zhu Y, Xu M. Involvement of abscisic acid in ozone–induced puerarin production of *Pueraria thomsnii* Benth. suspension cell cultures. Plant Cell Reports. 2012;31:179–185.

[166] Naik PM, Manohar SH, Praveen N, Murthy HN. Effects of sucrose and pH levels on *in vitro* shoot regeneration from leaf explants of *Bacopa monnieri* and accumulation of bacoside A in regenerated shoots. Plant Cell Tissue and Organ Culture 2010;100:235–239.

[167] Praveen N, Murthy HN. Synthesis of withanolide A depends on carbon source and medium pH in hairy root cultures of *Withania somnifera*. Industrial Crops and Products. 2012;35:241–243.

[168] Praveen N, Murthy HN. Establishment of cell suspension cultures of *Withania somnifera* for the production of withanolide A. Bioresource Technology. 2010;101:6735–6739.

[169] Sara–Alsadat R, Ahmad M, Mokhtar JJ. Paclitaxel production is enhanced in suspension–cultured hazel (*Corylus avellana* L.) cells by using a combination of sugar, precursor, and elicitor. Engineering in Life Sciences. 2015;15:234–242.

[170] Munish S, Ashok A, Rajinder G, Sharada M. Enhanced bacoside production in shoot cultures of *Bacopa monnieri* under the influence of abiotic elicitors. Natural Product Research. 2015;29(8):745–749.

[171] Izabela G, Halina W. The effect of methyl jasmonate on production of antioxidant compounds in shoot cultures of *Salvia officinalis* L. Herba Polonica. 2009;55:238–243.

[172] Firouzi A, Mohammadi SA, Khosrowchahli M, Movafeghi A, Hasanloo T. Enhancement of silymarin production in cell culture of *Silybum marianum* (L) Gaertn by elicitation and precursor feeding. Journal of Herbs, Spices and Medicinal Plants. 2013;19(3):262–274.

[173] Li B, Wang B, Li H, Peng L, Ru M, Liang Z, Yan X, Zhu Y. Establishment of *Salvia castanea* Diels f. *tomentosa* Stib. Hairy root cultures and the promotion of tanshinone accumulation and gene expression with Ag^{+}, methyl jasmonate, and yeast extract elicitation. Protoplasma. 2015; DOI: 10.1007/s00709–015–0790–9.

[174] Yuan Y, Huang L, Cui G, Mao Y, He X. Effect of gibberellins and its synthetic inhibitor on metabolism of tanshinones. Chinese Journal of Experimental Traditional Medical Formulae. 2008;6:002.

[175] Abbasi BH, Stiles AR, Saxena PK, Liu CZ. Gibberellic acid increases secondary metabolite production in *Echinacea purpurea* hairy roots. Applied Biochemistry and Biotechnology. 2012;168(7):2057–2066.

[176] Patil JG, Ahire ML, Nitnaware KM, Panda S, Bhatt VP, Kishor PB, Nikam TD. In vitro propagation and production of cardiotonic glycosides in shoot cultures of *Digitalis purpurea* L. by elicitation and precursor feeding. Applied Microbiology and Biotechnology. 2013;97:2379–2393.

[177] Coste A, Vlase L, Halmagyi A, Deliu C, Coldea G. Effects of plant growth regulators and elicitors on production of secondary metabolites in shoot cultures of *Hypericum hirsutum* and *Hypericum maculatum*. Plant Cell Tissue and Organ Culture. 2011;106:279–282.

[178] Ajungla L, Patil PP, Barmukh RB, Nikam TD. Influence of biotic and abiotic on accumulation of hyoscyamine and scopolamine in root cultures of *Datura metel* L. Indian Journal of Biotechnology. 2009;8(3): 317–322.

[179] Li J, Wang J, Li J, Li J, Liu S, Gao W. Salicylic acid induces the change in the adventitious root of *Glycyrrhiza uralensis* Fisch.: bioactive compounds and antioxidant enzymes. Research on Chemical Intermediates. 2015; DOI: 10.1007/s11164–015–2099–x.

[180] Krac˘un–Kolarevic´ M, Dmitrovic´ S, Filipovic´ B, Peric´ M, Mis˘ic´ D, Simonovic´ A, Todorovic S. Influence of sodium salicylate on rosmarinic acid, carnosol and carnosic acid accumulation by *Salvia officinalis* L. shoots grown in vitro. Biotechnology Letters. 2015;37(8):1693–1701.

[181] Purohit M, Pande D, Datta A, Srivastava PS. Enhanced xanthotoxin content in regenerating cultures of *Ammi majus* and micropropagtion. Planta Medica. 1995;61:481–482.

[182] Sonja GS, Oliver T, Stéphane M, Alain D, Claude J, Daniel H. Effects of polysaccharide elicitors on secondary metabolite production and antioxidant response in *Hypericum perforatum* L. shoot cultures. Scientific World Journal. 2014;11:609–649.

[183] Sonja GS, Oliver T, Ste´phane M, Christophe H, Alain D, Brigitte C, Fre´de´ric L, Eric L, Claude J, Daniel H. Fungal elicitor–mediated enhancement in phenylpropanoid and naphtodianthrone contents of *Hypericum perforatum* L. cell cultures. Plant Cell Tissue and Organ Culture. 2015;122:213–226.

[184] Qianliang M, Chunyan S, Chengjian Z, Min J, Qiaoyan Z, Hong Z, Khalid R, Ting H, Luping Q. Elicitors from the endophytic fungus Trichoderma atroviride promote Salvia miltiorrhiza hairy root growth and tanshinone biosynthesis. Journal of Experimental Botany. 2013;64(18):5687–5694.

5

Signalling Crosstalk of Plant Defence Responses to Xylem-Invading Pathogens

Sabina Berne and Branka Javornik

Abstract

Xylem is a plant vascular tissue that transports water and dissolved minerals from the roots to the rest of the plant. It consists of specialized water-conducting tracheary elements, supporting fibre cells and storage parenchyma cells. Certain plant pathogenic fungi, oomycetes and bacteria have evolved strategies to invade xylem vessels and cause highly destructive vascular wilt diseases that affect the crop production and forest ecosystems worldwide. In this chapter, we consider the molecular mechanisms of root-specific defence responses against vascular wilt pathogens, with an emphasis on the most important and well-studied fungal (*Verticillium* spp. and *Fusarium oxysporum*) and bacterial (*Xanthomonas* spp. and *Ralstonia solanacearum*) pathogens. In particular, we present the current understanding of plant immune responses, from invasion perception to signal transduction and termination. Furthermore, we address the role of specific transcription factors involved in plant immunity and their regulatory network. We also highlight the crucial roles of phytohormones as signalling molecules in local and systemic defence responses. Finally, we summarize the current knowledge of plant defence responses to xylem-invading pathogens to devise new strategies and methods for controlling these destructive plant pathogens.

Keywords: Vascular wilt pathogens, effectors, plant innate immunity, signal transduction, biotic stress

1. Introduction

The disease triangle concept, introduced in the 1960s by George McNew to predict plant disease outcomes, shows the complex interactions among the environment, the host and the infectious (or abiotic) agent [1]. Plants, continuously challenged by numerous abiotic stresses, potential pests and pathogens, have evolved efficient strategies to perceive and respond to such threats. Plants lack specialized immune cells and their survival relies upon a highly sophisticated innate immune system, in which each plant cell responds autonomously [2–5]. The first line of defence is a basal resistance response called pattern-triggered immunity (PTI). It is induced by recognition of exogenous microbe or pathogen-associated molecular patterns (MAMPs or PAMPs) or endogenous molecules released on pathogen perception or pathogen-induced cell damage (damage-associated molecular patterns, DAMPs) via pattern-recognition receptors (PRRs) in the plasma membrane [4]. Successful pathogens overcome PTI by secreting effectors, hydrolytic enzymes or toxins, which suppress or interfere with host defence molecules [6]. In an evolutionary arms race, plants have evolved a robust defence response network termed effector-triggered immunity (ETI) to intercept pathogen effectors through intracellular receptors, such as nucleotide-binding site/leucine-rich repeat (NLR) proteins [5,7,8]. An intricate network of signalling pathways transduces these incoming signals into a diverse array of immune responses activating reactive oxygen species (ROS) generation, MAP kinases, Ca^{2+} signalling, the production of phytohormones and extensive transcriptional reprogramming [9].

In the past, comprehensive research has been dedicated to understanding plant physiological and molecular responses to individual abiotic and biotic stresses under controlled laboratory conditions. Recent studies of plant responses to concurrent abiotic and biotic stress conditions [10–16] have demonstrated that plants perceive and respond to combined stresses in a specific and unique manner. Moreover, the underlying signalling pathways are carefully modulated [14,17,18] and coordinated to ensure that plant growth and fitness are not significantly retarded [19].

Vascular wilt pathogens are soil-borne bacteria, fungi and oomycetes that employ various infection strategies to invade plant roots at different infection sites [20]. They subsequently advance inter- or intracellularly through the root cortex and enter the xylem vessels, where they proliferate and spread passively with xylem sap to aerial plant parts [21]. The characteristic wilt symptoms develop as a consequence of obstructed transportation of water and minerals, either due to the physical blockage of vessels by the pathogen or indirectly due to the activation of plant physical defence responses (e.g. formation of tyloses, accumulation of pectin-rich gels and gums) that confine the further spread of the pathogen [22]. In addition to wilting, other disease symptoms include vein clearing, leaf epinasty, chlorosis, vascular browning, stunting, necrosis and eventually plant death [21–24].

Primarily due to the specific lifestyle of vascular wilt pathogens, relatively little is known about their interactions with host plants and root-specific defence responses on molecular and biochemical levels compared to foliar pathogens. This chapter, therefore, summarizes the currently available molecular, cellular and systems biology data gathered from studies of signalling networks in model plants and crops challenged by bacterial or fungal pathogens and applies this general knowledge to advance understanding of vascular wilt pathogenesis and implement all these findings into the design of new strategies for the protection of crops and forest ecosystems.

2. Plant immune signalling initiation

Several factors contribute to the complex regulatory mechanisms in the initiation of plant immune signalling: (i) as sessile organisms, plants need to respond promptly to danger signals, (ii) each plant cell reacts autonomously to different stimuli, but the response needs to be integrated at a higher organizational level to ensure the plant's survival and (iii) immune reactions are energy- and resource-demanding processes requiring the proper timing and amplitude of response [4]. Typically, immune responses occur on recognition of conserved microbe-, pathogen- or damage-associated molecular patterns or after perception of effector molecules that are species-, race-, or strain-specific and contribute to pathogen virulence [25]. However, not all microbial elicitors conform to the common distinction between PAMPs and effectors, and so Thomma et al. [26] proposed that plant immunity should be considered as a continuum, instead of a two-branched system composed of PAMP-triggered immunity (PTI) and effector-triggered immunity (ETI). An alternative perspective of plant innate immunity as a system that evolves to detect invasion has recently been extended into a so-called invasion model [27]. In this model, invasion patterns (IPs), externally encoded or modified-self ligands that signify invasion, are perceived by plant invasion pattern receptors (IPRs) and induce IP-triggered responses that do not result in immunity by default.

2.1. Plant PRRs convey danger signals to the intracellular immune signalling pathways

Recently emerging structural biology data on plant–pathogen interactions, [28] together with data obtained primarily from genetic and biochemical studies on the leaves of model plants and crops, have revealed that fine-tuning and coordination of immune responses are achieved within large protein complexes at the plasma membrane, where plant PRRs reside [4]. Plant PRRs are either receptor-like kinases (RLKs) or receptor-like proteins (RLPs) [4,6,25,29]. RLKs are modular proteins comprised of an extracellular domain involved in recognition of MAMPs/DAMPs, a single-pass transmembrane domain and a cytosolic serine/threonine kinase domain that transmits a signal to downstream signalling components. RLPs have a similar architecture, but their short cytoplasmic tail lacks kinase activity. RLPs, therefore, probably form heteromeric complexes with RLKs or other cytosolic kinases to relay downstream signalling. Extracellular domains of PRRs contain various motifs involved in recognition and binding of ligands. Leucine-rich repeat (LRR) motifs are widespread and serve as a scaffold for protein- or peptide–protein interactions [30]. Well-characterized examples of LRR–RLKs are *Arabidopsis* FLS2 (flagellin-sensitive 2) and EFR (elongation factor Tu receptor), which bind flagellin fragment flg22 and EF-Tu peptide elf18, respectively [31,32]. Lysine motifs (LysMs), lectin and epidermal growth factor (EGF)-like domains are found in PRRs that recognize carbohydrate moieties, such as fungal chitin [33,34] or bacterial peptidoglycans [35].

2.1.1. Perception of chitin

The chitin-responsive PRR system has been thoroughly investigated in both dicots and monocots. The perception of chitin in monocot plants is best described in rice (*Oryza sativa*). Chitin elicitor binding protein (OsCEBiP) is a receptor-like protein that specifically binds chitin oligomers [36]. In the absence of chitin, OsCEBiP exists as a homodimer [37]. On binding of chitin octamer, OsCEBiP associates with receptor-like kinase OsCERK1 and forms heterodimers [37,38]. This interaction activates the OsCERK1 kinase domain to become phosphorylated. Subsequently, active OsCERK1 phosphorylates a guanine nucleotide exchange factor OsRac-GEF, which activates a Rho-type small GTPase OsRac [39]. OsRac acts as a molecular switch in many plant signalling pathways and, among other things, regulates the production of ROS by the NADPH oxidase OsRbohB [40].

The model dicot plant *Arabidopsis thaliana* harbours chitin elicitor receptor kinase 1/LysM-containing receptor-like kinase1 (CERK1/LYK1), which perceives chitin (a polymer of N-acetyl-D-glucosamine, NAG) through its LysM motif [33]. Although the AtCERK1 extracellular domain contains three tandem LysMs, only LysM2 binds NAG_5 [41]. This interaction, however, fails to trigger immune responses. Downstream signalling has been observed only on binding of chitin octamer, which acts as a bivalent ligand and induces CERK1 dimerization [41]. Another LysM-containing cell surface receptor, AtLYK5, has recently been proposed as the primary chitin receptor, due to a significantly higher binding affinity for NAG_8 compared to AtCERK1 [42]. AtLYK5 exists as a homodimer in the absence of chitin. Binding of chitin to AtLYK5 homodimer promotes the association of AtLYK5 with AtCERK1. This leads to dimerization of AtCERK1 and activation of its kinase domain. The chitin signal is then transduced downstream to mitogen-activated protein kinases MPK3 and MPK6 [33].

2.1.2. Perception of flagellin

Extensive research of the model plant *Arabidopsis thaliana* has elucidated molecular mechanisms triggered in response to recognition of bacterial flagellin by evolutionary conserved LRR RLK flagellin-sensitive 2 (FLS2) [31]. The extracellular domain of FLS2 contains 28 LRR and binds the 22-amino acid long flagellin epitope flg22 [43]. Immediately after, FLS2 associates with co-receptor brassinosteroid insensitive 1 (BRI1)-associated receptor kinase 1 (BAK1)/ somatic embryogenesis receptor-like kinase 3 (SERK3) to form a heterodimer [44–46]. BAK1 is a key regulatory LRR RLK coordinating growth–defence trade-offs [47], since it is required for early defence responses in PTI [48] but also implicated in brassinosteroid hormone signalling [49]. BAK1 phosphorylates receptor-like cytoplasmic kinase *Botrytis*-induced kinase 1 (BIK1), which interacts and forms a complex with both BAK1 and FLS2 [50]. BIK1 is subsequently auto-phosphorylated at tyrosine and serine/threonine residues [51]. Activated BIK1 contributes to flg22-triggered calcium influx from apoplast [52] and phosphorylates NADPH oxidase RbohD involved in reactive oxygen species (ROS) production [53].

2.1.3. Perception of peptidoglycan

Peptidoglycan (PGN), a polymer of N-acetylglucosamine and N-acetylmuramic acid branched with a short peptide, is an essential component of a bacterial cell wall and is another well-studied MAMP. In *Arabidopsis,* the PGN perception system is comprised of two GPI-anchored LysM domain RLPs, LYM1 and LYM3, which bind PGN, and a transmembrane RLK AtCERK1, which probably relays the PGN signal downstream [35]. Rice has a similar PGN detection system, involving LysM RLK OsCERK1 [54] and two LysM-containing proteins, OsLYP4 and OsLYP6, which not only bind PGN but also associate with chitin oligomers [55].

2.1.4. Perception of DAMPs

Some PRRs respond to damage-associated molecular patterns (DAMPs), endogenous molecules such as cell wall fragments or peptides released on pathogen attack or various abiotic stresses [25,56]. Partial degradation of pectic polysaccharide homogalacturonan (HGA) by pathogen or plant polygalacturonases produces oligogalacturonides (OGs), oligomers of α-1,4–linked galacturonic acid [57,58]. In *Arabidopsis,* OGs are perceived by wall-associated kinase 1 (WAK1) Ser/Thr RLK kinase with an extracytoplasmic domain that contains several EGF-like repeats [59]. The signal is then relayed to the MAP kinase signalling pathway, where AtMPK3 and AtMPK6 become phosphorylated and induce expression of several defence genes [60]. Additional OGs-triggered defence responses include activation of NADPH oxidase AtRbohD involved in the generation of ROS, production of NO and deposition of callose in the plant cell walls [57,58].

The best studied peptides acting as DAMPs belong to the plant elicitor peptides (Peps) family. They are processed from precursor PROPEPs [61,62]. AtPep1, a 23 amino acid peptide released from the C-terminal of PROPEP1, was the first peptide elicitor isolated from *Arabidopsis* [61], but similar peptides were also later confirmed in other plants [63]. AtPep1 is recognized by two LRR RLK, PEPR1 and its paralog PEPR2 [64,65]. However, signalling is initiated only in complex with co-receptor LRR RLK BAK1 [66]. The active receptor complex consequently induces the expression of MAP kinase 3 (MPK3), WRKY transcription factors and defence-related genes such as *PR-1* and *PDF1.2* (encoding defensin) [63]. Moreover, the cytosolic kinase domain of PEPR1 has guanylyl cyclase activity, which generates cGMP from GTP [67]. An increased local concentration of cGMP has been proposed to open cyclic nucleotide-gated channels (CNGC2) in the plasma membrane and activate cytosolic Ca^{2+} signalling [67,68].

2.2. Intracellular immune receptors NLRs detect pathogen effectors

In addition to PRRs, plants have evolved a second class of immune receptors that intercept effectors in different parts of the cell [8]. These intracellular receptors, so called NLRs, are characterized as multi-domain proteins that have a conserved central nucleotide-binding (NB) domain and variable C-terminal leucine-rich repeats (LRR) domain [28]. In terms of their distinct N-terminal domains, NLRs are broadly divided into two groups: TNLs that harbour a Toll–interleukin 1 receptor (TIR) domain and CNLs that contain a coiled-coil (CC) domain [69–71]. NLRs belong to signal transduction ATPases with numerous domains (STAND) that operate as molecular switches cycling between an inactive closed ADP-bound state and active open state with bound GTP [72–74]. In the resting state, N-terminal TIR or CC and C-terminal LRR domains sterically inhibit the NB domain from ADP–ATP exchange. On pathogen recognition, a series of conformational changes occur that expose the NB domain, promote ADP–ATP exchange and initiate signal transduction [8]. Effector recognition by NLRs often, but not always, leads to a form of programmed cell death termed as a hypersensitive response [75]. NLRs are, therefore, under precise control by accessory proteins. NLRs interact with conserved Hsp90-Sgt1-RAR1 protein complexes for proper folding, accumulation and regulation [76,77]. Moreover, Sgt1 interacts with the suppressor of *rps4-RLD* (SRFR1) negatively to regulate NLRs accumulation and prevent autoimmune activation [78].

Various strategies of effector recognition by NLRs exist and are represented in different models. In the gene-for-gene model, originally described by Flor in 1971, NLRs can recognize effectors directly (e.g. *Pita-AvrPita* [79], *Pto-AvrPto* [80] and many others) acting like receptor and ligand pairs that initiate a cascade of reactions leading to resistance [3]. When considering pathogen effector diversity, it is more likely that a single NLR recognizes multiple effectors from diverse pathogens in the presence of other host proteins. This hypothesis is explained by the guard model, in which the NLR protein is assigned the role of a sentinel that is activated indirectly by detecting an effector-modified host protein and induces a defence response [81,82]. An example of such mechanism is given by *Arabidopsis* CNL receptors, RPM1 and RPS2, which constantly monitor host protein RIN4 (a negative regulator of basal resistance) for interference with *Pseudomonas syringae* effectors AvrB, AvrRpm1 and AvrRpt2 [83]. Another indirect effector recognition strategy is proposed in the decoy model [84]. As guarded effector targets are evolutionarily unstable, it is likely that the targeted host gene has duplicated and evolved into decoy proteins. These serve as bait to trap effectors that target structurally related proteins involved in basal defence. For example, in the inactive state, *Arabidopsis* CNL RPS5 interacts with its N-terminal CC domain with protein kinase PBS1, which has no function in basal resistance [85]. Cleavage of PBS1 by *Ps. syringae* effector AvrPphB activates RPS5 [86]. However, several PBS1-like kinases (including Botrytis-induced kinase 1 (BIK1)) are also cleaved by AvrPphB [87]. PBS1, therefore, acts as a decoy that prevents cleavage of BIK1, which is an important component of PRR signalling [53] and the key AvrPphB target. An integrated decoy model has recently been proposed [5]. In this model, the effector-targeted plant protein is duplicated and fused to one member of the NLR pair to act as bait that, on effector binding, triggers defence signalling by the second NLR.

Activated NLRs trigger a variety of immune responses, from the generation of ROS, elevation of intracellular Ca^{2+}, activation of MAPK cascades, transcriptional reprogramming to production of phytohormones [8]. Although effector-triggered responses are qualitatively similar to immune responses elicited by MAMPs/DAMPs, there are quantitative differences in the strength and duration of pathways, which result in different resistance responses and signalling networks [17].

2.3. Signal transduction cascades

Perception of MAMPs/DAMPs by their cognate receptors triggers an array of immune responses, comprising changes in intracellular calcium levels $[Ca^{2+}]_i$, membrane potential depolarization, extracellular alkalinization, production of ROS, NO and phosphatidic acid, activation of kinases (mitogen-activated protein kinases (MAPKs) or Ca^{2+}-dependent protein kinases (CDPKs)), transcriptional reprogramming and changes in plant hormone concentrations (e.g. ethylene, salicylic and jasmonic acid) [88,89].

2.3.1. Calcium and ROS signalling interconnection

Ca^{2+} is a ubiquitous second messenger released in response to various stresses and developmental processes. In *Arabidopsis,* various MAMPs/DAMPs induce distinct and sustained elevations of intracellular calcium concentration ($[Ca^{2+}]_i$), which differ in the lag phase and amplitude of response [89,90]. Moreover, changes in $[Ca^{2+}]_i$ are organ-specific and correlate with the expression patterns of the corresponding MAMP/DAMP receptors. Chitin octamer and Pep1 induce similar responses in seedling shoots and roots, while roots are insensitive to elf18 and show only a minor response to flg22 [90]. Furthermore, Ca^{2+} fluxes are generated from different sources; flg22/FLS2 signalling involves the release of Ca^{2+} from intracellular stores (e.g. endoplasmic reticulum and/or tonoplast) and inositol phosphate signalling, whereas Pep/PEPR signalling requires an influx of Ca^{2+} from the apoplast [68]. The identity of plant Ca^{2+} channels and pumps involved in the generation of Ca^{2+} signals is largely unknown, although some candidates (e.g. ionotropic glutamate receptor (iGluR)-like channels, cyclic nucleotide gated channels (CNGCs) and annexins in plasma membrane and two-pore-channel 1 (TPC1) in the tonoplast membrane) have been investigated [88,91]. Elevated $[Ca^{2+}]_i$ is detected by Ca^{2+}sensor proteins such as calmodulins (CaMs), calcium-dependent protein kinases (CDPKs), calcineurin B-like (CBL) proteins and CBL-interacting protein kinases (CIPKs) [88,91]. CaMs are highly conserved eukaryotic proteins that bind free Ca^{2+} with four EF-hand motifs and regulate the function of their interacting proteins, such as CaM-binding transcription factors [88]. CDPKs are unique proteins acting as sensors and decoders of Ca^{2+} signals and are suited for rapid responses to stimuli. Binding of Ca^{2+} via four EF-hand domain motifs in the C-terminus activates the CDPKs' N-terminal kinase domain and promotes transmission of a Ca^{2+} signal by phosphorylating different target proteins [91]. The functional specificity of CDPKs is achieved by targeting distinct membrane subdomains and involves specific lipid modifications (e.g. N-terminal myristoylation, S-acylation) [91]. In contrast to CDPKs, CBLs are Ca^{2+} sensors without enzymatic activity. They bind Ca^{2+} with four EF hands and then associate with CIPKs through the NAF motif in the kinase C-terminal regulatory domain. This interaction liberates kinase from auto-inhibition and enables conversion of the Ca^{2+} signal into phosphorylation events [91]. Like CDPKs, CBLs have different lipid modifications (e.g. N-terminal myristoylation, S-acylation) that determine their localization and, consequently, the site of action of CBL–CIPK complexes [91].

Rapid production of reactive oxygen species (ROS) in response to MAMPs/DAMPs depends primarily on respiratory burst oxidase homologues (RBOHs) [92,93]. These NADPH oxidases are integral membrane proteins that generate superoxide anions (O_2^-), which are rapidly converted into hydrogen peroxide (H_2O_2). RBOHs have cytosolic FAD- and NADPH-binding domains in the C-terminal region, six membrane-spanning domains and a cytosolic N-terminal extension that harbours two EF-hand motifs and multiple phosphorylation sites [94]. Recent studies in *Arabidopsis* have revealed different regulation mechanisms of RBOHD and RBOHF-dependent ROS production. RBOHF regulation involves direct binding of Ca^{2+} to its EF-hands and Ca^{2+}-dependent phosphorylation by CBL1/9–CIPK26 complexes [95]. Direct binding of $Ca^{2+}$to EF-hand motifs on MAMP-induced elevation of $[Ca^{2+}]_i$ causes conformational changes and activation of RBOHD [96]. Additionally, RBOHD is activated by protein phosphorylation at multiple sites via calcium-dependent protein kinase 5 (CPK5)[97,98] and MAMP-receptor-associated Botrytis-induced kinase 1 (BIK1) [53]. In addition to local defences, Ca^{2+} and

RBOHD-dependent ROS production is implicated in the systemic signal propagation required for long-distance signalling [98–100]. In accordance with the current model [98], the perception of MAMPs triggers a rapid rise of $[Ca^{2+}]_i$, causing the activation of CPK5 and subsequent phosphorylation of RBOHD and other CPK5 substrates. Apoplastic H_2O_2, generated after dismutation of the O_2^- produced by the RBOHD, probably represents the cell-permeable signal, which serves as the stimulus for further reiterations of calcium-dependent CPK5 activation and RBOHD phosphorylation, resulting in rapid propagation of the MAMP signal throughout the plant.

2.3.2. *MAPK signalling*

Plant mitogen-activated protein kinase (MAPK) cascades generally comprise MAPKK kinases (or MEKKs), which receive signals from receptors/sensors and phosphorylate downstream MAPK kinases (or MKKs) and which subsequently activate MAPKs (or MPKs) that control the activities and synthesis of a plethora of transcription factors (TFs), enzymes, hormones, peptides and antimicrobial chemicals [101,102]. In *Arabidopsis*, two kinase cascades, MKK4/MKK5–MPK3/MPK6 [103] and MEKK1–MKK1/MKK2--MPK4, [104] are activated after perception of MAMPs/DAMPs. The activation of MEKK1–MKK1/MKK2–MPK4 negatively regulates ROS and salicylic acid (SA) production [105,106], as well as repressing cell death and immune responses [107]. MKK4/MKK5–MPK3/MPK6 cascade positively regulates the expression of several defence–related genes [60,103] and promotes accumulation of camalexin via transcription factor WRKY33 [108,109]. Moreover, activation of MPK3/MPK6 is required for full priming of stress responses [110] and increases ethylene production via ACC synthases ACS2/ACS6 [111]. Given the essential nature of the MKK4/MKK5–MPK3/MPK6 cascade, its activation has to be precisely controlled, since inappropriate activation (e.g. constitutively activated MKK4/MKK5 [112] or over-expression of MPK3 [113]) may promote hypersensitive response (HR)-like cell death or be lethal to plants. MPK3 has also recently been indicated to be a negative regulator of defence gene expression, flg22-triggered SA accumulation and disease resistance to *Pseudomonas syringae* [114]. Another negative regulator of MAPK activities is MAPK phosphatase 2 (MKP2), which interacts with and dephosphorylates MPK3 and MPK6 [115]. Additionally, a Raf-like MAPKK kinase (EDR1) has been proposed to negatively regulate the MKK4/MKK5–MPK3/MPK6 cascade by physically interacting with MKK4 and MKK5 via its N-terminal domain [113].

2.4. Transcriptional reprogramming converges with complex phytohormone signalling networks

Transcription factors (TFs) involved in plant immunity reside in transcriptional complexes and, together with co-regulatory proteins, directly or indirectly recruit RNA polymerase II to the target promoters or release it from them [116]. TFs vital for plant immunity comprise members of the AP2/ERF, bHLH, bZIP, MYB, NAC and WRKY TF families and perform diverse roles [9]. For instance, certain members of apetala2/ethylene-response element binding factor (AP2/ERF) participate in the regulation of genes related to the jasmonic acid (JA) and ethylene hormone signalling pathways [117]. *At*MYC2/JAI1/JIN1 and closely related proteins *At*MYC3 and *At*MYC4 belong to basic-helix-loop-helix (bHLH) TFs and coordinate JA-mediated defence responses with other phytohormones (salicylic acid (SA), abscisic acid (ABA), gibberellins (GA) and auxin) [118]. TGA/basic domain leucine zipper (bZIP) family members are central players in SA-mediated resistance to biotrophic pathogens. Moreover, *At*TGA2, 5 and 6 TF have central roles in establishing systemic acquired resistance (SAR), regulate host detoxification pathways and are essential activators of certain ethylene-induced defence responses [119].

TF expression and activities are regulated in multiple ways. Certain nucleotide-binding site/leucine-rich repeat (NLR) proteins directly regulate transcription by physically interacting with TFs [120–123]. Several TFs are controlled by phosphorylation as downstream targets of activated MAPK cascades [109,124–127]. Another mechanism of TFs activation is carried out by Ca^{2+}sensors such as CaMs and CDPKs [128–133]. Additional factors (e.g. components of mediator complex [134,135], chromatin modifications [136–138]) and levels of regulation (e.g. ubiquitination [139], sumoylation [140], alternative mRNA splicing [141]) also contribute to the complexity of transcriptional networks and fine-tuning of immune responses.

Coordination of diverse stress responses and growth is resolved within complex phytohormone signalling networks, in which salicylic acid (SA), jasmonic acid (JA) and ethylene (ET) perform central roles, while other hormones merely modulate their responses [142]. SA is synthesized in chloroplasts from chorismate by isochorismate synthase [143] and exported to the cytosol [144]. In response to pathogens and various abiotic stresses, SA mediates expression of pathogenesis-related (PR) genes and the synthesis of antimicrobial compounds to provide basal defence and systemic acquired resistance (SAR) [145]. SA activates these defence responses through transcription cofactor nonexpresser of PR genes 1 (NPR1) [146] and transcription factors TGA2, TGA5 and TGA6 [119]. In the absence of SA, NPR1 is sequestered in the cytosol and forms oligomeric complexes stabilized by intermolecular disulphide bonds [147]. In response to activation of SA pathway, thioredoxins reduce these disulphide bonds, causing the release of NPR1 in monomeric form, which can translocate to the nucleus via a nuclear translocation signal (NLS) [148]. In addition, NPR1 protein levels oscillate through $CUL3^{NPR3}$- and $CUL3^{NPR4}$-mediated degradation in the nucleus, which is required for fine-tuning of immune responses [147,149,150].

Jasmonates (JAs) are plant hormones with essential roles in plant defence and development [118]. JAs are derived from α-linolenic acid liberated from membrane phospholipids by the action of phospholipase A and enzymatically converted in a series of steps in chloroplasts and peroxisomes, to be finally transformed into bioactive molecule JA-isoleucine (JA-Ile) in the cytosol [151]. JA-signalling is activated after repressor removal [152,153]. In unstimulated cells, jasmonate ZIM domain (JAZ) proteins repress transcription of JA signalling components, such as the basic-helix–loop-helix (bHLH) master transcription factor MYC2 and its close homologues MYC3 and MYC4 [154]. On JA signal perception by coronatine insensitive 1 (COI1), a component of the Skp1-Cul-F-box protein (SCF) E3 ligase complex, JAZ repressor proteins are targeted for proteasome-mediated degradation and MYC2 activates the transcription of several JA-responsive genes [154,155].

Ethylene (ET) is a gaseous hormone that often works synergistically with JA [156]. Important steps in ET biosynthesis are the conversion of *S*-AdoMet to 1-aminocyclopropane-1-carboxylic acid (ACC) by ACC synthase and oxidation of ACC by ACC oxidase to form ET [157]. In *Arabidopsis*, MPK3 and MPK6 phosphorylate ACS2 and ACS6 proteins to prevent rapid degradation of ACS2/ACS6 by the 26S proteasome pathway and enhance ET production in response to MAMP perception [111]. MPK3 and MPK6 also phosphorylate TF WRKY33, which subsequently binds to ACS promoters and regulates ET production [158]. Generated ET binds to its receptors, such as ethylene response 1 (ETR1) in the ER membrane. When ET is absent, active receptors ETR1 negatively regulate ethylene insensitive2 (EIN2) through phosphorylation via Raf-like protein kinase constitutive triple response1 (CTR1) [159]. At the same time, TFs ethylene insensitive3 (EIN3) and EIN3-like1 (EIL1) are recruited by two F-box proteins, EBF1 and EBF2, to 26S proteasomal degradation. On ET signal perception, the ETR1 receptors are inactivated and CTR1 repressed. Subsequently, the C-terminal part of EIN2 is cleaved and translocated to the nucleus [160]. This induces degradation of EBF1 and EBF2 and stabilizes EIN3 and EIL1, which regulate expression of ET-responsive genes (e.g. TF ERF1 and ORA59) [161,162].

Plant hormonal crosstalk is extensive and occurs in several combinations [163]. The molecular mechanism underlying SA-mediated reprogramming of the JA transcriptional network points to immune signalling antagonism and the involvement of transcriptional regulators NPR1, TGA, WRKY and ORA59 as signal integrators [164]. Phytohormones JA and ET synergistically regulate plant defence responses to necrotrophic fungi via JA-induced EIN3 and EIL1 activation and ET-induced EIN3 and EIL1 stabilization. In addition, antagonistic effects observed in JA and ET signalling are mediated by the interaction of JA-activated MYC2 TF and ET-stabilized TF EIN3 [165].

2.5. Inactivation of immune signalling pathways

Various negative regulation mechanisms ensure immune signalling activation is switched off when there is no danger. In the absence of ligand, several phosphatases interact with PRRs and their associated kinases to keep immune complexes inactive through dephosphorylation. For instance, Ser/Thr phosphatase type 2A (PP2A) constitutively associates with BRI1-associated kinase1 (BAK1) and negatively controls BAK1 activation in PRR immune receptor complexes [166]. Negative regulation exerted by downstream phosphatases is illustrated by *Arabidopsis* MAPK phosphatase 1 (MKP1) operating as a negative regulator of MPK6-mediated MAMP responses [167] and also observed with MAPK phosphatase 2 (MKP2), which acts as the key regulator of MPK3 and MPK6 networks controlling both abiotic and specific pathogen responses in plants [115].

Ubiquitination and proteasomal degradation are other mechanisms by which plant immune responses are attenuated. For example, two U-box E3 ubiquitin ligases, PUB12 and PUB13, are recruited to flagellin-induced FLS2/BAK1 receptor complex and phosphorylated by BAK1 to polyubiquitinate FLS2 and promote its degradation [168]. Downregulation of immune signalling can also be achieved by ligand-induced endocytosis and degradation [169]. This has been reported in localization studies of flg22-induced FLS2 receptors in *Arabidopsis* [170] and was recently proposed as a mechanism for desensitizing host cells to flg22 stimulus and in turning over ligand-bound FLS2 [171].

3. Plant defence responses to xylem-invading pathogens

3.1. General perception of MAMPs in roots

Despite the fact that roots are subjected to a rich microbial community, the perception of MAMPs and immune responses in roots are poorly understood. Millet et al. [172] studied immune responses in *Arabidopsis* roots after exposure to flg22, PGN and chitin. Flg22 and PGN initiated signalling only in association with LRR RLK BAK1. Furthermore, bacterial flg22 and PGN triggered a localized response in the elongation zone of the root tip, while chitin induced a response only in the mature zones of roots. It is thus likely plants have evolved tissue-specific MAMP-triggered immune responses, depending on the nature of the attacker [172,173]. While fungi and nematodes can directly penetrate the epidermal layer of roots, bacteria cannot and therefore exploit the weakest part of the roots as infection site. This hypothesis has been confirmed by recent FLS2 expression studies [174], which indicated that basal FLS2 promotor activity is restricted to the vascular cylinder and outgrowing lateral roots. Moreover, the FLS2 receptor system in roots is functional, since flg22 treatment induced rapid calcium influx and caused phosphorylation of MAPK [174]. Whole transcriptome expression analysis of flg22-elicited roots also revealed a set of genes specifically upregulated in roots, with functions in hormone and stress signalling, root and lateral root development, signalling and defence [174].

3.2. Perception of vascular wilt pathogens

At early stages of infection, vascular wilt pathogens are faced with preformed physical and chemical root defences and MAMP-induced immune responses that hinder their invasion [20]. Once they breach the rigid secondary xylem walls and enter the xylem vessels, vascular wilt pathogens are presumably recognized by specific extracellular receptors in the parenchyma cells surrounding the xylem vessels [21].

3.2.1. Perception of Verticillium spp.

In tomato, extracellular LRR RLP Ve1 [24,175,176] plays a role in xylem defence and provides resistance against race 1 strains of *V. dahliae* and *V. albo-atrum* [177,178]. In recent years, several other homologue genes have been reported in *Gossypium*, *Solanum* and *Mentha*. A functional *Ve1* orthologue has also been discovered in *Nicotiana glutinosa* [179]. Ve1 recognizes a small effector protein, Ave1, with a similarity to plant natriuretic peptides involved in regulation of water and ion homeostasis [180]. Phylogenetic analysis has indicated hundreds of Ave1 homologues in plants but only a few in fungi, suggesting *Verticillium* spp. acquired Ave1 through horizontal gene transfer [180].

Ve1 forms heterodimers with a tomato orthologue of the *Arabidopsis* RLK suppressor of BIR1-1/evershed (SOBIR1/EVR) in the absence of Ave1 [181]. However, Ve1-mediated signalling also requires other critical signalling components, such as SERK1 and SERK3/BAK1, to establish *Verticillium* resistance in tomato and *Arabidopsis* [177,178,182]. Additionally, Ve1-mediated signalling depends on ER-QC-assisted folding mediated by ER-resident chaperones HSP70 binding proteins (BiPs) and lectin-type calreticulins (CRTs) [183].

3.2.2. Perception of Fusarium oxysporum f. sp. lycopersici

Three I (immunity) genes have been identified in tomato [184] in a resistance response to *Fusarium oxysporum* f. sp. *lycopersici* (*Fol*) that involves callose deposition, accumulation of phenolics and formation of tyloses and gels [185]. The *I-2* gene encodes an intracellular CC-NB-LRR receptor protein that perceives *Fol* effector protein Avr2 (secreted in xylem 3; Six3) [186]. Avr2 is under the control of transcription factor Sge1 and is highly expressed in roots and xylem vessels [187]. Avr2 forms homodimers and requires nuclear localization to trigger I-2-mediated cell death [188], which can be strongly suppressed by *Fol* effector Six6 [189]. It has recently been shown that Six5 also contributes to the virulence of *Fol* in tomato plants that Six5 and Avr2 can interact and are together required for *I-2*-mediated resistance [190].

3.2.3. Perception of Xanthomonas oryzae pv. oryzae

The rice LRR RLK Xa21 that provides resistance to *Xanthomonas oryzae* pv. *oryzae* (*Xoo*) [191] recognizes sulphated peptide from the N-terminal part of the secreted quorum-sensing (QS) signal molecule activator of Xa21 (Ax21) [192,193]. In the absence of a signal, Xa21 associates with XB24, a protein with ATPase activity that enhances Xa21 autophosphorylation of Ser/Thr residues [194]. On Ax21 recognition, XB24 is released and Xa21 becomes activated to induce rice defence responses [193]. Subsequently, a protein phosphatase 2C (XB15) specifically interacts with activated Xa21, leading to dephosphorylation and inactivation of Xa21 [195]. In addition, several other proteins (e.g. RING finger ubiquitin ligase XB3 [196] and plant-specific ankyrin-repeat protein XB25 [197]) are associated with Xa21 and required for Xa21 accumulation and resistance to *Xoo* [193]. Moreover, Xa21 binds XB10, a WRKY62 transcription factor. When overexpressed, XB10 suppresses the activation of defence-related genes *OsPR1* and *OsPR10* and acts as a negative regulator of basal and Xa21-mediated immunity [198]. Xa21 also interacts with the endoplasmic reticulum (ER) chaperone BiP3, which regulates its stability and processing [199].

3.2.4. Perception of Ralstonia solanacearum

A pair of *Arabidopsis thaliana* TIR-NB-LRR proteins, RRS1 and RPS4, function together in disease resistance against *Colletotrichum higginsianum*, *Pseudomonas syringae* pv.*tomato* and *Ralstonia solanacearum* [200–202]. RRS1 and RPS4 proteins form an inactive heterodimer complex through the SH motif in their TIR domains [5,121]. RRS1 protein recognizes and, through its C-terminal WRKY domain, directly binds *R. solanacearum* effector PopP2 [203–205]. This leads to disruption of RRS1/RPS4 TIR heterodimer (but not full-length heterocomplex), allowing the formation of signalling active RPS4 TIR homodimer. PopP2 interacts with other WRKY domain-containing proteins and acetylates lysines to block DNA binding, suggesting that PopP2 interferes with WRKY TF-dependent defence [205,206].

3.3. Induced defence responses to vascular wilt pathogens

Recognition of vascular wilt pathogens by plant immune receptor complexes activates defence responses in the xylem vessels. Physical defence responses that confine pathogens from further spread comprise the formation of tyloses, accumulation of pectin-rich gels and gums, vascular coating and callose and secondary cell wall deposition [20,21]. An interesting adaptation to vascular wilt infection is vein clearing, a tissue-specific developmental programme leading to the formation of new xylem elements [207]. Furthermore, significant metabolic changes have been reported in response to xylem infection and involve the induction of pathogenesis-related (PR) proteins, peroxidases, proteases as well as the production of antimicrobial secondary metabolites such as phytoalexins, sulphur-containing compounds and phenolic compounds [20,21].

Studies of defence signalling in response to root pathogens have so far mainly focused on the leaves and have provided evidence that defence mechanisms involve similar signalling pathways (Ca^{2+}-signalling, induction of ROS and MAPK cascades, modulation of phytohormone signalling) [20]. Moreover, plant microarray and RNASeq studies have revealed that the interaction between vascular wilt pathogens and host plants involves transcriptional reprogramming of hundreds of genes [208–211]. Interestingly, in an incompatible interaction, only modest changes in gene and protein expression have been reported [210,212–214] and most of the differentially expressed genes have been repressed in roots rather than in leaves [211]. Moreover, genes implicated in photorespiration, hypoxia, glycoxylate metabolism and auxin signalling show inverse regulation on infection with the foliar pathogen *Cladosporium fulvum* or root pathogen *Verticillium dahliae* [210].

Genome-wide analyses on transcriptional and proteomic levels, together with functional characterization of individual genes, have revealed a convergence of signalling pathways in response to individual pathogens, in mostly controlled conditions. In the field, plants are simultaneously challenged by multiple stress factors, both biotic and abiotic. Even though signalling components of plant regulatory networks are partly shared in both and point to general stress response mechanisms, there is evidence of specific responses to combined stresses that are controlled by different signalling pathways and such studies may provide additional candidates for crop protection breeding [14–16,18].

Author details

Sabina Berne and Branka Javornik*

*Address all correspondence to: branka.javornik@bf.uni-lj.si

Department of Agronomy, Biotechnical Faculty, University of Ljubljana, Ljubljana, Slovenia

References

[1] Scholthof K-BG. The disease triangle: pathogens, the environment and society. Nat Rev Microbiol. Nature Publishing Group; 2006 Dec 27;5(2):152–6.

[2] Jones JDG, Dangl JL. The plant immune system. Nature. 2006 Nov 16;444(7117):323–9.

[3] Dodds PN, Rathjen JP. Plant immunity: towards an integrated view of plant-pathogen interactions. Nat Rev Genet. 2010 Aug;11(8):539–48.

[4] Macho AP, Zipfel C. Plant PRRs and the activation of innate immune signaling. Mol Cell. 2014 Apr 24;54(2):263–72.

[5] Cesari S, Bernoux M, Moncuquet P, Kroj T, Dodds PN. A novel conserved mechanism for plant NLR protein pairs: the "integrated decoy" hypothesis. Front Plant Sci. 2014 Jan;5:606.

[6] Trdá L, Boutrot F, Claverie J, Brulé D, Dorey S, Poinssot B. Perception of pathogenic or beneficial bacteria and their evasion of host immunity: pattern recognition receptors in the frontline. Front Plant Sci. Frontiers; 2015 Apr 8;6:219.

[7] Bonardi V, Cherkis K, Nishimura MT, Dangl JL. A new eye on NLR proteins: focused on clarity or diffused by complexity? Curr Opin Immunol. 2012 Feb;24(1):41–50.

[8] Cui H, Tsuda K, Parker JE. Effector-triggered immunity: from pathogen perception to robust defense. Annu Rev Plant Biol. 2015 Jan;66:487–511.

[9] Tsuda K, Somssich IE. Transcriptional networks in plant immunity. New Phytol. 2015 Jan 26;206(3):932–47.

[10] Atkinson NJ, Lilley CJ, Urwin PE. Identification of genes involved in the response of *Arabidopsis* to simultaneous biotic and abiotic stresses. Plant Physiol. 2013 Aug 1;162(4):2028–41.

[11] Rasmussen S, Barah P, Suarez-Rodriguez MC, Bressendorff S, Friis P, Costantino P, et al. Transcriptome responses to combinations of stresses in *Arabidopsis*. Plant Physiol. 2013 Apr;161(4):1783–94.

[12] Bostock RM, Pye MF, Roubtsova T V. Predisposition in plant disease: exploiting the nexus in abiotic and biotic stress perception and response. Annu Rev Phytopathol. 2014 Jan;52:517–49.

[13] Prasch CM, Sonnewald U. Simultaneous application of heat, drought, and virus to *Arabidopsis* plants reveals significant shifts in signaling networks. Plant Physiol. 2013 Aug;162(4):1849–66.

[14] Atkinson NJ, Jain R, Urwin PE. The Response of Plants to Simultaneous Biotic and Abiotic Stress. In: Mahalingam R, editor. Combined Stresses in Plants SE - 9. Springer International Publishing; 2015. pp. 181–201.

[15] Suzuki N, Rivero RM, Shulaev V, Blumwald E, Mittler R. Abiotic and biotic stress combinations. New Phytol. 2014 Jul 11;203(1):32–43.

[16] Kissoudis C, van de Wiel C, Visser RGF, van der Linden G. Enhancing crop resilience to combined abiotic and biotic stress through the dissection of physiological and molecular crosstalk. Front Plant Sci. 2014 Jan;5:207.

[17] Tsuda K, Katagiri F. Comparing signaling mechanisms engaged in pattern-triggered and effector-triggered immunity. Curr Opin Plant Biol. 2010 Aug;13(4):459–65.

[18] Atkinson NJ, Urwin PE. The interaction of plant biotic and abiotic stresses: from genes to the field. J Exp Bot. 2012 Jun 30;63(10):3523–43.

[19] Denancé N, Sánchez-Vallet A, Goffner D, Molina A. Disease resistance or growth: the role of plant hormones in balancing immune responses and fitness costs. Front Plant Sci. Frontiers; 2013 Jan 24;4:155.

[20] De Coninck B, Timmermans P, Vos C, Cammue BPA, Kazan K. What lies beneath: belowground defense strategies in plants. Trends Plant Sci. 2015 Oct;20(2):91–101.

[21] Yadeta KA, J Thomma BPH. The xylem as battleground for plant hosts and vascular wilt pathogens. Front Plant Sci. 2013 Jan;4:97.

[22] Doohan F. Fungal Pathogens of Plants. Fungi. John Wiley & Sons, Ltd; 2011. pp. 313–344.

[23] Pietro A Di, Madrid MP, Caracuel Z, Delgado-Jarana J, Roncero MIG. *Fusarium oxysporum*: exploring the molecular arsenal of a vascular wilt fungus. Mol Plant Pathol. 2003 Sep 1;4(5):315–25.

[24] Fradin EF, Thomma BPHJ. Physiology and molecular aspects of *Verticillium* wilt diseases caused by *V. dahliae* and *V. albo-atrum*. Mol Plant Pathol. Blackwell Publishing Ltd; 2006;7(2):71–86.

[25] Boller T, Felix G. A renaissance of elicitors: perception of microbe-associated molecular patterns and danger signals by pattern-recognition receptors. Annu Rev Plant Biol. 2009 Jan;60:379–406.

[26] Thomma BPHJ, Nürnberger T, Joosten MHAJ. Of PAMPs and effectors: the blurred PTI-ETI dichotomy. Plant Cell. 2011 Jan;23(1):4–15.

[27] Cook DE, Mesarich CH, Thomma BPHJ. Understanding Plant Immunity as a Surveillance System to Detect Invasion. Annu Rev Phytopathol. 2015 Jun 5;53:541–63.

[28] Wirthmueller L, Maqbool A, Banfield MJ. On the front line: structural insights into plant-pathogen interactions. Nat Rev Microbiol. Nature Publishing Group, a division of Macmillan Publishers Limited. All Rights Reserved.; 2013 Nov;11(11):761–76.

[29] Holton N, Nekrasov V, Ronald PC, Zipfel C. The phylogenetically-related pattern recognition receptors EFR and XA21 recruit similar immune signaling components in

monocots and dicots. PLoS Pathog. Public Library of Science; 2015 Jan 21;11(1):e1004602.

[30] Matsushima N, Miyashita H. Leucine-Rich Repeat (LRR) Domains Containing Intervening Motifs in Plants. Biomolecules. 2012 Jan;2(2):288–311.

[31] Gómez-Gómez L, Boller T. FLS2: an LRR receptor-like kinase involved in the perception of the bacterial elicitor flagellin in *Arabidopsis*. Mol Cell. 2000 Jun;5(6):1003–11.

[32] Zipfel C, Kunze G, Chinchilla D, Caniard A, Jones JDG, Boller T, et al. Perception of the bacterial PAMP EF-Tu by the receptor EFR restricts *Agrobacterium*-mediated transformation. Cell. 2006 May 19;125(4):749–60.

[33] Miya A, Albert P, Shinya T, Desaki Y, Ichimura K, Shirasu K, et al. CERK1, a LysM receptor kinase, is essential for chitin elicitor signaling in *Arabidopsis*. Proc Natl Acad Sci U S A. 2007 Dec 4;104(49):19613–8.

[34] Petutschnig EK, Jones AME, Serazetdinova L, Lipka U, Lipka V. The lysin motif receptor-like kinase (LysM-RLK) CERK1 is a major chitin-binding protein in *Arabidopsis thaliana* and subject to chitin-induced phosphorylation. J Biol Chem. 2010 Sep 10;285(37):28902–11.

[35] Willmann R, Lajunen HM, Erbs G, Newman M-A, Kolb D, Tsuda K, et al. *Arabidopsis* lysin-motif proteins LYM1 LYM3 CERK1 mediate bacterial peptidoglycan sensing and immunity to bacterial infection. Proc Natl Acad Sci U S A. 2011 Dec 6;108(49): 19824–9.

[36] Kaku H, Nishizawa Y, Ishii-Minami N, Akimoto-Tomiyama C, Dohmae N, Takio K, et al. Plant cells recognize chitin fragments for defense signaling through a plasma membrane receptor. Proc Natl Acad Sci U S A. 2006 Jul 18;103(29):11086–91.

[37] Shimizu T, Nakano T, Takamizawa D, Desaki Y, Ishii-Minami N, Nishizawa Y, et al. Two LysM receptor molecules, CEBiP and OsCERK1, cooperatively regulate chitin elicitor signaling in rice. Plant J. 2010 Oct;64(2):204–14.

[38] Hayafune M, Berisio R, Marchetti R, Silipo A, Kayama M, Desaki Y, et al. Chitin-induced activation of immune signaling by the rice receptor CEBiP relies on a unique sandwich-type dimerization. Proc Natl Acad Sci U S A. 2014 Jan 21;111(3):E404–13.

[39] Akamatsu A, Wong HL, Fujiwara M, Okuda J, Nishide K, Uno K, et al. An OsCEBiP/OsCERK1-OsRacGEF1-OsRac1 module is an essential early component of chitin-induced rice immunity. Cell Host Microbe. 2013 Apr 17;13(4):465–76.

[40] Kosami K, Ohki I, Nagano M, Furuita K, Sugiki T, Kawano Y, et al. The crystal structure of the plant small GTPase OsRac1 reveals its mode of binding to NADPH oxidase. J Biol Chem. 2014 Oct 10;289(41):28569–78.

[41] Liu T, Liu Z, Song C, Hu Y, Han Z, She J, et al. Chitin-induced dimerization activates a plant immune receptor. Science. 2012 Jun 1;336(6085):1160–4.

[42] Cao Y, Liang Y, Tanaka K, Nguyen CT, Jedrzejczak RP, Joachimiak A, et al. The kinase LYK5 is a major chitin receptor in *Arabidopsis* and forms a chitin-induced complex with related kinase CERK1. Elife. 2014 Jan;3:e03766.

[43] Chinchilla D, Bauer Z, Regenass M, Boller T, Felix G. The *Arabidopsis* receptor kinase FLS2 binds flg22 and determines the specificity of flagellin perception. Plant Cell. 2006 Feb;18(2):465–76.

[44] Chinchilla D, Zipfel C, Robatzek S, Kemmerling B, Nürnberger T, Jones JDG, et al. A flagellin-induced complex of the receptor FLS2 and BAK1 initiates plant defence. Nature. 2007 Jul 26;448(7152):497–500.

[45] Roux M, Schwessinger B, Albrecht C, Chinchilla D, Jones A, Holton N, et al. The *Arabidopsis* leucine-rich repeat receptor-like kinases BAK1/SERK3 and BKK1/SERK4 are required for innate immunity to hemibiotrophic and biotrophic pathogens. Plant Cell. 2011 Jun;23(6):2440–55.

[46] Sun Y, Li L, Macho AP, Han Z, Hu Z, Zipfel C, et al. Structural basis for flg22-induced activation of the *Arabidopsis* FLS2-BAK1 immune complex. Science. 2013 Nov 1;342(6158):624–8.

[47] Belkhadir Y, Yang L, Hetzel J, Dangl JL, Chory J. The growth-defense pivot: crisis management in plants mediated by LRR-RK surface receptors. Trends Biochem Sci. Elsevier; 2014 Oct 10;39(10):447–56.

[48] Heese A, Hann DR, Gimenez-Ibanez S, Jones AME, He K, Li J, et al. The receptor-like kinase SERK3/BAK1 is a central regulator of innate immunity in plants. Proc Natl Acad Sci U S A. 2007 Jul 17;104(29):12217–22.

[49] Li J, Wen J, Lease KA, Doke JT, Tax FE, Walker JC. BAK1, an *Arabidopsis* LRR receptor-like protein kinase, interacts with BRI1 and modulates brassinosteroid signaling. Cell. 2002 Jul 26;110(2):213–22.

[50] Lu D, Wu S, Gao X, Zhang Y, Shan L, He P. A receptor-like cytoplasmic kinase, BIK1, associates with a flagellin receptor complex to initiate plant innate immunity. Proc Natl Acad Sci U S A. 2010 Jan 5;107(1):496–501.

[51] Lin W, Li B, Lu D, Chen S, Zhu N, He P, et al. Tyrosine phosphorylation of protein kinase complex BAK1/BIK1 mediates *Arabidopsis* innate immunity. Proc Natl Acad Sci U S A. 2014 Mar 4;111(9):3632–7.

[52] Ranf S, Eschen-Lippold L, Fröhlich K, Westphal L, Scheel D, Lee J. Microbe-associated molecular pattern-induced calcium signaling requires the receptor-like cytoplasmic kinases, PBL1 and BIK1. BMC Plant Biol. 2014 Jan;14:374.

[53] Li L, Li M, Yu L, Zhou Z, Liang X, Liu Z, et al. The FLS2-associated kinase BIK1 directly phosphorylates the NADPH oxidase RbohD to control plant immunity. Cell Host Microbe. 2014 Mar 12;15(3):329–38.

[54] Ao Y, Li Z, Feng D, Xiong F, Liu J, Li J-F, et al. OsCERK1 and OsRLCK176 play important roles in peptidoglycan and chitin signaling in rice innate immunity. Plant J. 2014 Dec;80(6):1072–84.

[55] Liu B, Li J-F, Ao Y, Qu J, Li Z, Su J, et al. Lysin motif-containing proteins LYP4 and LYP6 play dual roles in peptidoglycan and chitin perception in rice innate immunity. Plant Cell. 2012 Aug;24(8):3406–19.

[56] Albert M. Peptides as triggers of plant defence. J Exp Bot. 2013 Dec 1;64(17):5269–79.

[57] Ferrari S, Savatin D V, Sicilia F, Gramegna G, Cervone F, Lorenzo G De. Oligogalacturonides: plant damage-associated molecular patterns and regulators of growth and development. Front Plant Sci. 2013 Jan;4:49.

[58] Benedetti M, Pontiggia D, Raggi S, Cheng Z, Scaloni F, Ferrari S, et al. Plant immunity triggered by engineered in vivo release of oligogalacturonides, damage-associated molecular patterns. Proc Natl Acad Sci U S A. 2015 Apr 28;112(17):5533–8.

[59] Brutus A, Sicilia F, Macone A, Cervone F, De Lorenzo G. A domain swap approach reveals a role of the plant wall-associated kinase 1 (WAK1) as a receptor of oligogalacturonides. Proc Natl Acad Sci U S A. 2010 May 18;107(20):9452–7.

[60] Galletti R, Ferrari S, De Lorenzo G. *Arabidopsis* MPK3 and MPK6 play different roles in basal and oligogalacturonide- or flagellin-induced resistance against *Botrytis cinerea*. Plant Physiol. 2011 Oct 1;157(2):804–14.

[61] Huffaker A, Pearce G, Ryan CA. An endogenous peptide signal in *Arabidopsis* activates components of the innate immune response. Proc Natl Acad Sci U S A. 2006 Jun 27;103(26):10098–103.

[62] Huffaker A, Ryan CA. Endogenous peptide defense signals in *Arabidopsis* differentially amplify signaling for the innate immune response. Proc Natl Acad Sci U S A. 2007 Jun 19;104(25):10732–6.

[63] Yamaguchi Y, Huffaker A. Endogenous peptide elicitors in higher plants. Curr Opin Plant Biol. 2011 Aug;14(4):351–7.

[64] Yamaguchi Y, Pearce G, Ryan CA. The cell surface leucine-rich repeat receptor for AtPep1, an endogenous peptide elicitor in *Arabidopsis,* is functional in transgenic tobacco cells. Proc Natl Acad Sci U S A. 2006 Jun 27;103(26):10104–9.

[65] Yamaguchi Y, Huffaker A, Bryan AC, Tax FE, Ryan CA. PEPR2 is a second receptor for the Pep1 and Pep2 peptides and contributes to defense responses in *Arabidopsis*. Plant Cell. 2010 Feb;22(2):508–22.

[66] Postel S, Küfner I, Beuter C, Mazzotta S, Schwedt A, Borlotti A, et al. The multifunctional leucine-rich repeat receptor kinase BAK1 is implicated in *Arabidopsis* development and immunity. Eur J Cell Biol. Jan;89(2-3):169–74.

[67] Qi Z, Verma R, Gehring C, Yamaguchi Y, Zhao Y, Ryan CA, et al. Ca2+ signaling by plant *Arabidopsis thaliana* Pep peptides depends on AtPepR1, a receptor with guanylyl cyclase activity, and cGMP-activated Ca2+ channels. Proc Natl Acad Sci U S A. 2010 Dec 7;107(49):21193–8.

[68] Ma Y, Walker RK, Zhao Y, Berkowitz GA. Linking ligand perception by PEPR pattern recognition receptors to cytosolic Ca2+ elevation and downstream immune signaling in plants. Proc Natl Acad Sci U S A. 2012 Nov 27;109(48):19852–7.

[69] Collier SM, Moffett P. NB-LRRs work a "bait and switch" on pathogens. Trends Plant Sci. 2009 Oct;14(10):521–9.

[70] Maekawa T, Kufer TA, Schulze-Lefert P. NLR functions in plant and animal immune systems: so far and yet so close. Nat Immunol. 2011 Sep;12(9):817–26.

[71] Griebel T, Maekawa T, Parker JE. NOD-like receptor cooperativity in effector-triggered immunity. Trends Immunol. 2014 Nov;35(11):562–70.

[72] Danot O, Marquenet E, Vidal-Ingigliardi D, Richet E. Wheel of Life, Wheel of Death: A Mechanistic Insight into Signaling by STAND Proteins. Structure. 2009 Mar 13;17(2):172–82.

[73] Takken FLW, Tameling WIL. To nibble at plant resistance proteins. Science. 2009 May 8;324(5928):744–6.

[74] Lukasik E, Takken FLW. STANDing strong, resistance proteins instigators of plant defence. Curr Opin Plant Biol. 2009 Aug;12(4):427–36.

[75] Coll NS, Epple P, Dangl JL. Programmed cell death in the plant immune system. Cell Death Differ. Macmillan Publishers Limited; 2011 Aug;18(8):1247–56.

[76] Elmore JM, Lin Z-JD, Coaker G. Plant NB-LRR signaling: upstreams and downstreams. Curr Opin Plant Biol. 2011 Aug;14(4):365–71.

[77] Zhang M, Kadota Y, Prodromou C, Shirasu K, Pearl LH. Structural basis for assembly of Hsp90-Sgt1-CHORD protein complexes: implications for chaperoning of NLR innate immunity receptors. Mol Cell. 2010 Jul 30;39(2):269–81.

[78] Li Y, Li S, Bi D, Cheng YT, Li X, Zhang Y. SRFR1 negatively regulates plant NB-LRR resistance protein accumulation to prevent autoimmunity. PLoS Pathog. 2010 Jan; 6(9):e1001111.

[79] Jia Y, McAdams SA, Bryan GT, Hershey HP, Valent B. Direct interaction of resistance gene and avirulence gene products confers rice blast resistance. EMBO J. 2000 Aug 1;19(15):4004–14.

[80] Tang X, Frederick R, Zhou J, Halterman D, Jia Y, Martin G. Initiation of Plant Disease Resistance by Physical Interaction of AvrPto and Pto Kinase. Science. 1996 Dec 20;274(5295):2060–3.

[81] Van Der Biezen EA, Jones JDG. Plant disease-resistance proteins and the gene-for-gene concept. Trends Biochem Sci. 1998 Dec;23(12):454–6.

[82] Van der Hoorn RAL, De Wit PJGM, Joosten MHAJ. Balancing selection favors guarding resistance proteins. Trends Plant Sci. 2002 Mar;7(2):67–71.

[83] Marathe R, Dinesh-Kumar SP. Plant defense: one post, multiple guards?! Mol Cell. 2003 Mar;11(2):284–6.

[84] Van der Hoorn RAL, Kamoun S. From Guard to Decoy: a new model for perception of plant pathogen effectors. Plant Cell. 2008 Aug;20(8):2009–17.

[85] Ade J, DeYoung BJ, Golstein C, Innes RW. Indirect activation of a plant nucleotide binding site-leucine-rich repeat protein by a bacterial protease. Proc Natl Acad Sci U S A. 2007 Mar 13;104(7):2531–6.

[86] Shao F, Golstein C, Ade J, Stoutemyer M, Dixon JE, Innes RW. Cleavage of *Arabidopsis* PBS1 by a bacterial type III effector. Science. 2003 Aug 29;301(5637):1230–3.

[87] Zhang J, Li W, Xiang T, Liu Z, Laluk K, Ding X, et al. Receptor-like cytoplasmic kinases integrate signaling from multiple plant immune receptors and are targeted by a *Pseudomonas syringae* effector. Cell Host Microbe. 2010 Apr 22;7(4):290–301.

[88] Seybold H, Trempel F, Ranf S, Scheel D, Romeis T, Lee J. Ca 2+ signalling in plant immune response: from pattern recognition receptors to Ca 2+ decoding mechanisms. New Phytol. 2014 Dec 24;204(4):782–90.

[89] Wu S, Shan L, He P. Microbial signature-triggered plant defense responses and early signaling mechanisms. Plant Sci. 2014 Nov;228:118–26.

[90] Ranf S, Eschen-Lippold L, Pecher P, Lee J, Scheel D. Interplay between calcium signalling and early signalling elements during defence responses to microbe- or damage-associated molecular patterns. Plant J. 2011 Oct;68(1):100–13.

[91] Steinhorst L, Kudla J. Calcium and reactive oxygen species rule the waves of signaling. Plant Physiol. 2013 Oct 1;163(2):471–85.

[92] Suzuki N, Miller G, Morales J, Shulaev V, Torres MA, Mittler R. Respiratory burst oxidases: the engines of ROS signaling. Curr Opin Plant Biol. 2011 Dec;14(6):691–9.

[93] Baxter A, Mittler R, Suzuki N. ROS as key players in plant stress signalling. J Exp Bot. 2014 Mar;65(5):1229–40.

[94] Glyan'ko AK, Ischenko AA. Structural and functional characteristics of plant NADPH oxidase: A review. Appl Biochem Microbiol. 2010 Sep 11;46(5):463–71.

[95] Drerup MM, Schlücking K, Hashimoto K, Manishankar P, Steinhorst L, Kuchitsu K, et al. The Calcineurin B-like calcium sensors CBL1 and CBL9 together with their interacting protein kinase CIPK26 regulate the *Arabidopsis* NADPH oxidase RBOHF. Mol Plant. 2013 Mar;6(2):559–69.

[96] Ogasawara Y, Kaya H, Hiraoka G, Yumoto F, Kimura S, Kadota Y, et al. Synergistic activation of the *Arabidopsis* NADPH oxidase AtrbohD by Ca2+ and phosphorylation. J Biol Chem. 2008 Apr 4;283(14):8885–92.

[97] Kimura S, Kaya H, Kawarazaki T, Hiraoka G, Senzaki E, Michikawa M, et al. Protein phosphorylation is a prerequisite for the Ca2+-dependent activation of *Arabidopsis* NADPH oxidases and may function as a trigger for the positive feedback regulation of Ca2+ and reactive oxygen species. Biochim Biophys Acta. 2012 Feb;1823(2):398–405.

[98] Dubiella U, Seybold H, Durian G, Komander E, Lassig R, Witte C-P, et al. Calcium-dependent protein kinase/NADPH oxidase activation circuit is required for rapid defense signal propagation. Proc Natl Acad Sci U S A. 2013 May 21;110(21):8744–9.

[99] Miller G, Schlauch K, Tam R, Cortes D, Torres MA, Shulaev V, et al. The plant NADPH oxidase RBOHD mediates rapid systemic signaling in response to diverse stimuli. Sci Signal. 2009 Jan 18;2(84):ra45.

[100] Mittler R, Vanderauwera S, Suzuki N, Miller G, Tognetti VB, Vandepoele K, et al. ROS signaling: the new wave? Trends Plant Sci. 2011 Jun;16(6):300–9.

[101] Tena G, Boudsocq M, Sheen J. Protein kinase signaling networks in plant innate immunity. Curr Opin Plant Biol. 2011 Oct;14(5):519–29.

[102] Popescu SC, Popescu G V, Bachan S, Zhang Z, Gerstein M, Snyder M, et al. MAPK target networks in *Arabidopsis thaliana* revealed using functional protein microarrays. Genes Dev. 2009 Jan 1;23(1):80–92.

[103] Asai T, Tena G, Plotnikova J, Willmann MR, Chiu W-L, Gomez-Gomez L, et al. MAP kinase signalling cascade in *Arabidopsis* innate immunity. Nature. 2002 Mar 28;415(6875):977–83.

[104] Gao M, Liu J, Bi D, Zhang Z, Cheng F, Chen S, et al. MEKK1, MKK1/MKK2 and MPK4 function together in a mitogen-activated protein kinase cascade to regulate innate immunity in plants. Cell Res. 2008 Dec;18(12):1190–8.

[105] Xu J, Xie J, Yan C, Zou X, Ren D, Zhang S. A chemical genetic approach demonstrates that MPK3/MPK6 activation and NADPH oxidase-mediated oxidative burst are two independent signaling events in plant immunity. Plant J. 2014 Jan;77(2):222–34.

[106] Rodriguez MCS, Petersen M, Mundy J. Mitogen-activated protein kinase signaling in plants. Annu Rev Plant Biol. Annual Reviews; 2010 Jan 4;61:621–49.

[107] Kong Q, Qu N, Gao M, Zhang Z, Ding X, Yang F, et al. The MEKK1-MKK1/MKK2-MPK4 kinase cascade negatively regulates immunity mediated by a mitogen-activated protein kinase kinase kinase in *Arabidopsis*. Plant Cell. 2012 May;24(5):2225–36.

[108] Ren D, Liu Y, Yang K-Y, Han L, Mao G, Glazebrook J, et al. A fungal-responsive MAPK cascade regulates phytoalexin biosynthesis in *Arabidopsis*. Proc Natl Acad Sci U S A. 2008 Apr 8;105(14):5638–43.

[109] Mao G, Meng X, Liu Y, Zheng Z, Chen Z, Zhang S. Phosphorylation of a WRKY transcription factor by two pathogen-responsive MAPKs drives phytoalexin biosynthesis in *Arabidopsis*. Plant Cell. 2011 Apr;23(4):1639–53.

[110] Beckers GJM, Jaskiewicz M, Liu Y, Underwood WR, He SY, Zhang S, et al. Mitogen-activated protein kinases 3 and 6 are required for full priming of stress responses in *Arabidopsis thaliana*. Plant Cell. 2009 Mar;21(3):944–53.

[111] Han L, Li G-J, Yang K-Y, Mao G, Wang R, Liu Y, et al. Mitogen-activated protein kinase 3 and 6 regulate *Botrytis cinerea*-induced ethylene production in *Arabidopsis*. Plant J. 2010 Oct;64(1):114–27.

[112] Ren D, Yang H, Zhang S. Cell death mediated by MAPK is associated with hydrogen peroxide production in *Arabidopsis*. J Biol Chem. 2002 Jan 4;277(1):559–65.

[113] Zhao C, Nie H, Shen Q, Zhang S, Lukowitz W, Tang D. EDR1 physically interacts with MKK4/MKK5 and negatively regulates a MAP kinase cascade to modulate plant innate immunity. PLoS Genet. 2014 Jan;10(5):e1004389.

[114] Frei dit Frey N, Garcia AV, Bigeard J, Zaag R, Bueso E, Garmier M, et al. Functional analysis of *Arabidopsis* immune-related MAPKs uncovers a role for MPK3 as negative regulator of inducible defences. Genome Biol. 2014 Jan;15(6):R87.

[115] Lumbreras V, Vilela B, Irar S, Solé M, Capellades M, Valls M, et al. MAPK phosphatase MKP2 mediates disease responses in *Arabidopsis* and functionally interacts with MPK3 and MPK6. Plant J. 2010 Sep;63(6):1017–30.

[116] Moore JW, Loake GJ, Spoel SH. Transcription dynamics in plant immunity. Plant Cell. 2011 Aug;23(8):2809–20.

[117] Licausi F, Ohme-Takagi M, Perata P. APETALA2/Ethylene Responsive Factor (AP2/ERF) transcription factors: mediators of stress responses and developmental programs. New Phytol. 2013 Aug 7;199(3):639–49.

[118] Kazan K, Manners JM. MYC2: the master in action. Mol Plant. 2013 May;6(3):686–703.

[119] Zander M, Thurow C, Gatz C. TGA Transcription Factors Activate the Salicylic Acid-Suppressible Branch of the Ethylene-Induced Defense Program by Regulating ORA59 Expression. Plant Physiol. 2014 Jul 2;165(4):1671–83.

[120] Chang C, Yu D, Jiao J, Jing S, Schulze-Lefert P, Shen Q-H. Barley MLA immune receptors directly interfere with antagonistically acting transcription factors to initiate disease resistance signaling. Plant Cell. 2013 Mar;25(3):1158–73.

[121] Williams SJ, Sohn KH, Wan L, Bernoux M, Sarris PF, Segonzac C, et al. Structural basis for assembly and function of a heterodimeric plant immune receptor. Science. 2014 Apr 18;344(6181):299–303.

[122] Padmanabhan MS, Ma S, Burch-Smith TM, Czymmek K, Huijser P, Dinesh-Kumar SP. Novel positive regulatory role for the SPL6 transcription factor in the N TIR-NB-LRR receptor-mediated plant innate immunity. PLoS Pathog. 2013 Mar;9(3):e1003235.

[123] Inoue H, Hayashi N, Matsushita A, Xinqiong L, Nakayama A, Sugano S, et al. Blast resistance of CC-NB-LRR protein Pb1 is mediated by WRKY45 through protein-protein interaction. Proc Natl Acad Sci U S A. 2013 Jun 4;110(23):9577–82.

[124] Ishihama N, Yoshioka H. Post-translational regulation of WRKY transcription factors in plant immunity. Curr Opin Plant Biol. 2012 Aug;15(4):431–7.

[125] Ishihama N, Yamada R, Yoshioka M, Katou S, Yoshioka H. Phosphorylation of the *Nicotiana benthamiana* WRKY8 transcription factor by MAPK functions in the defense response. Plant Cell. 2011 Mar;23(3):1153–70.

[126] Meng X, Xu J, He Y, Yang K-Y, Mordorski B, Liu Y, et al. Phosphorylation of an ERF transcription factor by *Arabidopsis* MPK3/MPK6 regulates plant defense gene induction and fungal resistance. Plant Cell. 2013 Mar;25(3):1126–42.

[127] Pecher P, Eschen-Lippold L, Herklotz S, Kuhle K, Naumann K, Bethke G, et al. The *Arabidopsis thaliana* mitogen-activated protein kinases MPK3 and MPK6 target a subclass of "VQ-motif"-containing proteins to regulate immune responses. New Phytol. 2014 Jul;203(2):592–606.

[128] Galon Y, Nave R, Boyce JM, Nachmias D, Knight MR, Fromm H. Calmodulin-binding transcription activator (CAMTA) 3 mediates biotic defense responses in *Arabidopsis*. FEBS Lett. 2008 Mar 19;582(6):943–8.

[129] Du L, Ali GS, Simons KA, Hou J, Yang T, Reddy ASN, et al. Ca^{2+}/calmodulin regulates salicylic-acid-mediated plant immunity. Nature. 2009 Jan 4;457(7233):1154–8.

[130] Wang L, Tsuda K, Truman W, Sato M, Nguyen L V, Katagiri F, et al. CBP60g and SARD1 play partially redundant critical roles in salicylic acid signaling. Plant J. 2011 Sep;67(6):1029–41.

[131] Pedrini N, Zhang S, Juárez MP, Keyhani NO. Molecular characterization and expression analysis of a suite of cytochrome P450 enzymes implicated in insect hydrocarbon degradation in the entomopathogenic fungus *Beauveria bassiana*. Microbiology. 2010 Aug 1;156(Pt 8):2549–57.

[132] Kim HS, Park HC, Kim KE, Jung MS, Han HJ, Kim SH, et al. A NAC transcription factor and SNI1 cooperatively suppress basal pathogen resistance in *Arabidopsis thaliana*. Nucleic Acids Res. 2012 Oct;40(18):9182–92.

[133] Gao X, Chen X, Lin W, Chen S, Lu D, Niu Y, et al. Bifurcation of *Arabidopsis* NLR immune signaling via Ca^{2+}-dependent protein kinases. PLoS Pathog. 2013 Jan; 9(1):e1003127.

[134] Wathugala DL, Hemsley PA, Moffat CS, Cremelie P, Knight MR, Knight H. The Mediator subunit SFR6/MED16 controls defence gene expression mediated by salicylic acid and jasmonate responsive pathways. New Phytol. 2012 Jul;195(1):217–30.

[135] Çevik V, Kidd BN, Zhang P, Hill C, Kiddle S, Denby KJ, et al. MEDIATOR25 acts as an integrative hub for the regulation of jasmonate-responsive gene expression in *Arabidopsis*. Plant Physiol. 2012 Sep;160(1):541–55.

[136] Berr A, Ménard R, Heitz T, Shen W-H. Chromatin modification and remodelling: a regulatory landscape for the control of *Arabidopsis* defence responses upon pathogen attack. Cell Microbiol. 2012 Jun;14(6):829–39.

[137] Kim K-C, Lai Z, Fan B, Chen Z. *Arabidopsis* WRKY38 and WRKY62 transcription factors interact with histone deacetylase 19 in basal defense. Plant Cell. 2008 Sep;20(9):2357–71.

[138] Alvarez ME, Nota F, Cambiagno DA. Epigenetic control of plant immunity. Mol Plant Pathol. 2010 Jul;11(4):563–76.

[139] Marino D, Peeters N, Rivas S. Ubiquitination during plant immune signaling. Plant Physiol. 2012 Sep;160(1):15–27.

[140] Park HJ, Kim W-Y, Park HC, Lee SY, Bohnert HJ, Yun D-J. SUMO and SUMOylation in plants. Mol Cells. 2011 Oct;32(4):305–16.

[141] Yang S, Tang F, Zhu H. Alternative splicing in plant immunity. Int J Mol Sci. 2014 Jan;15(6):10424–45.

[142] Pieterse CMJ, Van der Does D, Zamioudis C, Leon-Reyes A, Van Wees SCM. Hormonal modulation of plant immunity. Annu Rev Cell Dev Biol. 2012 Jan;28:489–521.

[143] Wildermuth MC, Dewdney J, Wu G, Ausubel FM. Isochorismate synthase is required to synthesize salicylic acid for plant defence. Nature. 2001 Nov 29;414(6863):562–5.

[144] Serrano M, Wang B, Aryal B, Garcion C, Abou-Mansour E, Heck S, et al. Export of salicylic acid from the chloroplast requires the multidrug and toxin extrusion-like transporter EDS5. Plant Physiol. 2013 Aug;162(4):1815–21.

[145] Zhang Y, Xu S, Ding P, Wang D, Cheng YT, He J, et al. Control of salicylic acid synthesis and systemic acquired resistance by two members of a plant-specific family of transcription factors. Proc Natl Acad Sci U S A. 2010 Oct 19;107(42):18220–5.

[146] Zhang Y, Fan W, Kinkema M, Li X, Dong X. Interaction of NPR1 with basic leucine zipper protein transcription factors that bind sequences required for salicylic acid induction of the PR-1 gene. Proc Natl Acad Sci U S A. 1999 May 25;96(11):6523–8.

[147] Pajerowska-Mukhtar KM, Emerine DK, Mukhtar MS. Tell me more: roles of NPRs in plant immunity. Trends Plant Sci. 2013 Jul;18(7):402–11.

[148] Tada Y, Spoel SH, Pajerowska-Mukhtar K, Mou Z, Song J, Wang C, et al. Plant immunity requires conformational changes [corrected] of NPR1 via S-nitrosylation and thioredoxins. Science. 2008 Aug 15;321(5891):952–6.

[149] Fu ZQ, Yan S, Saleh A, Wang W, Ruble J, Oka N, et al. NPR3 and NPR4 are receptors for the immune signal salicylic acid in plants. Nature. 2012 Jun 14;486(7402):228–32.

[150] Yan S, Dong X. Perception of the plant immune signal salicylic acid. Curr Opin Plant Biol. 2014 Aug;20:64–8.

[151] Kazan K, Manners JM. Jasmonate signaling: toward an integrated view. Plant Physiol. 2008 Apr 1;146(4):1459–68.

[152] Chini A, Fonseca S, Fernández G, Adie B, Chico JM, Lorenzo O, et al. The JAZ family of repressors is the missing link in jasmonate signalling. Nature. 2007 Aug 9;448(7154):666–71.

[153] Kazan K, Manners JM. JAZ repressors and the orchestration of phytohormone crosstalk. Trends Plant Sci. 2012 Jan;17(1):22–31.

[154] Lyons R, Manners JM, Kazan K. Jasmonate biosynthesis and signaling in monocots: a comparative overview. Plant Cell Rep. 2013 Jun;32(6):815–27.

[155] Antico CJ, Colon C, Banks T, Ramonell KM. Insights into the role of jasmonic acid-mediated defenses against necrotrophic and biotrophic fungal pathogens. Front Biol (Beijing). 2012 Jan 8;7(1):48–56.

[156] Kazan K, Lyons R. Intervention of Phytohormone Pathways by Pathogen Effectors. Plant Cell. 2014 Jun 10;26(6):2285–309.

[157] Wang KL-C, Li H, Ecker JR. Ethylene biosynthesis and signaling networks. Plant Cell. 2002 Jan;14 Suppl:S131–51.

[158] Li G, Meng X, Wang R, Mao G, Han L, Liu Y, et al. Dual-level regulation of ACC synthase activity by MPK3/MPK6 cascade and its downstream WRKY transcription factor during ethylene induction in *Arabidopsis*. PLoS Genet. 2012 Jun;8(6):e1002767.

[159] Ju C, Yoon GM, Shemansky JM, Lin DY, Ying ZI, Chang J, et al. CTR1 phosphorylates the central regulator EIN2 to control ethylene hormone signaling from the ER membrane to the nucleus in *Arabidopsis*. Proc Natl Acad Sci U S A. 2012 Nov 20;109(47):19486–91.

[160] Wen X, Zhang C, Ji Y, Zhao Q, He W, An F, et al. Activation of ethylene signaling is mediated by nuclear translocation of the cleaved EIN2 carboxyl terminus. Cell Res. 2012 Nov;22(11):1613–6.

[161] Zander M, Chen S, Imkampe J, Thurow C, Gatz C. Repression of the *Arabidopsis thaliana* jasmonic acid/ethylene-induced defense pathway by TGA-interacting glutaredoxins depends on their C-terminal ALWL motif. Mol Plant. 2012 Jul;5(4):831–40.

[162] Qiao H, Shen Z, Huang SC, Schmitz RJ, Urich MA, Briggs SP, et al. Processing and subcellular trafficking of ER-tethered EIN2 control response to ethylene gas. Science. 2012 Oct 19;338(6105):390–3.

[163] De Vleesschauwer D, Xu J, Höfte M. Making sense of hormone-mediated defense networking: from rice to *Arabidopsis*. Front Plant Sci. 2014 Jan;5:611.

[164] Caarls L, Pieterse CMJ, Van Wees SCM. How salicylic acid takes transcriptional control over jasmonic acid signaling. Front Plant Sci. 2015 Jan;6:170.

[165] Song S, Huang H, Gao H, Wang J, Wu D, Liu X, et al. Interaction between MYC2 and ETHYLENE INSENSITIVE3 modulates antagonism between jasmonate and ethylene signaling in *Arabidopsis*. Plant Cell. 2014 Jan;26(1):263–79.

[166] Segonzac C, Macho AP, Sanmartín M, Ntoukakis V, Sánchez-Serrano JJ, Zipfel C. Negative control of BAK1 by protein phosphatase 2A during plant innate immunity. EMBO J. 2014 Sep 17;33(18):2069–79.

[167] Anderson JC, Bartels S, González Besteiro MA, Shahollari B, Ulm R, Peck SC. *Arabidopsis* MAP Kinase Phosphatase 1 (AtMKP1) negatively regulates MPK6-mediated PAMP responses and resistance against bacteria. Plant J. 2011 Jul;67(2):258–68.

[168] Lu D, Lin W, Gao X, Wu S, Cheng C, Avila J, et al. Direct ubiquitination of pattern recognition receptor FLS2 attenuates plant innate immunity. Science. 2011 Jun 17;332(6036):1439–42.

[169] Nathalie Leborgne-Castel, Bouhidel K. Plasma membrane protein trafficking in plant-microbe interactions: a plant cell point of view. Front Plant Sci. Frontiers; 2014 Jan 22;5:735.

[170] Robatzek S, Chinchilla D, Boller T. Ligand-induced endocytosis of the pattern recognition receptor FLS2 in *Arabidopsis*. Genes Dev. 2006 Mar 1;20(5):537–42.

[171] Smith JM, Salamango DJ, Leslie ME, Collins CA, Heese A. Sensitivity to Flg22 is modulated by ligand-induced degradation and de novo synthesis of the endogenous flagellin-receptor FLAGELLIN-SENSING2. Plant Physiol. 2014 Jan;164(1):440–54.

[172] Millet YA, Danna CH, Clay NK, Songnuan W, Simon MD, Werck-Reichhart D, et al. Innate immune responses activated in *Arabidopsis* roots by microbe-associated molecular patterns. Plant Cell. 2010 Mar;22(3):973–90.

[173] Faulkner C, Robatzek S. Plants and pathogens: putting infection strategies and defence mechanisms on the map. Curr Opin Plant Biol. 2012 Dec;15(6):699–707.

[174] Beck M, Wyrsch I, Strutt J, Wimalasekera R, Webb A, Boller T, et al. Expression patterns of flagellin sensing 2 map to bacterial entry sites in plant shoots and roots. J Exp Bot. 2014 Dec;65(22):6487–98.

[175] Kawchuk LM, Hachey J, Lynch DR, Kulcsar F, van Rooijen G, Waterer DR, et al. Tomato Ve disease resistance genes encode cell surface-like receptors. Proc Natl Acad Sci U S A. 2001;98(11):6511–5.

[176] Fradin EF, Zhang Z, Rovenich H, Song Y, Liebrand TWH, Masini L, et al. Functional analysis of the tomato immune receptor Ve1 through domain swaps with its non-functional homolog Ve2. PLoS One. 2014 Jan;9(2):e88208.

[177] Fradin EF, Zhang Z, Juarez Ayala JC, Castroverde CDM, Nazar RN, Robb J, et al. Genetic Dissection of *Verticillium* Wilt Resistance Mediated by Tomato Ve1. Plant Physiol. 2009;150(1):320–32.

[178] Fradin EF, Abd-El-Haliem A, Masini L, van den Berg GCM, Joosten MHAJ, Thomma BPHJ. Interfamily Transfer of Tomato Ve1 Mediates *Verticillium* Resistance in *Arabidopsis*. Plant Physiol. 2011;156(4):2255–65.

[179] Zhang Z, Fradin E, de Jonge R, van Esse HP, Smit P, Liu C-M, et al. Optimized agro-infiltration and virus-induced gene silencing to study Ve1-mediated *Verticillium* resistance in tobacco. Mol Plant Microbe Interact. 2013 Mar;26(2):182–90.

[180] De Jonge R, van Esse HP, Maruthachalam K, Bolton MD, Santhanam P, Saber MK, et al. Tomato immune receptor Ve1 recognizes effector of multiple fungal pathogens uncovered by genome and RNA sequencing. Proc Natl Acad Sci. National Acad Sciences; 2012;109(13):5110–5.

[181] Liebrand TWH, van den Berg GCM, Zhang Z, Smit P, Cordewener JHG, America AHP, et al. Receptor-like kinase SOBIR1/EVR interacts with receptor-like proteins in plant immunity against fungal infection. Proc Natl Acad Sci U S A. 2013 Jun 11;110(24):10010–5.

[182] Liebrand TWH, van den Burg HA, Joosten MHAJ. Two for all: receptor-associated kinases SOBIR1 and BAK1. Trends Plant Sci. 2014 Feb;19(2):123–32.

[183] Liebrand TWH, Kombrink A, Zhang Z, Sklenar J, Jones AME, Robatzek S, et al. Chaperones of the endoplasmic reticulum are required for Ve1-mediated resistance to *Verticillium*. Mol Plant Pathol. 2014 Jan;15(1):109–17.

[184] Huang C-C, Lindhout P. Screening for resistance in wild *Lycopersicon* species to *Fusarium oxysporum* f.sp. *lycopersici* race 1 and race 2. Euphytica. Kluwer Academic Publishers; 93(2):145–53.

[185] Beckman CH. Phenolic-storing cells: keys to programmed cell death and periderm formation in wilt disease resistance and in general defence responses in plants? Physiol Mol Plant Pathol. 2000 Sep;57(3):101–10.

[186] Houterman PM, Ma L, van Ooijen G, de Vroomen MJ, Cornelissen BJC, Takken FLW, et al. The effector protein Avr2 of the xylem-colonizing fungus *Fusarium oxysporum* activates the tomato resistance protein I-2 intracellularly. Plant J. 2009 Jul;58(6):970–8.

[187] Michielse CB, van Wijk R, Reijnen L, Cornelissen BJC, Rep M. Insight into the molecular requirements for pathogenicity of *Fusarium oxysporum* f. sp. lycopersici through large-scale insertional mutagenesis. Genome Biol. 2009 Jan;10(1):R4.

[188] Ma L, Cornelissen BJC, Takken FLW. A nuclear localization for Avr2 from *Fusarium oxysporum* is required to activate the tomato resistance protein I-2. Front Plant Sci. 2013 Jan;4:94.

[189] Gawehns F, Houterman PM, Ichou FA, Michielse CB, Hijdra M, Cornelissen BJC, et al. The *Fusarium oxysporum* effector Six6 contributes to virulence and suppresses I-2-mediated cell death. Mol Plant Microbe Interact. The American Phytopathological Society; 2014 May 4;27(4):336–48.

[190] Ma L, Houterman PM, Gawehns F, Cao L, Sillo F, Richter H, et al. The AVR2-SIX5 gene pair is required to activate I-2-mediated immunity in tomato. New Phytol. 2015 May 13.

[191] Song WY, Wang GL, Chen LL, Kim HS, Pi LY, Holsten T, et al. A receptor kinase-like protein encoded by the rice disease resistance gene, Xa21. Science. 1995 Dec 15;270(5243):1804–6.

[192] Lee S-W, Han S-W, Sririyanum M, Park C-J, Seo Y-S, Ronald PC. A type I-secreted, sulfated peptide triggers XA21-mediated innate immunity. Science. 2009 Dec 6;326(5954):850–3.

[193] Park C-J, Han S-W, Chen X, Ronald PC. Elucidation of XA21-mediated innate immunity. Cell Microbiol. 2010 Aug;12(8):1017–25.

[194] Chen X, Chern M, Canlas PE, Ruan D, Jiang C, Ronald PC. An ATPase promotes autophosphorylation of the pattern recognition receptor XA21 and inhibits XA21-mediated immunity. Proc Natl Acad Sci U S A. 2010 May 27;107(17):8029–34.

[195] Park C-J, Peng Y, Chen X, Dardick C, Ruan D, Bart R, et al. Rice XB15, a protein phosphatase 2C, negatively regulates cell death and XA21-mediated innate immunity. PLoS Biol. 2008 Oct 23;6(9):e231.

[196] Wang Y-S, Pi L-Y, Chen X, Chakrabarty PK, Jiang J, De Leon AL, et al. Rice XA21 binding protein 3 is a ubiquitin ligase required for full Xa21-mediated disease resistance. Plant Cell. 2006 Dec;18(12):3635–46.

[197] Jiang Y, Chen X, Ding X, Wang Y, Chen Q, Song W-Y. The XA21 binding protein XB25 is required for maintaining XA21-mediated disease resistance. Plant J. 2013 Mar;73(5):814–23.

[198] Peng Y, Bartley LE, Chen X, Dardick C, Chern M, Ruan R, et al. OsWRKY62 is a negative regulator of basal and Xa21-mediated defense against *Xanthomonas oryzae* pv. *oryzae* in rice. Mol Plant. 2008 May;1(3):446–58.

[199] Park C-J, Bart R, Chern M, Canlas PE, Bai W, Ronald PC. Overexpression of the endoplasmic reticulum chaperone BiP3 regulates XA21-mediated innate immunity in rice. PLoS One. 2010 Jan;5(2):e9262.

[200] Narusaka M, Shirasu K, Noutoshi Y, Kubo Y, Shiraishi T, Iwabuchi M, et al. RRS1 and RPS4 provide a dual Resistance-gene system against fungal and bacterial pathogens. Plant J. 2009 Oct;60(2):218–26.

[201] Narusaka M, Hatakeyama K, Shirasu K, Narusaka Y. *Arabidopsis* dual resistance proteins, both RPS4 and RRS1, are required for resistance to bacterial wilt in transgenic *Brassica* crops. Plant Signal Behav. 2014 Jan;9(7):e29130.

[202] Saucet SB, Ma Y, Sarris PF, Furzer OJ, Sohn KH, Jones JDG. Two linked pairs of *Arabidopsis* TNL resistance genes independently confer recognition of bacterial effector AvrRps4. Nat Commun. 2015 Jan;6:6338.

[203] Deslandes L, Olivier J, Theulieres F, Hirsch J, Feng DX, Bittner-Eddy P, et al. Resistance to *Ralstonia solanacearum* in *Arabidopsis thaliana* is conferred by the recessive RRS1-R gene, a member of a novel family of resistance genes. Proc Natl Acad Sci U S A. 2002 Mar 19;99(4):2404–9.

[204] Deslandes L, Olivier J, Peeters N, Feng DX, Khounlotham M, Boucher C, et al. Physical interaction between RRS1-R, a protein conferring resistance to bacterial wilt, and PopP2, a type III effector targeted to the plant nucleus. Proc Natl Acad Sci U S A. 2003 Jul 24;100(13):8024–9.

[205] Sarris PF, Duxbury Z, Huh SU, Ma Y, Segonzac C, Sklenar J, et al. A Plant Immune Receptor Detects Pathogen Effectors that Target WRKY Transcription Factors. Cell. 2015 May 21;161(5):1089–100.

[206] Le Roux C, Huet G, Jauneau A, Camborde L, Trémousaygue D, Kraut A, et al. A Receptor Pair with an Integrated Decoy Converts Pathogen Disabling of Transcription Factors to Immunity. Cell. 2015 May 21;161(5):1074–88.

[207] Reusche M, Thole K, Janz D, Truskina J, Rindfleisch S, Drübert C, et al. *Verticillium* infection triggers VASCULAR-RELATED NAC DOMAIN7-dependent de novo xylem formation and enhances drought tolerance in *Arabidopsis*. Plant Cell. 2012 Sep; 24(9):3823–37.

[208] Hu J, Barlet X, Deslandes L, Hirsch J, Feng DX, Somssich I, et al. Transcriptional responses of *Arabidopsis thaliana* during wilt disease caused by the soil-borne phytopathogenic bacterium, *Ralstonia solanacearum*. PLoS One. 2008 Jan;3(7):e2589.

[209] Zhang Y, Wang XF, Ding ZG, Ma Q, Zhang GR, Zhang SL, et al. Transcriptome profiling of *Gossypium barbadense* inoculated with *Verticillium dahliae* provides a resource for cotton improvement. BMC Genomics. BioMed Central Ltd; 2013;14(1):637.

[210] Van Esse HP, Fradin EF, de Groot PJ, de Wit PJGM, Thomma BPHJ. Tomato transcriptional responses to a foliar and a vascular fungal pathogen are distinct. Mol Plant Microbe Interact. 2009 Mar;22(3):245–58.

[211] Chen YC, Wong CL, Muzzi F, Vlaardingerbroek I, Kidd BN, Schenk PM. Root defense analysis against *Fusarium oxysporum* reveals new regulators to confer resistance. Sci Rep. Nature Publishing Group; 2014 Jan 7;4:5584.

[212] Robb J, Shittu H, Soman K V, Kurosky A, Nazar RN. Arsenal of elevated defense proteins fails to protect tomato against *Verticillium dahliae*. Planta. 2012;236(2):623–33.

[213] Mandelc S, Timperman I, Radišek S, Devreese B, Samyn B, Javornik B. Comparative proteomic profiling in compatible and incompatible interactions between hop roots and *Verticillium albo-atrum*. Plant Physiol Biochem. 2013 Jul;68:23–31.

6

Molecules and Methods for the Control of Biotic Stress Especially the Insect Pests — Present Scenario and Future Perspective

Santosh Kumar Upadhyay and Sudhir P. Singh

Abstract

Continuous increase in population has unbalanced the demand and supply of agricultural produce. In this scenario, food security in a sustainable manner is being challenged due to several factors. Insect pests are considered as one of the major factors, which accounts for 35–100% crop damage, worldwide. Synthetic insecticides contributed significantly, but several safety concerns are associated with them. Transgenic crops with enhanced biotic or abiotic stress tolerance have shown promising contribution in achieving greater crop productivity. Transgenic cotton expressing Cry toxin of *Bacillus thuringiensis* has tremendously increased the production as well as the societal status of farmers in our country. However, a concomitant increase in the population of minor pests like aphids, whiteflies and others has demanded certain new approaches. Researchers have isolated several other toxic proteins like lectins, protease inhibitors, amylase inhibitors, chitinases, and tried various novel approaches like gene pyramiding, tissue specific expression and modulation in metabolites expression to combat emerging problems of insect pests. Conversely, the emergence of a new type of crop insect pests demands more specific effort for each insect. Besides this, there are several safety and ethical concerns that are associated with the use of genetically modified crops, which also need to be resolved as per demand. Development of a dedicated scientific forum for the proper demonstration of advantages and disadvantages of genetically modified crops to the citizens at ground level might be useful in resolving the societal and ethical concerns in our country.

Keywords: Insecticidal proteins, Cry toxins, Lectins, Protease inhibitors, Chinitase, Transgenic crops

1. Introduction

World population is projected to increase over 1,000 million in the next four decades. An immediate priority for agriculture industry is to achieve maximum production in an environmentally sustainable and cost-effective manner. Food security is on high agenda at the political and social level [1]. Our progeny can face a severe shortage of food supply due to the over demand of continuously increasing population. Jaques Diouf, the Director General FAO, stated (2011) *"The silent hunger crisis, affecting one sixth of all humanity, poses a serious risk for world peace and security"*. The current challenge is to increase primary crop production in agricultural sustainability manner. In order to achieve these goals, it is important to identify and address the major limitations of productivity. Crop damage caused by insect pests is one of the major confinements, which is estimated to be around 35–100%, globally [2]. Synthetic insecticides have made significant contributions in food production, but they are also responsible for environmental and health hazards.

Transgenic crops with enhanced biotic or abiotic stress tolerance have shown promising contribution in achieving greater food security. A milestone was established about 25 years ago with the development of genetically engineered tobacco expressing the entomotoxic Cry protein from the bacterium *Bacillus thuringiensis* (Bt) [3, 4]. Presently, a number of Bt-Cry protein containing products are in the market of the United States (www.aphis.usda.gov) and some other countries. In March 2002, the Government of India permitted the release of transgenic cotton (Bollgard) expressing Bt toxins-Cry1Ac for commercial cultivation, which conferred resistance against bollworms [5]. Subsequently, BollgardII was released, which expresses Cry2Ab toxin along with Cry1Ac. Bt-Cry proteins have increased the productivity of crops substantially by controlling the major insects of order Lepidoptera and Coleoptera [6]. However, a concomitant increase in the population of minor pests (like whiteflies, aphids, leafhoppers and others) has threatened the success of Bt-transgenic crops [7–9]. An alternative strategy is to take advantage of the plant's own defense mechanisms, either by maneuvering the expression of the endogenous defense proteins or by introduction of an insect toxic gene (like lectins) derived from another plant. Several insecticidal proteins encoding genes have now been isolated from different sources and introduced into crop genomes to combat the issue of various groups of insect pests [10, 11]. Simultaneously, a new approach based on RNA interference is also reported for the control of crop insects [12–14]. Present chapter briefs about the insecticidal proteins and transgenic strategies for the control of crop insects.

2. Insecticidal proteins

2.1. Cry toxins of *Bacillus thuringiensis*

Introduction of Bt-Cry toxins revolutionized the area of insect control through transgenic technology. These are toxic to the insects of orders Lepidoptera, Diptera, Hymenoptera, Coleoptera and also to nematodes. These are produced as parasporal crystalline inclusions in *B. thuringiensis*. More than 500 Cry proteins/genes have been discovered till date, which are classified into 67 groups (Cry1–Cry67) on the basis of the primary structure [15, 16]. The genes are further divided into four phylogenetically unrelated protein families with different modes of action. These are: (1) three domain Cry toxins (3D) family, (2) mosquitocidal Cry toxins (Mtx) family, (3) binary-like (Bin) family and (4) the Cyt family of toxins [17]. Some Bt strains produce an additional insecticidal toxin called as VIP (vegetative insecticidal proteins) during the vegetative growth phase. Three VIP toxins: VIP1/VIP2, a binary toxin, and VIP3 have been characterized till date [18, 19].

Several insect-resistant transgenic crops have been developed by expressing Bt-Cry proteins, among which corn, cotton, soybean and canola are the most important crops. These transgenic crops are mostly expressing the Cry1Ac and Cry2Ab to control the chewing pests like *H. armigera, H. zea* and *Pectinophora gossypiella, Heliothis virescens* and *Ostrinia nubilalis* [20]. Some other cry toxin based products are also commercialized, which express Cry1A, Cry1F, Cry1EC, Cry34Ab/Cry35Ab binary toxin, Cry1Ab and Cry3Bb for the control of lepidopteran insect *Spodoptera frugiperda, S. litura* and coleopteran insect *Diabrotica virgifera* [6, 20, 21]. Further detail about the commercialized insecticidal crops are given in a later section.

2.2. Lectins

Lectins are carbohydrate-binding proteins, which possess at least one non-catalytic domain for specific and reversible binding to mono- or oligosaccharides [22, 23]. A typical lectin is multivalent in nature, therefore agglutinate cells. Lectins are extensively distributed in nature from prokaryotes to eukaryotes. The specific interaction with glycoconjugates makes them valuable in biomedical sciences and biotechnology [24]. Carbohydrates present in viruses, microorganisms, fungi, nematodes or phytophagous insects interact with plant lectins [25, 26]. In the past decades, many plant lectins are reported to be toxic to several economically important insect pests of various orders [27–29]. To analyze the insecticidal properties under natural conditions, many transgenic plants expressing lectins have been developed. The toxic effects of different lectins have been demonstrated on several insect species; these effects range from a severe delay in development to high mortality in insects [11].

2.2.1. GNA-related lectins

Galanthus nivalis agglutinin (GNA) purified from snowdrop bulbs is the best studied plant lectin for insecticidal properties. The snowdrop lectin specifically binds to terminal mannose residues in high-mannose-N-glycans, which occur very frequently on insect glycoproteins [30]. Toxicity of GNA has been shown for a wide range of insects; but homopteran insects are highly sensitive to GNA. Several GNA-related lectins have been isolated from different *Allium* species which have shown the potential for insect control [11, 29]. Further, accumulation of some lectins like *Allium porrum* agglutinin in the phloem sap in natural situations support the defensive role of lectins against sap-sucking insects [31].

GNA and related lectins have been successfully expressed for resistance against insect pests into a variety of crops [32]. Transgenic rice expressing ASAL caused significant mortality in nymph of hemipteran insect pests [33]. Onion (*Allium cepa*) lectin has shown more potential against mustard aphid (*Lipaphis erysimi*) in comparison to GNA and ASAL (*Allium sativum* leaf agglutinin) [34]. Transgenic rice expressing ASAL exhibited protection against tungro disease also, after infestation with the *N. virescens* [35]. Vajhala et al. [36] recently demonstrated significant protection in ASAL expressing transgenic cotton against jassid and whitefly. ASAL is also reported to be toxic to chewing insects like *Helicoverpa armigera* and *Spodoptera litura* [27] and several other sucking insects like *Nephotettix virescens* and *Nilaparvata lugens* [37]. Studies related to the mechanism of toxicity showed that ASAL shares the similar receptors with Bt-Cry toxin [28], but both the proteins interact at different positions without steric hindrance and increased the toxicity of each other [29]. Therefore, they can be pyramided together for broad-range insect resistance.

2.2.2. Legume lectins

Legume lectins are purified from seeds and bind to carbohydrate structures like Thomsen-nouveau (Tn) antigen or complex N-glycan with terminal galactose and sialic acid residues. Pea lectin (*Pisum sativum* agglutinin, PSA) expressed in transgenic oilseed rape (*Brassica napus*) shows growth retardation of the pollen beetle larvae (*Meligethes aeneus*) [38] and no effect on the adult beetles [39]. A legume lectin known as Gleheda purified from ground ivy (*Glechoma hederacea*) exhibits high insecticidal activity against the Colorado potato beetle larvae (*Leptinotarsa decem-lineata*) [40]. GS-II lectin isolated from the seed of *Griffonia simplicifolia* shows toxicity to Cowpea weevil (*Callosobruchus maculates*) [41]. A mannose-binding legume lectin concanavalin A (ConA) from jackbean has shown toxicity to the hemipteran pea aphid (*Acyrthosiphon pisum*) [42, 43] and tara plant hopper (*Tarophagous proserpina*) [44].

2.2.3. Hevein-related lectins

Hevein-related plant lectins exhibit specificity for chitin (chitin forms endo- and exo-arthropods, nematodes and fungi). These are also studied for insecticidal properties [45]. Due to the absence of chitin in mammals, hevein-related lectins are considered safe for the usage in genetically modified crops. Wheat germ agglutinin (WGA) has shown a negative effect on the development of the cowpea weevil (*Callosobruchus maculatus*) larvae and southern corn root worm (*Diabrotica undecimpunctata*) [46, 47]. WGA is active against lepidopteran insect larvae also [47, 48].

2.2.4. Other insecticidal lectins

Several other plant lectins have shown insecticidal property. Transgenic tobacco plants expressing tobacco leaf lectin (NICTABA) is detrimental to the cotton leafworm (*S. littoralis*) and the tobacco hornworm (*M. sexta*) [49]. Another protein, *phloem protein 2* (PP2) belonging to the NICTABA family, also possesses insecticidal activity [50, 51]. The amaranthins and the jacalin-related lectins have also shown the potential for insect control, especially against sap-sucking insect pests. Transgenic cotton expressing *Amaranthus caudatus* agglutinin (ACA) under the control of a phloem-specific promoter shows a strong resistance against nymphs of the cotton aphid (*Aphis gossypii*) [52]. Transgenic tobacco expressing Heltuba, a jacalin-related lectin from the *Helianthus tuberosus*, showed reduced development and fecundity of the peach–

potato aphid (*M. persicae*) [53]. Another promising jacalin-related lectin HFR1 is produced in resistant varieties of wheat (*T. aestivum*) during infestation by the Hessian fly larvae. Although HFR1 has not shown any toxicity against Hessian fly, it shows the strong insecticidal activity to the larvae of fruit fly (*D. melanogaster*) [54].

2.3. Proteinase inhibitors

Proteinase inhibitors (PIs) are small molecular weight proteins which affect several metabolic pathways. They are the major components in seeds and storage organs of crops. Mickel and Standish [55] demonstrated the role of PIs in plant defense for the first time and noticed the abnormality in the development of larvae of certain insects fed on soybean products. The feature was attributed to trypsin inhibitors, and it was found to be toxic to the larvae of flour beetle (*Tribolium confusum*) [56].

PIs inhibit the digestion of proteins in midgut and cause mortality of insects due to nutritional imbalance [57, 58]. PIs also interfere with several metabolic processes (like moulting) by blocking the proteolytic activation of enzymes [59]. They affect growth and development, multiplication rate and insect life span [60–62]. PIs have been expressed in several transgenic plants for resistance against insect pests of several classes [63–65]. Pea and soybean trypsin–chymotrypsin inhibitors (PsTI-2, SbBBI) belonging to the Bowman–Birk family [66] and mustard-type trypsin–chymotrypsin variant Chy8 [67] cause significant mortality of pea aphid *A. pisum*. Plant-derived PIs have been used for the development of insect-resistant transgenic plants and projected as an alternative to Bt-Cry proteins [68, 69].

The majority of plant PIs originate from three main families, namely Solanaceae, Leguminosae and Gramineae [70]. Plant PIs can be grouped into four classes: serine, thiol, metallo and aspartyl. Most plant PIs are inhibitors of microbial and animal serine proteases, such as chymotrypsin, trypsin, elastase and subtilisin [71]. Specificity of protease inhibitor families is mainly based on the amino acid residues present in the active site [72].

2.3.1. Serine (Serpin) protease inhibitors

It is found in almost all kingdoms of organisms [73–76]. Several serine PIs have been purified from plants and characterized [77, 78]. Plant serine PIs show inconsistent and varied specificities towards plant proteases [79]. *Hordeum vulgare* serine PI inhibits trypsin, chymotrypsin [80], thrombin, plasma kallikrein, Factor VIIa and Factor Xa [81]. *Triticum aestivum* serine PI inhibits chymotrypsin and cathepsin G [82]. Serine protease inhibitors have been used most commonly for the development of transgenic plants for the control of insect pests [83–85].

2.3.2. Cysteine protease inhibitors

An inhibitor of cysteine proteinases was first described in egg white by Sen and Whitaker [86] and was later named cystatin [87]. Cysteine proteinases inhibitors are widely distributed in plants, animals and microorganisms [88]. Their role in defense has been explored by *in vitro* analysis on inhibition of digestive proteinases from insect pests and nematodes [89–91]. First plant cystatin was isolated from rice seeds and as of now, more than 80 members of different plant species have been characterized [92, 93]. Barley cystatin in artificial diets hampered the life cycle of two aphid species and also in transgenic *Arabidopsis* [94]. Expression of such inhibitors in maize enhanced the resistance against phytophagous mites [95]. Inhibition of these proteases provides a promising control on insects and therefore PIs can be employed as a potential source of defense in plants against insect pests.

2.3.3. *Aspartyl protease inhibitors*

It is relatively less studied class, due to the rare occurrence [91]. Potato tubers possess cathepsin D, an aspartic proteinase inhibitor which showed substantial amino acid sequence similarity with the soybean trypsin inhibitor [96]. Aspartic proteases have been found in coleoptera species, such as *Callosobruchus maculatus* [97] and *H. hampei* [98], in which the acidic pH in midgut provides a favourable condition for these proteases [58].

2.3.4. *Metallo-proteases inhibitors*

The metallo carboxypeptidase inhibitors (MCPIs) have been identified in solanaceaous plants tomato and potato [99]. The MCPIs are 38–39 amino acid residues long polypeptide [100, 101]. Plants have evolved at least two families of metalloproteinase inhibitors, the metallo-carboxypeptidase inhibitor family in potato and tomato [102] and a cathepsin D inhibitor family in potato [103]. The inhibitor is produced in potato tubers and accumulates with potato inhibitor I and II families (serine proteinase inhibitors) during tuber development. The inhibitor also accumulates in potato leaf with inhibitor I and II in response to wounding and have the potential to inhibit all the major digestive enzymes (like trypsin, chymotrypsin, elastase, carboxypeptidase A and carboxypeptidase B) of higher animals and many insects [104].

2.4. α-Amylase inhibitors

α-Amylases (α-1,4-glucan-4-glucanohydrolases) are hydrolytic enzymes, which catalyze the hydrolysis of α-1,4-glycosydic bonds in polysaccharides. They are present in microorganisms, animals and plants [105–107]. They are the most important digestive enzymes of many insects which feed exclusively on seed products. Inhibition of α-amylase impairs the digestion in an organism and causes shortage of free sugar for energy. α-Amylase inhibitors (α-AI) are found in many plants as a part of the defense system and abundant in cereals and legumes [108–111].

α-AI of *Phaseolus vulgaris* is the most studied amylase inhibitor and have shown toxic effects to several insect pests [110, 111]. Like lectins, they possess carbohydrate-binding property. There are at least four types of *Phaseolus* amylase inhibitors on the basis of α-AIs: AI-1, AI-2, AI-3 and the null type [112]. AI-1 is present in the most cultivated common bean varieties and inhibits mammalian α-amylases. It also inhibits α-amylases in insects like *C. chinensis, C. maculatus* and *B. pisorum* [106]. AI-2 is 78% homologous to AI-1 and found in few wild accessions. It inhibits the *Z. subfasciatus* larval α-amylase and pea bruchid α-amylase [106, 111, 113]. This inhibitor is a good example of co-evolution of insect digestive enzymes and plant defense proteins.

They are potential molecules for the development of insect-resistant transgenic plants [114, 115]. Seeds of transgenic pea and azuki, expressing α-AI-1 inhibitor of *P. vulgaris,* shows resistance against pea weevil (*Bruchus pisorum*), cowpea weevil (*C. maculatus*) and azuki bean weevil (*Callosobruchus chinensis*) [110, 113, 116].

2.5. Chitinase

Chitinases are being employed in plant defense in many ways. It has been used in controlling the growth of fungi and insects. Expression of poplar chitinase in tomato leads to growth inhibition in Colorado potato beetle [117]. Secretome analysis of tobacco cell suspension represents chitinase as the major defense protein [118]. A chitinase-like domain containing 56-kDa defense protein (MLX56) provides strong resistance against cabbage armyworm, *Mamestra brassicae,* and Eri silkworm, *Samia ricini* [119]. Two chitinase like proteins LA-a and LA-b (latex abundant) from Mulberry (*Morus* sp.) latex are found to be toxic against *Drosophila melanogaster* [120].

Chitinases have also been isolated from insects and found to be equally promising in plant defense. Transgenic tobacco plants expressing chitinase of tobacco hornworm (*Manduca sexta*) shows resistance to tobacco budworm *Heliothis virescens* [121]. Hornworm chitinase expressing transgenic plants are also resistant against fungal infection [122]. Further, a recombinant baculovirus expressing chitinase of hard tick (*Haemaphysalias longicornis*) has been shown as bio-acaricide for tick control [123].

3. Insect-resistant transgenic crops

Development of many transgenic crops has been reported for insect resistance. Both private and public sector organizations are involved in the process and they used δ-endotoxins of *Bacillus thuringiensis* to achieve resistance against insects. Among transgenic plants, cotton and maize were the most successful and released for commercial cultivation. These crops are being adopted annually at very high rates. In other words, area under Bt-crops are increasing day-by-day. Successful deployment of these crops has decreased the pesticide usage. However, the sustainability and durability of pest resistance are still a matter of discussion. It is also important to focus on next-generation insect-resistant transgenic crops.

3.1. First-generation insect-resistant transgenic crops

Insect-resistant transgenic crops have not only increased the economy but also the environmental and health benefits [69, 124]. Six transgenic crops (canola, corn, cotton, papaya, squash and soybean) were planted in 2003 in the USA alone. These crops increased farm income by US$ 1.9 billion by producing an additional 2.4 million tonnes of food and fiber and reduced the use of pesticides by 21,000 tonnes.

In 2009, China government approved the cultivation of Bt-rice (the country has been growing Bt-cotton since 1997). Farm surveys of randomly selected households cultivating Bt-rice varieties have been performed. The benefit of Bt-rice has been acknowledged to the level of small and poor farmers, it is due to the lesser crop damage by the insects and therefore higher crop yields and less use of pesticides. An improved health has also been observed in Bt-rice cultivating farmers compared to non-Bt rice cultivating farmers [126]. Government of India approved the cultivation of Bt-cotton in 2003, which resulted in a 70% reduction in insecticide applications. This saves up to US$ 30 per ha in insecticide costs and results 80–87% increase in cotton yield [127]. A spectacular decrease in pesticide usage in Bt cotton fields has also been reported from China. The pesticide poisoning to the farmers reduced from 22% to 4.7% [128].

To assess probable hazards of Bt toxins on non-target insects, field evaluation was performed in Spain [129]. Bt-maize did not show negative impact on non-target pests. Similar numbers of cutworms and wireworms were present in Bt versus non-Bt fields. Surprisingly, higher numbers of aphids and leafhoppers were observed in Bt field.

3.2. Strategies for next-generation insect resistance

3.2.1. Engineering of Cry toxin by domains swapping

Most of the Cry toxins share common three-domain structure in activated form [130]. Domain I gets inserted into the target membrane and forms pore; domain II is associated with receptor binding and thus determines specificity, and domain III is also involved in receptor-binding specificity. It has been demonstrated in a couple of studies that hybrid Cry toxins exhibit higher toxicity. Domain III of Cry1Ac increased the efficacy of various other Cry1 proteins in Cry1–Cry1Ac hybrid [131]. Similarly, Singh et al. (2004) developed a hybrid toxin against *Spodoptera litura*. They replaced a region in domain III of Cry1Ea toxin by 70 amino acid homologous region of Cry1Ca. Transgenic tobacco and cotton expressing hybrid gene are highly effective/toxic to all stages of larvae of *S. litura*. Another hybrid *Bt* gene was developed by replacing part of domain II of Cry1Ba with that of Cry2a [132]. The transgenic potato expressing the hybrid toxin showed resistance against Colorado potato beetle, potato tuber moth and European corn borer. The strategy provides new opportunities for resistance management as the target receptor recognition of hybrid toxins is expected to be different from currently used Cry toxins.

3.2.2. Plant-derived insecticidal lectins and protease inhibitors

Detail about lectins and protease inhibitors have been discussed in earlier section. Some other insecticidal roles are summarized here. Besides insecticidal potential, GNA and ASAL also serve as a carrier protein for other insecticidal peptides and proteins to the haemolymph of lepidopteran larvae. It has been demonstrated by feeding GNA-allatostatin and GNA-SFI1 fusions to the tomato moth *Lacanobia oleracea* [133–135]. SFI1 is a neurotoxin isolated from the spider *Segestria florentina*. The individual toxin did not cause toxicity through oral delivery; however, the fusion proteins with GNA were toxic.

Lectins are reported to be insecticidal towards sap-sucking insects, where Bt-toxins are not effective. Transgenic tobacco expressing garlic (*Allium sativum*) leaf lectin showed substantial control over peach potato aphids [136]. Fusion of galactose-binding domain of the non-toxic ricin B-chain with Cry1Ac provides additional binding domains, which increases interactions with the gut receptors in target insects. Transgenic rice and maize expressing the fusion protein show high toxicity in comparison to the Bt-toxin alone [137].

Protease inhibitors (PIs) expressing transgenic plants are not as effective as Bt and insecticidal lectin expressing plants. This is due to the adaptation in gut proteases in phytophagous insects. High genetic diversity in gut proteases and low potency of protease inhibitors is responsible for such adaptation. The combination of inhibitors (potato PI–II and carboxypeptidase) is not enough to avoid the compensatory adaptation [68]. However, inhibitors like barley trypsin inhibitor [65], equistatin from sea anemone [138], other cystatins [139, 140] or use of multiple inhibitors [141] or combination of inhibitors and lectins [142] might also be useful to provide resistance against insects in transgenic plants.

3.2.3. *Multiple insecticidal proteins containing transgenic crop*

Second-generation Bt transgenic cotton [Bollgard II (Cry1Ac + Cry 2Ab) and Widestrike (Cry1Ac + Cry1F)] are developed to increase the level of resistance against cotton bollworm [143, 144]. It has also been demonstrated that the expression of three insecticidal proteins (Cry1Ac, Cry2A and GNA) into Indica rice control three major pests, rice leaf folder (*Cnaphalocrocis medinalis*), yellow stem borer (*Scirpophaga incertulas*) and the brown plant hopper (*Nilaparvata lugens*) [145]. Cry proteins target the leaf folder and the stem borer, and GNA targets the plant hopper. Comparison of three different Bt-cotton lines (either single Cry1Ac or Cry2Ab, or both genes) for insect damage showed that the lines containing two Bt genes performed better [144]. Broccoli expressing both Cry1Ac and Cry1C exhibited increased resistance to diamondback moths and delayed the resistance development [146, 147]. Similarly, transgenic tobacco expressing Cry1Ac and cowpea trypsin inhibitor (CpT 1) delayed resistance development in *H. armigera* [148]. Recently, Bharathi et al. [149] pyramided two lectin genes ASAL and GNA and showed increased resistance against brown plant hopper, green leaf hopper and white backed plant hopper, as compared to their parental lines expressing single lectin. The performance of transgenic plant pyramided with genes has shown that the insecticidal functions of most of the toxins are non-overlapping and non-competitive.

3.2.4. *Tissue-specific or regulated expression*

Insecticidal proteins are usually expressed under constitutive promoter for higher accumulation of the proteins. Although the constitutive expression has some advantages, tissue-specific or inducible expression is desirable under certain circumstances. Insect attacks epidermal cells first and therefore the expression of insecticidal proteins under epidermal cell-specific promoters can be a useful strategy. For example, CER6 is an epidermal cell-specific promoter responsible for the expression of an enzyme for cuticular wax production [150]. Similarly, phloem-feeding insects can be targeted by using phloem-specific promoter like PP2 promoter of pumpkin [151], rice sucrose synthase Rss promoter [152] and root phloem-specific promoter AAP3 [153]. Tissue-specific expression of several insecticidal proteins has demonstrated as a good potential for insect control in several studies. Phloem-specific expression of ASAL under promoter Asus1 protects tobacco against aphid, *Myzus nicotianae* [154]. Transgenic chickpea expressing ASAL under rolC promoter showed effective control over *A. craccivora* [155] and transgenic Indian mustard (*Brassica juncea*) expressing ASAL under Rss I promoter showed resistance against aphid *Lipaphis erysimi* [136]. Researchers are also working on sap-sucking pest inducible phloem-specific promoters, which are not only insect-inducible but also insect-specific in nature [156]. Another strategy is temporal expression of insecticidal proteins as some insects infest a crop in a particular phase only. For example, pink bollworm (*Pectinophora gossypiella*) attacks and feeds on the cotton bolls only. At this stage, cotton plants are mature; the expression of Cry toxins goes down and becomes insufficient for effective control.

3.2.5. Strategies to over express secondary metabolites

Secondary metabolites synthesized by the plants participate in a number of physiological and biochemical processes. Our group demonstrated that the over-expression of pectin methylesterase of *Arabidopsis thaliana* and *Aspergillus niger* in transgenic tobacco plants enhances methanol production, which in turn provided resistance against sap-sucking as well as chewing insect pests [157]. Similarly, transgenic tobacco expressing AtMYB-12 gene showed enhanced production of rutin in leaves and callus, which confers resistance against *H. armigera* and *S. litura* larvae [158, 159]. WsSGTL1, a sterol glycosyltransferases isolated from *Withania somnifera*, was expressed and functionally characterized in transgenic tobacco plants, which showed significant resistance towards *S. litura* [160]. Tobacco plants were transformed by a multigene transfer vector containing three coffee N-methyltransferases genes CaMXT1, CaMXMT1 and CaDXMT1 responsible for producing caffeine in transgenic plants which showed tolerance to *S. litura* [161]. Dixit et al [162] demonstrated the insect resistance by altering the amino acid composition in sap.

4. Conclusions and perspectives

Transgenic technology (especially Bt crops) has contributed significantly in increasing the crop production worldwide. The crops are protected from being damaged by insect pests. Certainly, this methodology provides an environmentally safe alternative for the synthetic pesticides. Further, it has also been proven to be useful in enhancing nutritional values of crops, improvement of stress tolerance and production of pharmaceutical proteins. Introduction of Bt cotton varieties in India has tremendously increased the yields of cotton and thereby profits to the farmers. Bt proteins are able to control the damage caused by Lepidopteran and Coleopteran insects, but not effective against sap-sucking Homopteran pests [8, 9]. Therefore, an unusual increase in the population of homopteran pests like whiteflies, aphids and leafhoppers on transgenic cotton has been reported [7]. Further, development of resistance in insects against toxins is also going to be a major point of concern, which might ultimately challenge the future of Bt crops. Some defense-related proteins like plant lectins, PIs and chitinases are reported to be toxic to various homopteran insect pests. However, several safety and societal concerns are raised from time to time. Further, there is non-availability of an effective and safe protein against several important and emerging insects, which need an *ab initio* approach to resolve this issue. A promising and biosafe strategy to defeat the above problem can be: (a) exploration of the plant's own defense mechanisms and manipulation of their expressions or (b) by introducing a gene for insect control derived from other plants, especially derived from non-host plants, and (c) pyramiding of insecticidal proteins for the control of multiple insect pests. Exploration of RNAi mediated insect control by targeting high expressing and/or important vital genes can also be an effective approach (12-14, 163, 164). Besides this, in our country, we need a dedicated forum to popularize the use of genetically modified crops and convince the government as well as citizens at ethical issues.

Acknowledgements

SKU acknowledges Department of Science and Technology-INSPIRE faculty fellowship.

Author details

Santosh Kumar Upadhyay[1*] and Sudhir P. Singh[2*]

*Address all correspondence to:

*Address all correspondence to: skupadhyay@pu.ac.in; pratapsudhir17@gmail.com

1 Department of Botany, Panjab University, Chandigarh, India

2 Center of Innovative and Applied Bioprocessing, Mohali, India

References

[1] Anon (2010) Food 2030, London UK: Department for Food and Rural Affairs.

[2] Ferry N, Gatehouse AMR (2010) Transgenic crop plants for resistance to biotic stress, In Transgenic crop plants: utilization and biosafety 2 (eds C Kole, CH Michler, AG Abbott & TC Hall) pp. 1–66, Germany: Springer-Verlag.

[3] Andrews RE, Faust RM, Wabiko H, Raymond KC, Bulla LA (1987) The biotechnology of Bacillus thuringiensis, Crit Rev Biotech 6 163–232.

[4] Vaeck M, Reynaerts A, Hofte H, Jansens S, de Beuckleer M, Dean C, Zabeau M, Van Montagu M, Leemans J (1987) Transgenic plants protected from insect attack, Nature 328 33–37.

[5] Xie R, Zhuang M, Ross LS, Gomez I, Oltean DI, Bravo A, Soberon M, Gill S (2005) Single amino acid mutation in the cadherin recepter from Heliothis virescens affect its toxin binding ability to cry1A toxins, J Biol Chem 280 8416–8425.

[6] Sanahuja G, Banakar R, Twyman RM, Capell T, Christou P (2011) Bacillus thuringiensis: a century of research development and commercial applications, Plant Biotechnol J 9 283–300.

[7] Abro GH, Syed TS, Tunio GM, and Khuhro MA (2004) Performance of transgenic Bt cotton against insect pest infestation, Biotechnology 3 75–81.

[8] Dutt U (2007) Mealy bug infestation in Punjab: Bt cotton falls flat, Environment News Service 21 August (countercurrents.org).

[9] Virla EG, Casuso M, Frias E A (2010) A preliminary study on the effects of a transgenic corn event on the non targrt pest Dalbulus Maiid (Hemitera: Cicadeliidae), Crop protection 29 635–638.

[10] Carlini CR, Grossi-de-Sa MF (2002) Plant toxic proteins with insecticidal properties. A review on their potentialities as bioinsecticides, Toxicon 40 1515–1539.

[11] Vandenborre G, Smagghe G, and VanDamme EJM (2011) Plant lectins as defense proteins against phytophagous insects, Phytochemistry 72 1538-50.

[12] Upadhyay SK, Chandrashekar K, Thakur N, Verma PC, Singh PK, Tuli R (2011) RNA interference (RNAi) for the control of whitefly (*Bemisia tobaci*), Journal of Biosciences 36 153–161.

[13] Upadhyay SK, Dixit S, Sharma S, Singh H, Kumar J, Verma PC, Chandrashekar K (2013) siRNA machinery in whitefly (*Bemisia tabaci*), Plos ONE 8(12) e83692 doi: 10.1371/journal.pone.0083692.

[14] Thakur N, Upadhyay SK, Verma PC, Chandrashekar K, Tuli R, Singh PK (2014) Enhanced whitefly resistance in transgenic tobacco plants expressing double stranded RNA of v-ATPase A gene, PLoS ONE 9 e87235.

[15] Crickmore N, Zeigler DR, Schnepf E, Van Rie J, Lereclus D, Baum J, Bravo A, Dean DH (2010) Bacillus thuringiensis toxin nomenclature. http://www.biols.susx.ac.uk/Home/Neil Crickmore/Bt/index.html.

[16] Upadhyay SK, Singh PK (2011) Role of alkaline phosphatase in insecticidal action of Cry1Ac against Helicoverpa armigera larvae, Biotechnology lettters 33 2027-2036.

[17] Bravo A, Gill SS, Soberón M (2005) Bacillus thuringiensis mechanisms and use, In: Gilbert LI, Iatrou K, Gill SS, (Eds) Comprehensive Molecluar Insect Science, Elsevier BV ISBN 0-4 4-451516-X 175–206.

[18] Estruch JJ, Warren GW, Mullins MA, Nye GJ, Craig JA, Koziel MG (1996) Vip3A a novel Bacillus thuringiensis vegetative insecticidal protein with a wide spectrum of activities against lepidopteran insects, Proc Natl Acad Sci USA 93 5389–5394.

[19] Warren G (1997) Vegetative insecticidal proteins: novel proteins for control of corn pests, In: Carozzi N, Koziel M, (Eds) Advances in Insect Control: The Role of Transgenic Plants, Taylor & Francis Ltd p. 109.

[20] Christou P, Capell T, Kohli A, Gatehouse JA, Gatehouse AM (2006) Recent developments and future prospects in insect pest control in transgenic crops, Trends Plant Sci 11 302–308.

[21] Singh PK, Kumar M, Chaturvedi CP, Yadav D, Tuli R (2004) Development of hybrid delta endotoxin and its expression in tobacco and cotton for control of a polyphagous pest Spodoptera litura, Transgenic Res 13 397–410.

[22] Vijayan M, Chandra N (1999) Lectins, Curr Opin Struct Biol 9 707–714.

[23] Upadhyay SK, Saurabh S, Singh R, Rai P, Dubey NK, Chandrashekar K, Negi KS, Tuli R, Singh PK, (2011) Purification and Characterization of a Lectin with High Hemagglutination Property Isolated from *Allium altaicum*, Protein J 30 374–83.

[24] VanDriessche E, Fischer J, Beeckmans S, Bog-Hanse TC (1996) Lectins Biology Biochemistry Clinical Biochemistry Textop Hellerup Denmark 11 215–219.

[25] Upadhyay SK, Singh S, Chandrashekar K, Singh PK, Tuli R (2012) Compatibility of garlic (*Allium sativum* L.) leaf agglutinin and Cry1Ac δ-endotoxin for gene pyramiding, Applied microbiology and Biotechnology 93 2365–2375.

[26] Ripoll C, Favery B, Lecompte P, Van Damme EJM, Peumans WJ, Abad P, Jouanin L (2003) Evaluation of the ability of lectin from snowdrop (*Galanthus nivalis*) to protect plants against root-knot nematodes, Plant Sci 164 517–523.

[27] Upadhyay SK, Saurabh S, Rai P, Singh R, Chandrashekar K, Verma PC, Singh PK, Tuli R (2010a) SUMO fusion facilitates expression and purification of garlic leaf lectin but modifies some of its properties, J Biotechnol 146 1–8.

[28] Upadhyay SK, Mishra M, Singh H, Ranjan A, Chandrashekar K, Verma PC, Singh PK, Tuli R (2010b) Interaction of Allium sativum leaf agglutinin (ASAL) with midgut brush border membrane vesicle proteins and its stability in *Helicoverpa armigera*, Proteomics 10 4431–4440.

[29] Upadhyay SK, Singh PK (2012) Receptors of garlic (*Allium sativum*) lectins and their role in insecticidal action, Protein J 31(6) 439–446.

[30] Schachter H (2009) Paucimannose N-glycans in Caenorhabditis elegans and Drosophila melanogaster, Carbohydr Res 344 1391–1396.

[31] Peumans WJ, Smeets K, VanNerum K, VanLeuven F, VanDamme EJM (1997) Lectin and alliinase are the predominant proteins in nectar from leek (Allium porrum L,) flowers, Planta 201 298–302.

[32] Wang Z, Zhang K, Sun X, Tang K, Zhang J (2005) Enhancement of resistance to aphids by introducing the snowdrop lectin gene gna into maize plants, J Biosci 30 627–638.

[33] Yarasi B, Sadumpati V, Immanni CP, Vudem DR, Khareedu VR (2008) Transgenic rice expressing Allium sativum leaf agglutinin (ASAL) exhibits high-level resistance against major sap-sucking pests, BMC Plant Biology 8 102.

[34] Hossain MA, Maiti MK, Basu A, Sen S, Ghosh AK, Sen SK (2006) Transgenic expression of onion leaf lectin gene in Indian mustard offers protection against aphid colonization, Crop Sci 46 2022–2032.

[35] Saha P, Dasgupta I, Das S (2006) A novel approach for developing resistance in rice against phloem limited viruses by antagonizing the phloem feeding hemipteran vectors, Plant Mol Biol 62 735–752.

[36] Vajhala CSK, Sadumpati VK, Nunna HR, Puligundla SK, Vudem DR, et al. (2013) Development of transgenic cotton lines expressing Allium sativum agglutinin (ASAL) for enhanced resistance against major sap-sucking pests. PLoS ONE 8 e72542.

[37] Bala A, Roy A, Behura N, Hess D, Das S (2013) Insight to the mode of action of Allium sativum leaf agglutinin (ASAL) expressing in T3 rice lines on brown plant hopper, Am J Plant Sci 4 400–407.

[38] Melander M, Ahman I, Kamnert I, and Strömdahl AC (2003) Pea lectin expressed transgenically in oilseed rape reduces growth rate of pollen beetle larvae, Transgenic Res 12 555–567.

[39] Lehrman A, Ahman I, Ekbom B (2007) Influence of pea lectin expressed transgenically in oilseed rape (Brassica napus) on adult pollen beetle (Meligethes aeneus), J Appl Entomol 131 319–325.

[40] Wang W, Hause B, Peumans WJ, Smagghe G, Mackie A, Fraser R, VanDamme EJM (2003) The Tn antigen-specific lectin from ground ivy is an insecticidal protein with an unusual physiology, Plant Physiol 132 1322–1334.

[41] Zhu K, Huesing JE, Shade RE, Bressan RA, Hasegawa PM, Murdock LL (1996) An insecticidal N-acetylglucosamine-specific lectin gene from Griffonia simplicifolia (Leguminosae), Plant Physiol 110 195–202.

[42] Sauvion N, Charles H, Febvay G, Rahbé Y (2004a) Effects of jackbean lectin (ConA) on the feeding behavior and kinetics of intoxication of the pea aphid Acyrthosiphon pisum, Entomol Exp Appl 10 34–44.

[43] Sauvion N, Nerdon C, Febvay G, Gatehouse AMR, Rahbé Y (2004b) Binding of the insecticidal lectin Concanavalin A in pea aphid Acyrthosiphon pisum (Harris) and induced effects on the structure of midgut epithelial cells, J Insect Physiol 50 1137–1150.

[44] Powell KS (2001) Antimetabolic effects of plant lectins towards nymphal stages of the planthoppers Tarophagous proserpina and Nilaparvata lugens, Entomol Exp Appl 99 71–78.

[45] Merzendorfer H (2006) Insect chitin synthases: a review, J Comp Physiol 176 1–15.

[46] Huesing JE, Murdock LL, Shade RE (1991) Effect of wheat germ isolectins on the development of the cowpea weevil, Phytochemistry 30 785–788.

[47] Czapla TH, Lang BA (1990) Effect of plant lectins on the larval development of the European corn borer (Lepidoptera: Pyralidae) and the Southern corn rootworm (Coleoptera: Chrysomelidae), J Econ Entomol 83 2480–2485.

[48] Hopkins TL, Harper MS (2001) Lepidopteran peritrophic membranes and the effect of dietary wheat germ agglutinin on their formation and structure, Archiv Insect Biochem Physiol 47 100–109.

[49] Vandenborre G, Groten K, Smagghe G, Lannoo N, Baldwin IT, VanDamme EJM (2010a) Nicotiana tabacum agglutinin is active against Lepidopteran pest insects, J Exp Bot 61 1003–1014.

[50] Beneteau J, Renard D, Marché L, Douville E, Lavenant L, Rahbé Y, Dupont D, Vilaine F, Dinant S, (2010) Binding properties of the N -acetylglucosamine and high-mannose N-glycan PP2–A1 phloem lectin in Arabidopsis, Plant Physiol 153 1345–1361.

[51] Dinant S, Clark AM, Zhu Y, Vilaine F, Palauqui JC, Kusiak C, Thompson GA (2003) Diversity of the superfamily of phloem lectins (phloem protein 2) in angiosperms, Plant Physiol 131 114–128.

[52] Wu J, Luo X, Guo H, Xiao J, Tian Y (2006) Transgenic cotton expressing Amaranthus caudatus agglutinin confers enhanced resistance to aphids, Plant Breeding 125 390–394.

[53] Chang T, Chen L, Chen S, Cai H, Liu X, Xiao G, Zhu Z (2003) Transformation of tobacco with genes encoding Helianthus tuberosus agglutinin (HTA) confers resistance to peach-potato aphid (Myzus persicae), Transgenic Res 12 607–614.

[54] Subramanyam S, Smith DF, Clemens JC, Webb MA, Sardesai N, Williams CE (2008) Functional characterization of HFR-1 a high-mannoseN-glycan-specific wheat lectin induced by Hessian fly larvae, Plant Physiol 147 1412–1426.

[55] Mickel CE, Standish J (1947) Susceptibility of processed soy flour and soy grits in storage to attack by Tribolium castaneum, University of Minnesota, Agricultural Experimental Station, Technical Bulletin, 178 1–20,

[56] Lipke H, Fraenkel GS, Liener I (1954) Effect of soybean inhibitors on growth of Tribolium confusum, J Sci Food Agri 2 410–415.

[57] Broadway R, Duffey S (1986) Plant proteinase inhibitors: mechanism of action and effect on the growth and digestive physiology of larval Heliothis zea and Spodoptera exigua, J Insect Physiol 32 827–833.

[58] Ryan CA (1990) Proteinase inhibitors in plants: genes for improving defenses against insects and pathogens, Annu Rev Phytopathol 28 425–449.

[59] Hilder VA, Gatehouse AMR, Sheerman SE, Barker RF, Boulter D (1987) A novel mechanism of insect resistance engineered into tobacco, Nature 330 160–163.

[60] Gatehouse AM, Norton E, Davison GM, Babbé SM, Newell CA, Gatehouse JA (1999) Digestive proteolytic activity in larvae of tomato moth Lacanobia oleracea; effects of plant protease inhibitors in vitro and in vivo, J Insect Physiol 45 545–558.

[61] Annadana S, Peters J, Gruden K, Schipper A, Outchkourov NS, Beekwilder MJ, Udayakumar M, Jongsma MA (2002) Effects of cysteine protease inhibitors on oviposition rate of the western flower thrips Frankliniella occidentalis, J Insect Physiol 48 701–706.

[62] Oppert B, Morgan TD, Hartzer K, Lenarcic B, Galesa K, Brzin J, Turk V, Yoza K, Ohtsubo K, Kramer KJ (2003) Effects of proteinase inhibitors on digestive proteinases and growth of the red flour beetle Tribolium castaneum (Herbst) (Coleoptera: Tenebrionidae), Comp Biochem Physiol Toxicol Pharmacol 134 481–90.

[63] De Leo F, Ceci LR, Jouanin L, Gallerani R (2001) Analysis of mustard trypsin inhibitor-2 gene expression in response to developmental or environmental induction, Planta 212 710–717.

[64] Upadhyay SK, Chandrashekar K (2012) Interaction of salivary and midgut proteins of Helicoverpa armigera with soybean trypsin inhibitor. Protein J 31 259–64.

[65] Alfonso-Rubi J, Ortego F, Castanera P, Carbonero P, Diaz I (2003) Transgenic expression of trypsin inhibitor Cme from barley in indica and japonica rice confers resistance to the rice weevil Sitophilus oryzae, Transgenic Res 12 23–31.

[66] Rahbe Y, Ferrasson E, Rabesona H, Quillien L (2003a) Toxicity to the pea aphid Acyrthosiphon pisum of anti-chymotrypsin isoforms and fragments of Bowman-Birk protease inhibitors from pea seeds, Insect Biochem Molecul Biol 33 299–306.

[67] Ceci LR, Volpicella M, Rahbe Y, Gallerani R, Beekwilder J, Jongsma MA (2003) Selection by phage display of a variant mustard trypsin inhibitor toxic against aphids, The Plant J 33 557–566.

[68] Abdeen A, Virgós A, Olivella E, Villanueva J, Avilés X, Gabarra R, Prat S (2005) Multiple insect resistance in transgenic tomato plants over-expressing two families of plant proteinase inhibitors, Plant Mol Biol 57 189–202.

[69] Ferry N, Edwards M, Gatehouse JA, Capell T, Christou P, Gatehouse AMR (2006) Transgenic plants for insect pest control: a forward looking scientific perspective, Transgen Res 15 13–19.

[70] Richardson MJ (1991) Seed storage proteins: The enzyme inhibitors, In Methods in Plant Biochemistry, Edited by: Richardson MJ. New York: Academic Press; pp. 259-305.

[71] Michaud D (2000) Recombinant Protease Inhibitors in Plants. Eurekah Georgetown.

[72] Koiwa H, Bressan RA, Hasegawa PM (1997) Regulation of protease inhibitors and plant defense, Trends in Plant Sci 2 379–384.

[73] Gettins PG (2002) Serpin structure mechanism and function, Chem Rev 102 4751–4804.

[74] Christeller J, Liang W (2005) Plant serine protease inhibitors, Protein Peptide Lett 12 439–447.

[75] Law R, Zhang Q, McGowan S, Buckle AM, Silverman GA, Wong W, Rosado CJ, Langendorf CG, Bird PI, Whisstock JC (2006) An overview of the serpin superfamily, Genome Biol 7 216.

[76] Irving JA, Pike RN, Lesk AM, Whisstock JC (2000) Phylogeny of the serpin superfamily: implications of patterns of amino acid conservation for structure and function, Genome Res 10 1845–1864.

[77] Yoo BC, Aoki K, Xiang Y, Campbell LR, Hull RJ, Xoconostle-Cazares B, Monzer J, Lee JY, Ullman DE, Lucas WJ (2000) Characterization of Cucurbita maxima phloem serpin-1 (CmPS-1), A developmentally regulated elastase inhibitor, J Biol Chem 275 35122–35128.

[78] Tsybina T, Dunaevsky Y, Musolyamov A, Egorov T, Larionova N, Popykina N, Belozersky M (2004) New protease inhibitors from buckwheat seeds: properties partial amino acid sequences and possible biological role, Biol Chem 385 429–434.

[79] Al-Khunaizi M, Luke CJ, Cataltepe S, Miller D, Mills DR, Tsu C, Brömme D, Irving JA, Whisstock JC, Silverman GA (2002) The serpin SON-5 is a dual mechanistic class inhibitor of serine and cysteine proteinases, Biochemistry 41 3189–3199.

[80] Dahl SW, Rasmussen SK, Hejgaard J (1996a) Heterologous expression of three plant serpins with distinct inhibitory specificities, J Biol Chem 271 25083–25088.

[81] Dahl SW, Rasmussen SK, Petersen LC, Hejgaard J (1996b) Inhibition of coagulation factors by recombinant barley serpin BSZX, FEBS Lett 394 165–168.

[82] Roberts TH, Marttila S, Rasmussen SK, Hejgaard J (2003) Differentia gene expression for suicide-substrate serine proteinase inhibitor (serpins) in vegetative and grain tissues of barley, J Exp Bot 54 2251–2263.

[83] Savić JM, Smigocki AC (2012) Beta vulgaris L. serine proteinase inhibitor gene expression in insect resistant sugar beet, Euphytica 188 187–198.

[84] Schlüter U, Benchabane M, Munger A, Kiggundu A, Vorster J, Goulet MC, Cloutier C, Michaud D (2010) Recombinant protease inhibitors for herbivore pest control: a multitrophic perspective, J Exp Bot 61 4169–4183.

[85] Smigocki AC, Ivic-Haymes S, Li H, Savić J (2013) Pest protection conferred by a beta vulgaris serine proteinase inhibitor gene, PLoS ONE 8 e57303.

[86] Sen LC, Whitaker JR (1973) Some properties of a ficin-papain inhibitor from avian egg white, Arch Biochem Biophys 158 623–632.

[87] Anastasi A, Brown A, Kembhavi AA, Nicklin MJH, Sayers CA, Sunter DC, Barrett AJ (1983) Cystatin a protein inhibitor of cysteine proteinases, Improved purification from egg white characterization and detection in chicken serum, Biochem J 211 129–138.

[88] Oliveira AS, Filho JX, Sales MP (2003) Cysteine proteinases cystatins, Braz Arch Biol Technol 46 91–104.

[89] Pernas M, Sanchez M,R, Gomez L, Salcedo G (1998) A chestnut seed cystatin differentially effective against cysteine proteinases from closely related pests, Plant Mol Biol 38 1235–1242.

[90] Siqueira-Junior CL, Fernandes KVS, Machado OLT, Cunha M, Gomes VM, Moura D, Jacinto T (2002) 87 kDa tomato cystatin exhibits properties of a defence protein and forms crystals in prosystemin over-expressing transgenic plants, Plant Physiol Biochem 40 247–254.

[91] Haq SK, Atif SM, Khan RH (2004) Protein proteinase inhibitor genes in combat against insects pests and pathogens: natural and engineered phytoprotection, Archiv Biochem Biophys 431 145–159.

[92] Abe K, Emori Y, Kondo H, Suzuki K, Arai S (1987a) Molecular cloning of a cysteine proteinase inhibitor of rice (oryzacystatin): homology with animal cystatins and transient expression in the ripening process of rice seeds, J Biol Chem 262 16793–16797.

[93] Abe K, Kondo H, Arai S (1987b) Purification and characterization of a rice cysteine proteinase inhibitor, Agric Biol Chem 51 2763–2768.

[94] Carrillo L, Martinez M, Alvarez-Alfageme F, Castañera P, Smagghe G, Diaz I, Ortego F (2011a) A barley cysteine proteinase inhibitor reduces the performance of two aphid species in artificial diets and transgenic Arabidopsis plants, Transgenic Res 20 305–319.

[95] Carrillo L, Martinez M, Ramessar K, Cambra I, Castañera P, Ortego F, Díaz I (2011b) Expression of a barley cystatin gene in maize enhances resistance against phytophagous mites by altering their cysteine-proteases, Plant Cell Rep 30 101–112.

[96] Mares M, Meloun B, Pavlik M, Kostka V, Baudys M (1989) Primary structure of Cathepsin-D inhibitor from potatoes and its structural relationship to trypsin inhibitor family, FEBS Lett 251 94–98.

[97] Silva CP, Xavier-Filho J (1991) Comparison between the levels of aspartic and cysteine proteinases of the larval midguts of Callosobruchus maculatus (F.) and Zabrotes subfasciatus (Boh.) (Coleoptera: Bruchidae), Comp Biochem Physiol 99B 529–533.

[98] Preciado DPR, Valencia AJ (2000) Partial characterization of digestive proteinases from coffee berry borer adults (Hypothenemus hampei), Insect physiology neurosciences immunity and cell biology symposium and Poster Session abstract book II – XXI-International Congress of Entomology Brazil August 20–26.

[99] Habib H, Fazili KM (2007) Plant protease inhibitors: a defense strategy in plants, Biotechnol Molecul Biol Rev 2 068–085.

[100] Hass GM, Nau H, Biemann K, Grahn DT, Ericsson LH, Neurath H (1975) The amino acid sequence of a carboxypeptidase inhibitor from potatoes, Biochem 14 1334–1342.

[101] Hass GM, Hermodson MA (1981) Amino acid sequence of a carboxy-peptidase inhibitor from tomato fruit, Biochemistry 20 2256–2260.

[102] Rancour JM, Ryan CA (1968) Isolation of a carboxypeptidase B inhibitor from potatoes, Arch Biochem Biophys 125 380–382.

[103] Keilova H, Tomasek V (1976) Isolation and properties of cathepsin D inhibitor from potatoes, Collect Czech Chem Commun 41 489–497.

[104] Hollander-Czytko H, Andersen JL, Graham JS, Ryan CA (1997) Accumulation of metallocarboxy-peptidase inhibitor in leaves of wounded potato plants, Biochem Biophys Res Commun 101 1164–1170.

[105] Qian MX, Haser R, Payan F (1993) Structure and molecular-model refinement of pig pancreatic alpha-amylase at 2,1 angstrom resolution, J Mol Biol 231 785–799.

[106] Grossi-de-Sa MF, Chrispeels MJ (1997) Molecular cloning of bruchid (Zabrotes subfasciatus) alpha-amylase cDNA and interactions of the expressed enzyme with bean amylase inhibitors, Insect Biochem Mol Biol 27 271–281.

[107] Strobl S, Maskos K, Betz M, Wiegand G, Huber R, Gomis-Ru°th FX, Glockshuber R (1998) Crystal structure of yellow meal worm a-amylase at 1,64 A° resolution, J Mol Biol 278 617–628.

[108] Franco OL, Rigden DJ, Melo FR, Bloch C Jr, Silva CP, Grossi-de-Sa′ MF (2000) Activity of wheat α–amylase inhibitors towards bruchid α-amylases and structural explanation of observed specificities, Eur J Biochem 267 2166–2173.

[109] Iulek J, Franco OL, Silva M, Slivinski CT, Bloch C Jr, Rigden DJ, Grossi-de-Sa′ MF (2000) Purification biochemi-cal characterisation and partial primary structure of a

new a -amylase inhibitor from Secale cereale (Rye), Int J Biochem Cell Physiol 32 1195–1204.

[110] Ishimoto M, Chrispeels MJ (1996) Protective mechanism of the Mexican bean weevil against high levels of α-amylase inhibitor in the common bean, Plant Physiol 111 393–401.

[111] Grossi-de-Sa MF, Mirkov TE, Ishimoto M, Colucci G, Bateman KS, Chrispeels MJ (1997) Molecular characterization of a bean α-amylase inhibitor that inhibits the α-amylase of the Mexican bean weevil Zabrotes subfasciatus, Planta 203 295–303.

[112] Suzuki K, Ishimoto M (1999) Characterization of the third alpha-amylase inhibitor alpha AI-3 in the common bean (Phaseolus vulgaris L.), Breed Sci 49 275–280.

[113] Morton RL, Schroeder HE, Bateman KS, Chrispeels MJ, Armstrong E, Higgins TJV (2000) Bean α-amylase inhibitor-I in transgenic peas (Pisum sativum) provides complete protection from pea weevil (Bruchus pisorum) under field conditions, Proc Natl Acad Sci USA 97 3820–3825.

[114] Gatehouse AMR, Gatehouse JA (1998) Identifying proteins with insecticidal activity: use of encoding genes to produce insect-resistant transgenic crops, Pest Sci 52 165–175.

[115] Valencia A, Bustillo AE, Ossa GE, Chrispeels MJ (2000) α-Amylases of the coffee berry borer (Hypothenemus hampei) and their inhibition by two plant amylase inhibitors, Insect Biochem Mol Biol 30 207–213.

[116] Chrispeels MJ (1996) Transfer of bruchid resistance from the common bean to other starchy grain legumes by genetic engineering with the a-amylase inhibitor gene, In: Carozzi N, Koziel M, (Eds,) Transgenic Plants for Control of Insect Pests, Taylor & Francis London pp. 1–10.

[117] Lawrence SD, Novak NG (2006) Expression of poplar chitinase in tomato leads to inhibition of development in colorado potato beetle, Biotechnol Lett 28 593–599.

[118] Lipmann R, Kaspar S, Rutten T, Melzer M, Kumlehn J, Matros A, Mock HP (2009) Protein and metabolite analysis reveals permanent induction of stress defense and cell regeneration processes in a tobacco cell suspension culture, Int J Mol Sci 10 3012–3032.

[119] Wasano N, Konno K, Nakamura M, Hirayama C, Hattori M, Tateishi K (2009) A unique latex protein MLX56 defends mulberry trees from insects, Phytochemistry 70 880–888.

[120] Kitajima S, Kamei K, Taketani S, Yamaguchi M, Kawai F, Komatsu A, Inukai Y (2010) Two chitinase-like proteins abundantly accumulated in latex of mulberry show insecticidal activity, BMC Biochemistry 11 6.

[121] Ding X, Gopalakrishnan B, Johnson LB, White FF, Wang X, Morgan TD, Kramer KJ, Muthukrishnan S (1998) Insect resistance of transgenic tobacco expressing an insect chitinase gene, Transgenic Res 7 77–84.

[122] Kramer KJ, Muthukrishnan S (1997) Insect chitinases: Molecular biology and potential use as biopesticides, Insect Biochem Mol Bio 27 887–900.

[123] Assenga SP, You M, Shy CH, Yamagishi J, Sakaguchi T, Zhou J, Kibe MK, Xuan X, Fujisaki K (2006) The use of a recombinant baculovirus expressing a chitinase from the hard tick Haemaphysalias longicornis and its potential application as a bioacaricide for tick control, Parasitol Res 98 111–118.

[124] James C (2005) Preview: Global Status of Commercialized Biotech/ GM Crops: 2005. ISAAA Briefs No 34, ISAAA (http://www.africabio.com/pdf/Briefs2034. pdf).

[125] Sankula S, Marmon G, Blumenthal E (2005). Biotechnology-derived crops planted in 2004 – impacts on US agriculture. National Center for Food and Agricultural Policy (http://www.ncfap.org/whatwedo/pdf/2004biotechimpacts.pdf).

[126] High SM, Cohen MB, Shu QY, Altosaar I (2004) Achieving successful deployment of Bt rice, Trends Plant Sci 9 286–292.

[127] Qaim M, Zilberman D (2003) Yield effects of genetically modified crops in developing countries, Science 299 900–902.

[128] Huang J, Rozelle S, Pray C, and Wang Q (2002) Plant biotechnology in China, Science 295 674–677.

[129] Eizaguirre M, Albajes R, López C, Eras J, Lumbierres B, Pons X (2006) Six years after the commercial introduction of Bt maize in Spain: field evaluation impact and future prospects, Transgenic Res 15 1–12.

[130] De Maagd RA, Bravo A, Crickmore N (2001) How Bacillus thuringiensis has evolved specific toxins to colonize the insect world, Trends Genet 17 193–199.

[131] Karlova R, Weemen WMJ, Naimov S, Ceron J, Dukiandjiev S, Maagd RA de (2005) Bacillus thuringiensis delta endotoxin Cry1Ac domain III enhances activity against Heliothis virescens in some but not all Cry1-Cry1Ac hybrids, J Invertebr Pathol 88 169–172.

[132] Naimov S, Dukiandjiev S, De Maagd RA (2003) A hybrid Bacillus thuringiensis delta endotoxin gives resistance against a coleopteran and a lepidopteran pest in transgenic potato, Plant Biotechnol J 1 51–57.

[133] Fitches E, Audsley N, Gatehouse JA, Edwards JP (2002) Fusion proteins containing neuropeptides as novel insect control agent s: snowdrop lectin delivers fused allatostatin to insect haemolymph following oral ingestion, Insect Biochem Mol Biol 32 1653–1661.

[134] Fitches E, Edwards MG, Mee C, Grishin E, Gatehouse AM, Edwards JP, Gatehouse JA (2004) Fusion proteins containing insect-specific toxins as pest control agents:

snowdrop lectin delivers fused insecticidal spider venom toxin to insect haemolymph following oral ingestion, J Insect Physiol 50 61–71.

[135] Fitches E, Wiles D, Douglas EA, Hinchliffe G, Audsley N, and Gatehouse JA (2008) The insecticidal activity of recombinant garlic lectins towards aphids, Insect Biochem Mol Biol 38 905–915.

[136] Dutta I, Saha P, Majumder P, Sarkar A, Chakraborti D, Banerjee S, Das S (2005a) The efficacy of a novel insecticidal protein Allium sativum leaf lectin (ASAL) against homopteran insects monitored in transgenic tobacco, Plant Biotech J 3 601–611.

[137] Mehlo L, Gahakwa D, Nghia PT, Loc NT, Capell T, Gatehouse A, Gatehouse AMR, Christou P (2005) An alternative stra tegy for sustainable pest resistance in genetically enhanced crops, Proc Natl Acad Sci USA 102 7812–7816.

[138] Gruden K, Strukelj B, Popovic T, Lenarcic B, Bevec T, Brzin J, Kregar I, Herzog-Velikonja J, Stiekema WJ, Bosch D, Jongsma MA (1998) The cysteine protease activity of Colorado potato beetle (Leptinotarsa decemlineata Say) guts which is insensitive to potato protease inhibitors is inhibited by thyroglobulin type-1 domain inhibitors, Insect Biochem M ol Biol 28 549–560.

[139] Martínez M, Abraham Z, Carbonero P, and Díaz I (2005a) Comparative phylogenetic analysis of cystatin gene families from Arabidopsis rice and barley, Mol Genet Genomics 273 423–432.

[140] Martínez M, Rubio-Somoza I, Fuentes R, Lara P, Carbonero P, Díaz I, (2005b) The barley cystatin gene (Icy) is regulated by DOF transcription factors in aleur one cells upon germination, J Exp Bot 56 547–556.

[141] Outchkourov NS, de Kogel WJ, Wiegers GL, Abrahamson M, Jongsma MA (2004) Engineered multidomain cysteine protease inhibitors yield resistance against western flower thrips (Franklinielia occidentalis) in greenhouse trials, Plant Biotechnol J 2 449–458.

[142] Zhu-Salzman K, Ahn JE, Salzman R, A, Koiwa H, Shade RE, Balfe S (2003) Fusion of a soybean cysteine protease inhibitor and a legume lectin enhances anti-insect activity synergistically, Agric Amd Forest Entomol 5 317–323.

[143] Gahan LJ, Ma YT, Cobble MLM, Gould F, Moar WJ, Heckel DG (2005) Genetic basis of resistance to Cry1Ac and Cry2Aa in Heliothis virescens (Lepidoptera:Noctuidae), J Econ Entomol 98 1357–1368.

[144] Jackson RE, Bradley Jr JR, VanDuyn JW (2004) Performance of feral and Cry1Ac-selected Helicoverpa zea (Lepidoptera: Noctuidae) strains on transgenic cottons expressing one or two Bacillus thuringiensis kurstaki proteins under greenhouse conditions, J Entomol Sci 39 46–55.

[145] Bano-Maqbool S, Riazuddin S, Loc NT, Gatehouse AMR, Gatehouse JA, Christou P (2001) Expression of multiple insecticidal genes confers broad resistance against arange of different rice pests, Mol Breed 7 85–93.

[146] Cao J, Zhao JZ, Tang JD, Shelton AM, Earle ED (2002) Broccoli plants with pyramided Cry1Ac and Cry1C Bt genes control diamondback moths resistant to Cry1A and Cry1C proteins Theor Appl Genet, 105 258–264.

[147] Zhao JZ, Cao J, Li Y, Collins HL, Roush R, Earle ED, Shelton AM (2003) Transgenic plants expressing two Bacillus thuringiensis toxins delay insect resistance evolution, Nat Biotechnol 21 1493–1497.

[148] Zhao JZ, Fan YL, Fan XL, Shi XP, Lu MG (1999) Evaluation of transgenic tobacco expressing two insecticidal genes to delay resistance development of Helicoverpa armigera, Chin Sci Bull 44 1871–1874.

[149] Bharathi Y, VijayaKumar S, Pasalu I,C, Balachandran S,M, Reddy VD, Rao KV (2011) Pyramided rice lines harbouring Allium sativum (asal) and Galanthus nivalis (gna) lectin genes impart enhanced resistance against major sap-sucking pests, J Biotechnol 152 63–71.

[150] Hooker TS, Millar AA, Kunst L (2002) Significance of the expression of the CER6 condensing enzyme for cuticular wax production in Arabidopsis, Plant Physiol 129 1568–1580.

[151] Guo HN, Chen X, Zhang H, Fang R, Yuan Z, Zhang Z, Tian Y (2004) Characterization and activity enhancement of the phloem-specific pumpkin PP2 gene promoter, Transgenic Res 13 559–566.

[152] Nagadhara D, Ramesh S, Pasalu IC, Rao YK, Sarma NP, Reddy VD, Rao KV (2004) Transgenic rice plants expressing the snow drop lectin gene (gna) exhibit high-level resistance to the white backed plant hopper (Sogatella furcifera), Theor Appl Genet 109 1399– 1405.

[153] Okumoto S, Koch W, Tegeder M, Fischer WN, Biehl A, Leister D, Stierhof YD, and Frommer WB (2004) Root phloem-specific expression of the plasma membrane amino acid proton cotransporter AAP3, J Exp Bot 55 2155–2168.

[154] Sadeghi A, Broeders S, De Greve H, Hernalsteens JP, Peumans WJ, VanDamme EJ, Smagghe G (2007) Expression of garlic leaf lectin under the control of the phloem-specific promoter Asus1 from Arabidopsis thaliana protects tobacco plants against the tobacco aphid (Myzus nicotianae), Pest Manag Sci 63 1215–1223.

[155] Chakraborti D, Sarkar A, Mondal HA, Das S (2009) Tissue specific expression of potent insecticidal Allium sativum leaf agglutinin (ASAL) in important pulse crop chickpea (Cicer arietinum L.) to resist the phloem feeding Aphis craccivora, Transgenic Res 18 529–544.

[156] Dubey N (2012) Identification and characterization of phloem specific promoter induced by sap sucking pests. Ph. D. Thesis, Banaras Hindu University, India.

[157] Dixit S, Upadhyay SK, Singh H, Sidhu OP, Verma PC, Chandrashekar K (2013) Enhanced methanol production in plants provides broad spectrum insect resistance, PLoS ONE 8 e79664.

[158] Mishra P, Pandey A, Tiwari M, Chandrashekar K, Sidhu OP, Asif MH, Chakrabarty D, Singh PK, Trivedi PK, Nath P, Tuli R (2010) Modulation of transcriptome and metabolome of tobacco by Arabidopsis transcription factor, AtMYB12, leads to insect resistance, Plant Physiol 152 2258–2268.

[159] Pandey A, Misra P, Chandrashekar K, Trivedi PK (2012) Development of AtMYB12-expressing transgenic tobacco callus culture for production of rutin with biopesticidal potential, Plant Cell Rep 31 1867–1876.

[160] Pandey V, Niranjan A, Atri N, Chandrashekhar K, Mishra MK, Trivedi PK, Misra P (2014) WsSGTL1 gene from Withania somnifera, modulates glycosylation profile, antioxidant system and confers biotic and salt stress tolerance in transgenic tobacco, Planta doi 10.1007/s00425-014-2046-x.

[161] Kim YS, Uefuji H, Ogita S, Sano H (2006) Transgenic tobacco plants producing caffeine: a potential new strategy for insect pest control, Transgenic Res 15 667–672.

[162] Dixit S, Yadav S, Upadhyay SK, Verma PC, Chandrashekar K (2013) A method to produce insect resistance in plant by altering amino acid content in sap, Int J Biotech Res 3 13-20.

[163] Upadhyay SK, Sharma S, Singh H, Dixit S, Kumar J, Verma PC, Chandrashekar K (2015) Whitefly genome expression reveals host-symbiont interaction in amino acid biosynthesis. Plos ONE 10(5) e0126751 DOI: 10.1371/journal.pone.0126751.

[164] Thakur N, Mundey JK, SK Upadhyay (2015) RNAi: Implications in Entomological Research and Pest Control. ISBN 979-953-307-112-0: InTech - Open Access Publisher.

7

Drought Adaptation in Millets

Zerihun Tadele

Abstract

Millets are major food and feed sources in the developing world especially in the semi-arid tropical regions of Africa and Asia. The most widely cultivated millets are pearl millet [*Pennisetum glaucum* (L.) R. Br.], finger millet [*Eleusine coracana* (L.) Gaertn], foxtail millet [*Setaria italica* (L.) P. Beauvois], Japanese barnyard millet [*Echinochloa esculneta* (A. Braun) H. Scholz], Indian Barnyard millet [*Echinochloa frumetacea* Link], kodo millet [*Paspalum scrobiculatum* L.], little millet [*Panicum sumatrense* Roth.ex.Roem. & Schult.], proso millet [*Panicum miliaceum* L.], tef [*Eragrostis tef* (Zucc.) Trotter] and fonio or acha [*Digitaria exilis* (Kippist) Stapf and *D. iburua* Stapf]. Millets are resilient to extreme environmental conditions especially to inadequate moisture and are rich in nutrients. Millets are also considered to be a healthy food, mainly due to the lack of gluten (a substance that causes coeliac disease) in their grain. Despite these agronomic, nutritional and health-related benefits, millets produce very low yield compared to major cereals such as wheat and rice. This extremely low productivity is related to the challenging environment in which they are extensively cultivated and to the little research investment in these crops. Recently, several national and international initiatives have begun to support the improvement of diverse millet types.

Keywords: Abiotic stress, drought avoidance, drought escape, drought tolerance, millet

1. Introduction

Millets are among the major cereal crops in the developing world especially in the semi-arid tropical regions of Africa and Asia where they are used both as human food and livestock feed. Millets represent small grain crops that are mainly cultivated in marginal environments. Exceptional to this definition is pearl millet [*Pennisetum glaucum* (L.) R. Br.] that has a large seed size. Among the widely cultivated millets, those traditionally considered as millet are pearl millet, finger millet [*Eleusine coracana* (L.) Gaertn], foxtail millet [*Setaria italica* (L.) P. Beauvois], Japanese barnyard millet [*Echinochloa esculneta* (A. Braun) H. Scholz], Indian

Barnyard millet [*Echinochloa frumetacea* Link], kodo millet [*Paspalum scrobiculatum* L.], little millet [*Panicum sumatrense* Roth ex Roem. & Schult.] and proso millet [*Panicum miliaceum* L.] (Table 1). Tef [*Eragrostis tef* (Zucc.) Trotter] was included in the millet group at the First Small Millets Workshop held 30 years ago in Bangalore, India [1], while both tef and fonio or acha [*Digitaria exilis* (Kippist) Stapf and *D. iburua* Stapf] were grouped under small millets by international agricultural organizations in the mid-1990s [2]. The inclusion of tef and fonio to the millet family is justifiable due to the close relationship of the two species with other millets. The genetic difference between some traditional millets is as large as that between tef or fonio and other millets. Due to this substantial variability among themselves, millets are grouped into two subfamilies, namely Panicoideae, which includes pearl millet, foxtail millets, Japanese barnyard millet and Indian millet, and to Chloridoideae, which includes finger millet and tef, and eight genera (Table 1). This indicates that finger millet which is normally grouped under millet is more closely related to tef than to other millets [3]. The divergence among traditional millets is also exhibited in the chromosome number and ploidy level which range from the diploid pearl millet ($2n = 2x = 14$) to the hexaploid fonio ($2n = 6x = 54$) [4, 5]. Millets are also divergent in the size and colour of seeds, seed weight, plant stature and shape of their panicles (Table 1). The geographical distributions of small millets were recently summarized by Goron and Raizada [6]. Except for finger millet, which is extensively cultivated in Africa and Asia, other small millets are mainly grown in Asia.

Common name	Traditional millets				
	Pearl millet	Foxtail millet	Japanese Barnyard millet	Indian Barnyard millet	Kodo millet
Other names	Bulrush millet	Italian millet	Japanese millet	Billion dollar grass	Koda millet
Botanical names	*Pennisetum glaucum*	*Setaria italica*	*Echinochloa esculneta*	*Echinochloa frumetacea*	*Paspalum scrobiculatum*
Subfamily	Panicoideae	Panicoideae	Panicoideae	Panicoideae	Panicoideae
Tribe	Paniceae	Paniceae	Paniceae	Paniceae	Paniceae
Distribution			Japan, Korea, China	India, Pakistan, Nepal	
Ploidy level	Diploid	Diploid	Hexaploid	Hexaploid	Tetraploid
Chromosome number	2n = 2x = 14	2n = 2x = 18	2n = 6x = 36	2n = 6x = 36	2n = 4x = 40
Purpose	Food, feed	Food, biofuel	Food, feed	Food	Food, feed
Agronomic benefits	Drought & heat tolerance	Drought tolerance	Early maturity, anti-fungal	Early maturity	Drought tolerance
Nutritional benefits	High protein, starch & minerals		anti-diabetic	High protein content	High-quality protein
Health benefits	No gluten	No gluten	No gluten	No gluten	Low glycaemic index, anti-oxidant
Reference	[4, 20, 27]	[16, 27]	[5, 18, 27]	[6]	[27, 36]

Common name	Traditional millets			Other millets	
	Little millet	Proso millet	Finger millet	Tef	Fonio
Other names		Common millet	Ragi, African millet	Teff, lovegrass	Acha
Botanical names	*Panicum sumatrense*	*Panicum miliaceum*	*Eleusine coracana*	*Eragrostis tef*	*Digitaria exilis, D. Iburua*
Subfamily	Panicoideae	Panicoideae	Chloridoideae	Chloridoideae	Panicoideae
Tribe	Paniceae	Paniceae	Eragrostideae	Eragrostideae	Paniceae
Distribution					
Ploidy level	Tetraploid	Tetraploid	Tetraploid	Tetraploid	Diploid or hexaploid
Chromosome number	$2n = 4x = 36$	$2n = 4x = 36$	$2n = 4x = 36$	$2n = 4x = 40$	$2n = 2x = 30$ or $2n = 6x = 54$
Purpose	Food	Food, feed	Food, feed	Food, feed	Food, feed
Agronomic benefits	Abiotic stress tolerance	Drought tolerance, early maturity	Drought and salt tolerance	Water-logging tolerance, storage pest tolerance	Drought tolerance
Nutritional or	High in phytochemicals, fibre	Rich in amino acids	Rich in calcium, methionine and tryptophan	Rich in protein	Rich in amino acids
Health benefits	Anti-diabetic	Anti-cancer	Low glycaemic index, anti-oxidant	No gluten	
Reference	[38–40]	[6, 23, 27]	[6, 20, 27, 35, 37]	[33, 47]	[20, 24, 25]

Table 1. Description and benefits of millets

2. Importance of millets in global agriculture

2.1. Economic benefits

Millets play a key role in the economy of the developing world especially in countries with extensive areas of marginal land used for crop cultivation. In 2013, the global area under millet cultivation was 34.9 million hectares, corresponding to 4.7 % of the global area for all cereals including wheat, maize and rice [7] (Table 2). On the other hand, the global production of millets in the same year was estimated to be 36.7 million tons, which contributes only 1.2% to the total cereal production. This lower production was due to the inferior average yield of millets (only 0.9 t ha^{-1}) compared to other cereals (3.8 t ha^{-1}). However, the contribution of India to global millet production is significant. In 2013, India produced over 30% of the global millet yield from only 25% of the global millet area, mainly due to improved productivity. In the same year, while the mean seed yield of millet in India was 1.2 t ha^{-1}, it was only 0.8 t ha^{-1} for other countries. This 50% production advantage in India over other countries especially African countries was due to the widespread use of improved varieties and techniques. A decade ago, the rate of adoption of improved pearl millet cultivars by farmers was 65% in India but below 10% in some African countries [8].

Country	Traditional millets[a]	Tef	Fonio (Acha)	Total
Production (ton)				
India	10,910,000			10,910,000
Ethiopia	848,956[b]	4,418,642		5,267,598
Nigeria	5,000,000		90,000	5,090,000
Niger	2,995,000		6,000	3,001,000
China	1,746,000			1,746,000
Mali	1,152,331		22,090	1,174,421
Burkina Faso	1,078,570		19,887	1,098,457
Sudan (former)	1,090,000			1,090,000
Guinea	215,000		429,000	644,000
Chad	582,000			582,000
Senegal	572,155		1,030	573,185
Russia	418,844			418,844
USA	418,145			418,145
Tanzania	322,731			322,731
Pakistan	310,000			310,000
Nepal	305,588			305,588
Uganda	228,000			228,000
Myanmar	185,000			185,000
Ghana	155,131			155,131
Cameroon	97,000			97,000
others	1,233,696		19,000	1,252,696
Total production	29,864,147	4,418,642[c]	587,007	34,869,796
Total area (ha)	33,118,792	3,016,521	554,451	36,689,764
Yield (ton ha^{-1})[d]	0.90	1.47	1.06	0.95
Reference	[7,9]	[9]	[7]	

[a] Traditional millets include finger millet, foxtail millet, Indian barnyard millet, Japanese barnyard millet, kodo millet, little millet, pearl millet and proso millet.

[b] Only for finger millet.

[c] Only for Ethiopia.

[d] Average global yield except for tef where it is the national average yield for Ethiopia.

Table 2. The top 20 millet-producing countries in the world in 2013.

Tef and fonio are exclusively cultivated in Africa. While fonio is cultivated on a total of half a million hectares in West Africa mainly in Guinea, Guinea-Bissau and Côte d'Ivoire [7], tef is grown in the Horn of Africa especially in Ethiopia where it is annually cultivated on three million hectares of land and is a staple food for about 50 million people [9]. In the last two decades, the productivity of tef was raised by 100%, from just 0.7 t ha^{-1} in 1994 to 1.4 t ha^{-1} in 2013 mainly due to an increase in the use of improved cultivars.

In general, millets play a key role in food security in Asia and Africa. Together with sorghum, millets account for about half of the total cereal production in Africa [10]. Millets, are therefore considered as a poor man's crop due to their significant contributions to the diet of resource-limited farmers and consumers.

2.2. Agronomic benefits

Millets are resilient to the extreme climatic and soil conditions prevalent in the semi-arid regions of Asia and Africa. The similarities of millets are that they are all grown under extreme environmental conditions, especially those of inadequate moisture and poor soil fertility which are poorly suited to the major crops of the world [11] (Table 1). Proso millet is considered to have been domesticated before rice in China, based on the extreme resistance of this millet to drought [12, 13]. In addition to its resistance to drought, proso millet escapes the terminal drought that normally occurs late in the growing season since it matures in only three months; hence, proso millet is considered to be a millet with low water requirements [6].

Similar to maize and sorghum, millets possess a C4 photosynthesis system [14, 15]; hence, they prevent photorespiration and, as a consequence, efficiently utilize the scarce moisture present in the semi-arid regions. Since C4 plants are able to close their stomata for long periods, they can significantly reduce moisture loss through the leaves. In addition to its tolerance to drought, tef is tolerant to waterlogging especially in poorly drained soils where other crops such as maize and wheat could not survive. Foxtail millet is also considered to be a model plant for biofuel studies [16]. A novel peptide isolated from foxtail millet and barnyard millet has shown strong antifungal properties as has one from finger millet which is especially effective and works against four fungus species, namely *Alternaria, Trichoderma, Botrytis* and *Fusarium* [17, 18].

2.3. Nutritional benefits

Millets are rich sources of nutrients for both humans and animals. Saleh et al. [19] have compiled detailed information on the nutritional advantages of several millets. The grains of most millets possess levels of protein comparable to those of wheat but higher than those of rice [20] (Table 1). In addition, the seeds of finger millet contain valuable amino acids especially methionine [20], which is lacking in the diets of hundreds of millions of the poor who live on starchy staples such as cassava. Other reports indicate that finger millet is rich in lysine, threonine and valine [21, 22] while proso millet has plentiful leucine, isoleucine and methionine [23]. The seeds of fonio are also nutritious, especially in amino acids such as leucine, methionine and valine [24, 25]. Since proso millet is rich in essential amino acids including leucine, isoleucine and methionine, the protein quality of the grain is higher than that of wheat [23].

The grains of extensively cultivated pearl millet contain high amounts of starch, fibres and minerals [26, 27]. In general, millets have high amounts of vitamins, calcium, iron, potassium, magnesium and zinc [28].

The straws and crop residues of millets are also the main source of livestock feed for farmers in developing countries. In Ethiopia, compared to the straw from other cereals, the straw of tef is the most palatable to livestock and fetches the highest price [29].

2.4. Health-related benefits

In addition to being nutritious, millets are also considered to be a healthy food. Two recent reviews examined the health-related benefits associated with millets [19, 6]. A number of leading newspapers and media have recently indicated the potential of millets particularly tef as a global lifestyle crop [30–32]. This is particularly due to the lack of gluten in the grain of tef [33] (Table 1). Gluten is a substance present in wheat and other grains that causes celiac disease or other forms of allergies. Similar to tef, several other millets, particularly foxtail millet, do not contain gluten.

Six millet species (namely kodo, finger, proso, foxtail, little and pearl millets) were shown to have an anti-proliferative property and might have a potential in the prevention of cancer initiation [34, 35]. The anti-proliferative property of these millets is associated with the presence of phenolic extracts. Among the first four millets indicated above, the maximum phenolic content was obtained in kodo millet while the minimum was in foxtail millet [36].

Finger millet is also a popular food among diabetic patients because of its low glycaemic index and slow digestion due to high fibre content [37]. The glycaemic index of little millet was also lower than that of rice, wheat and sorghum; hence, it is considered to be an anti-diabetic grain [38]. The composition of useful antioxidants and related products could be enhanced through processing the grain. A study in little millet showed that the levels of phenolics, flavonoids and tannins were substantially increased by germinating, steaming and roasting soaked grains [39].

3. Drought: A major challenge to millet cultivation

Biotic stresses such as insect pests and diseases are a cause for substantial yield losses to diverse types of millets. However, abiotic stresses are the biggest contributor to losses every year. Although, in general, millets perform better than cereals such as wheat and rice in semi-arid environments, these challenging climatic and soil conditions are by no means an optimum environment for millet cultivation. In semi-arid and arid environments where millets are the dominant crop, drought or inadequate moisture is the major abiotic stress affecting productivity. Studies in pearl millet showed that drought impacts include growth, yield, membrane integrity, pigment, osmotic adjustment, water relations and photosynthetic activity [40].

3.1. Prevalence of drought

Drought is defined as a temporary reduction in moisture availability in which the amount of available water is significantly below normal for a specified period. In general, drought can be explained as meteorological, hydrological or agricultural drought [41]. Agricultural drought occurs when there is not enough soil moisture to meet the needs of a particular crop at a particular time. Drought is also commonly expressed as a shortage or absence of rainfall causing a loss in rain-fed agriculture. For example, the decline in the level of rainfall during severe drought years in Ethiopia was accompanied by serious reductions in rain-fed agricultural outputs; this is because a 10% drop in rainfall (below the long-term national averages) results in an average drop of 4.2% in cereal yields [42].

As indicated above, millets are crops of dry land areas of the world. According to the United Nations, dry lands, which cover 40% of the world's land area or one-third of the global arable land, support two billion people, of which 90% live in the developing world [43]. Dry lands are classified into four, namely hyper-arid deserts, arid, semi-arid and dry subhumid. Millets are extensively cultivated in the semi-arid region, which is characterized by low and erratic rainfall and periodic drought. Climate change is expected to worsen the situation in this part of the world by reducing the grassland productivity by 49–90% by 2020 [43]. The Sahel Region in Africa, covering over three million km^2 in 10 countries (namely northern Senegal, southern Mauritania, central Mali, northern Burkina Faso, the extreme south of Algeria, Niger, the extreme north of Nigeria, central Chad, central and southern Sudan and northern Eritrea) is the typical semi-arid region situated between the Sahara desert in the north and the tropical or savanna climate in the south [44].

The frequency and intensity of drought has increased in recent times. In Ethiopia, severe droughts used to occur periodically every 6–8 years [45], but recently, they have happened every 1–2 years especially in the south of the country [46].

Similar to other millets, drought is implicated among the major yield limiting factors in tef production [47]. Although tef grows in a wide variety of agro-ecological conditions ranging from semi-arid areas with low rainfall to areas with high rainfall, the rainfall pattern in most tef growing regions is not consistent enough to support the normal growth of the crop during the crop cycle. In most tef growing regions, greater rainfall variability exists over the growing period than over the year-cycle [48, 49] which results in poor agricultural outputs. The Water Requirement Satisfaction Index (WRSI), a crop-specific performance indicator taking rainfall and soil characteristics into account, indicates extreme and increasing variability in Ethiopia. A recent study also confirmed that climate will have a negative impact on the acreage and productivity of tef unless urgent interventions are implemented which favours mitigation and adaptation strategies [50].

3.2. Yield losses due to drought

Various yield loss studies made for millets treated with drought conditions are summarized in Table 3. Using polyvinylchloride (PVC) tubes filled with sandy soil, Matsuura and colleagues [51] investigated the effect of moisture deficit before and after flowering on four millets, namely proso millet, little millet, foxtail millet and wild millet [*Setaria glauca* (L.) Beauv.]. Compared to the well-watered plants, a significant yield reduction was obtained in all four millets when the drought treatment was implemented at early developmental stage, that is, before flowering (or heading). However, terminal drought, which occurs from the flowering stage to the harvesting of the crop, contributed to a significant yield loss only in proso and little millets while the effect on foxtail and wild millets was negligible.

Millet type	Yield loss (%)			Critical stage	Reference
	Early drought[a]	terminal drought[b]	Long-term drought[c]		
Proso millet	30.1*	34.6*	64.0*	Before and after heading	[51]
Little millet	62.6*	80.1*	80.5*		[51]
Foxtail millet	19.2*	3.4[NS]	20.3*	Before heading	[51]
Wild millet (*Setaria gluaca*)	27.3*	15.3[NS]	30.1*		[51]
	Mid-season stress[d]	Terminal stress[e]			
Pearl millet	6.6	60.1		Flowering	[53]
Finger millet		109.8*[f]		Flowering	[54]
	Prior to flowering	Beginning flowering	End of flowering		
Pearl millet	72	61	Insignificant	From four weeks to flowering	[52]
	Early stress[g]				
Tef	69–77				[55]

[a] Early drought: water stress from 25 days after sowing till flowering.

*Indicates statistically significant difference from the well-watered samples.

[b] Terminal drought: water stress from flowering till harvesting.

[c] Long-term drought: water stress from 25 days after sowing till harvesting.

[d] Mid-season stress: water stress for 30 days from floral initiation to flowering.

[e]Terminal stress: water stress at flowering.

[f] Water stress from 28 days after sowing to harvest.

[g] Early stress: water stress from two weeks after emergence until symptoms of stress observed.

Table 3. The magnitude of yield loss due to moisture scarcity in millets.

A study by Winkel et al. [52] in Niger where the annual rainfall is around 200 mm investigated the impact of water deficit at three stages of pearl millet development. The three stages were prior to flowering, at flowering and at the end of flowering. According to the findings of the work, the grain yield of pearl millet was severely reduced when moisture was limited prior to and at the flowering stage but not at the end of flowering. On the other hand, in pearl millet, terminal drought in which irrigation was terminated from the flowering until crop maturity, was severe, as it resulted in 60% yield loss [53]. The mid-season stress, which occurred from one month before flower initiation to full flowering, resulted in only 7% yield loss.

The study in two landraces of finger millet in which a drought treatment was imposed four weeks after sowing, resulted in 100% yield loss and over 30% biomass damage [54]. Similarly, yield loss reached up to 77% when the tef plant experienced drought at the flowering stage [55].

Although yield loss studies were not exhaustively made for most millets as they are considered drought tolerant, substantial damage occurs to these crops depending on the severity of drought. However, millets produce at least some grain and straw even in bad years unlike drought-intolerant cereals such as wheat and rice which completely fail to produce any yield.

4. Adaptation of millets to drought

4.1. Strategies to drought adaptation or tolerance

Plants cope with drought using three main strategies, namely, drought escape, drought avoidance and drought tolerance, although a fourth strategy, known as drought recovery, has also been identified [56–60].

Drought escape: Drought escape refers to the condition in which plants reach maturity before the drought occurs. Traits associated with drought escape are rapid growth, early flowering, high leaf nitrogen level and high photosynthetic capacity [58]. The study in West Africa indicated that pearl millet matches its phenology to the mean distribution of the rainfall where precipitation is limited and erratic [61]. In this case, the development of the main panicle coincided with an increasing period of rain, thus reducing the risks associated with drought events occurring prior to or at the beginning of flowering.

Drought avoidance: Drought avoidance refers to the ability of the plant to maintain a favourable water balance under moisture stress in order to avoid water deficit in the plant tissue. Two types of drought avoidance mechanisms have been identified: (i) those that reduce water loss through transpiration (e.g. low stomata conductance and reduced leaf) and (ii) those that maintain water uptake during drought period (e.g. high root-to-shoot ratio) [56, 58, 62].

Drought tolerance: Drought tolerance refers to the ability of the plant to produce some yield by withstanding low water potential [62]. Traits associated with drought tolerance are increased osmoprotectants (or compatible solutes such as betaines and amino acids), and osmotic adjustment (i.e. reducing osmotic potential through accumulation of organic and inorganic substances) [58, 60].

Drought recovery: Drought recovery refers to a condition in which plants recover from the adverse effects of drought in order to provide some yield and/or biomass. Desiccation-tolerant or resurrection plants particularly the wild *Eragrostis nindensis* is the typical example of drought recovery since it stabilizes its cells or membranes at desiccated state [63].

These strategies which are devised by plants to cope with drought are manifested through changes in some phenotypic traits. In a recent review, Kooyers [58] showed for each strategy the path followed by plants in terms of life cycle, altered phenotypes and to the type of drought the plant fits itself. This indicates that the strategies and mechanisms of drought tolerance are interrelated.

4.2. Mechanism of drought tolerance

Table 4 summarizes various mechanisms of drought tolerance in diverse millet types. These inherent properties of plants which include agronomical, morphological and physiological traits are briefly discussed below.

Parameter	Millet type	Response to drought	Reference
Agronomy-related traits			
Seed number and biomass	Pearl millet	Unaffected under drought	[64]
Seed yield	Pearl millet	High for drought-tolerant genotypes	[65]
Flowering time	Pearl millet	Adjust phenology to rainfall pattern	[53]
Morphology-related traits			
Shoot length	Little millet	Decreased under drought	[40]
Root length	Little millet	Increased under drought	[40]
Leaf tensile strength	Tef	Increased in drought-tolerant plants	[68]
Physiology-related traits			
Water extraction	Pearl millet	Less extraction before flowering; more extraction after flowering	[65]
Chlorophyll content	Little millet	Decreased under drought	[40]
Biochemical-related traits			
Anti-oxidants	Little millet	Accumulated under drought	[40]
ROS scavenging enzymes	Little millet, tef	Accumulated under drought	[40, 71]
Free proline	Tef, little millet	Increased concentration	[40, 71]
GB (glycine betaine)	Little millet	Accumulated under drought	[40]
Superoxide	Little millet	Accumulated under drought	[40]
AP (ascorbate peroxidase)	Tef, little millet	Increased specific activity	[40, 71]
CAT (catalase)	Little millet	Accumulated under drought	[40]
GR (glutathione reductase)	Tef	Increased concentration	[71]
MDAR (monodehydro-ascorbate reductase)	Tef	Increased concentration	[71]
Total free amino acid	Little millet	Increased concentration	[40]

Table 4. Traits associated to diverse drought tolerance mechanisms in millets.

Agronomy-related traits: These refer to the traits that are commonly known as yield and yield components. Among these, number of tillers, number and size of panicle, seed and biomass yield, seed weight and harvest index are the major ones. However, conclusions could not be made from the two studies using drought-tolerant pearl millet cultivars since drought did not affect the shoot biomass in the first case [64] while it boosted the seed yield in the second case [65].

Morphology-related traits: Morphological or anatomical traits which play important roles in drought tolerance include root- and shoot length and leaf area [66]. However, changes in the morphological and biochemical properties of the flag leaf play a key role in drought tolerance as flag leaves are the primary source of photosynthesis [67]. Mechanical properties of the plant also affect drought tolerance in millets. Balsamo et al. [68] studied the leaf tensile strength or also known as force to tear in three *Eragrostis* species with different levels of tolerance to drought. According to their findings, drought-tolerant *E. curvula* had higher tensile strength values than the moderately drought-tolerant *E. tef*, which in turn had higher values than the drought-susceptible *E. capensis*, indicating a positive correlation between drought tolerance and leaf tensile strength [68]. Structural investigations of leaves from the three species revealed the presence of extensive lignification of bundle sheath extensions in *E. tef* and *E. curvula* unlike in *E. capensis*. A study in maize indicated that lignification of the midrib parenchyma and epidermis was directly correlated with increased tensile strength [69].

Physiology-related traits: Among the several physiological traits that are differentially regulated during moisture deficit, osmotic adjustment is a major mechanism that increases drought avoidance to enable the plant produce some yield. Osmotic adjustment, which refers to the lowering of the osmotic potential in the cytoplasm due to the accumulation of compatible solutes such as proline, glycine betaine and organic acids, contributes to turgor maintenance of shoots and roots [40]. In little millet, drought stress increased the amount of proline and glycine betaine in both the root and leaf [40]. According to the authors, the accumulation of free amino acids in this millet during drought might be related to the disruption in protein synthesis, induced proteolysis or its partial hydrolysis [40].Water-use efficiency of the plant is also important as moisture is mostly limited in the areas where millets are extensively cultivated. The experiment using drought-sensitive and drought-tolerant pearl millet genotypes showed that under moisture deficit conditions, the total amount of water extracted by both genotypes was comparable [65]. However, compared to susceptible genotypes, tolerant genotypes extracted less water prior to flowering and more water after flowering, enabling these genotypes to support the tillers and maintain the stay-green phenotype.

Biochemical-related traits: Reactive oxygen species (ROS) are chemically reactive molecules that are useful in cell signalling at low concentrations but are damaging to cells when present at high concentrations. The main causes for the high production of ROS are environmental stresses such as drought and salinity [70]. In order to reduce the damaging effects of ROS, plants produce antioxidants, which include glutathione, ascorbate and carotenoids and ROS-scavenging enzymes which include superoxide dismutase (SOD), peroxidase (POD), catalase (CAT) and ascorbate peroxidase (AP or APX) [40]. In little millet, the activity of SOD, POD and CAT were elevated under drought conditions to enable the plant cope with unfavourable ROS accumulation [40]. Similarly, the activity of AP and monodehydro-ascorbate reductase (MDAR) increased in tef plants treated with drought compared to control plants receiving normal watering [71].

4.3. Genes involved in drought tolerance

The sequence of the genome and transcriptome of plants provides information important to the understanding of the types of genes involved in the regulation of drought tolerance, particularly in plants with increased resistant to moisture scarcity. So far, the genome of foxtail millet [72, 73] and tef [3] has been sequenced.

Transcriptome sequencing of millets after exposure to moisture-deficit condition provides information on genes differentially regulated under exposure to abiotic stresses particularly to drought. A transcriptome-wide study of finger millet plants exposed to drought obtained 2824 genes that were differentially expressed under these conditions [74].

Genes known to be involved in drought response and/or tolerance of selected millets are presented in Table 5. Wang et al. [75] indicated that the overexpression of SiLEA14, a type of LEA gene from foxtail millet, increased the tolerance of *Arabidopsis* plants to salt and osmotic stress. Parvathi et al. [76] reported the induction of several genes when finger millet was exposed to drought. The up-regulated genes include metallothionein, farnesylated protein ATFP6, Farnesyl pyrophosphate synthase and protein phosphatase 2A.

Gene name	Source of the gene	Test organism (type)	Reference
SiLEA	Foxtail millet	Overexpression in foxtail millet and Arabidopsis increased drought tolerance	[75]
SiARDP	Foxtail millet	Overexpression in foxtail millet and Arabidopsis increased drought tolerance	[83]
EcDehydrin7	Finger millet	Overexpression of EcDehydrin7	[80]
Ec-apx1	Finger millet	Expression increased under drought	[82]
Mt1D	bacteria	Finger millet expressing mt1D had better osmotic adjustment and chlorophyll retention under drought	[81]
Metallothionein,	Finger millet	Induced under drought	[76]
Farnesylated protein ATFP6	Finger millet	Induced under drought	[76]
Farnesyl pyrophosphate synthase	Finger millet	Induced under drought	[76]
Protein phosphatase 2A	Finger millet	Induced under drought	[76]
RISBZ4	Finger millet	Induced under drought	[76]
β-carbonic anhydrase (*Pg*CA)	Pearl millet	Up-regulated when exposed to drought	[79]

Table 5. Differentially regulated drought-related genes in millets.

Traits associated with drought tolerance were investigated using a genome scan and association mapping methods [77, 78]. A single gene known as β-carbonic anhydrase (*Pg*CA) was consistently up-regulated in pearl millet exposed to multiple abiotic stresses including drought, salinity and heat [79]. Hence, this particular gene is useful in adapting the plant to diverse abiotic stresses. Other genes known to be involved in drought response or tolerance in millets were EcDehydrin 7 [80], mt1D [81] and Ec-apx1 [82] from finger millet, and SiARDP [83] from wild foxtail millet.

Although not yet reported for millets, the suppression of two genes, namely, SAL1 and ERA1, increased the drought tolerance of the model plant *Arabidopsisthaliana* [84, 85]. The *era1* mutants develop tolerance to drought through a mechanism involving closing of the stomata [85].

5. Breeding millets for extreme drought tolerance

5.1. Germplasm acquisition and utilization

National and international efforts have been made to collect and maintain landraces of various millets types. The recent review by Goron and Naizanda [6] indicates the institutions involved in the preservation efforts and the amount of germplasm available at each institution. In general, India and China dominate the collections of millets. While institutions in India maintain 67% of the total of 33650 finger millet accessions, a single institution in China called the Chinese National Gene Bank preserves 61.2% of the total of 43,580 foxtail collections. Similarly, in the Ethiopian Institute of Biodiversity (EIB), over 5000 tef landraces collected from various tef-growing regions in the country are available [86]. Although these germplasm collections might not be exhaustive, they can play a key role in improving the productivity of respective crops. Further, large-scale expeditions need to be made for other millets in order to fully survey and bank the existing diversity in millets.

5.2. Breeding for drought tolerance

Breeding for drought tolerance is the major objective of many crop-breeding programmes due to the widespread prevalence of the moisture-deficit problem in global agriculture. A number of crops with drought tolerance have been developed. There are two options for the management of crops in water-limiting environments: the genetic and agronomic [87]. The genetic approach requires robust and reproducible screening methods for the identification of traits of drought tolerance in germplasm and breeding materials, and incorporation of the same into high-yielding varieties using conventional and biotechnological tools.

Crop breeding has relied for many years on conventional and ancient techniques such as selection and hybridization. Mutation breeding, the process of using chemicals or radiation to generate mutant plants with desirable traits, has also been used for several decades and has been a key in the release of over 2000 crop varieties to the farming community among which drought-tolerant cultivars are included [88]. Crop improvement techniques that apply modern genetic and omics (genomics, transcriptomics, proteomics and metabolomics) tools include the following: (i) *marker-assisted selection* (MAS) which refers to the utilization of molecular markers located near genes of interest to breed for traits that are difficult to observe, (ii) *TILLING* (targeting induced local lesions in genomes) [89] or EcoTILLING [90], the high-throughput and non-transgenic techniques which rapidly detect point mutations in mutagenized populations, and (iii) *Gene* targeting that relies on the following three tools to increase the efficiency of gene targeting: *zinc-finger nucleases* [91, 92], *TALEN* (transcription activator-like effector nuclease) [93] and *CRISPR/Cas* (clustered regularly interspaced short palindromic repeats)/(CRISPR associated), type II prokaryotic adaptive immune system [94, 95].

5.3. Improved crop management

The wise use of crop management practices which include the time of planting, frequency of tillage and the rate and time of fertilizer application is important particularly in the semi-arid regions where moisture is scarce. Flexibility to change from late maturing crops to early maturing crops when the rainfall arrives late in the season is important. In the central semi-arid regions of Ethiopia farmers start their season by planting sorghum in April. When sorghum fails due to late arrival of rain, they sow wheat in June. However, if the rain is still late or not enough for wheat plant establishment, farmers sow tef in July or early August as the last option. Compared to sorghum and wheat, tef requires less moisture and matures early.

Suggestions have been earlier given on the type of technologies to be adopted in the semi-arid regions of Southern Africa [96] and West Africa [97]. According to Mir and colleagues, these technologies should include genomics, physiology and breeding [98].

5.4. Agricultural inputs and insurance

Access to agricultural inputs such as improved seeds, fertilizer and chemicals as well as credit and markets is important for farmers. In semi-arid areas where millets are dominantly cultivated, the amount and pattern of rainfall is erratic. Due to this, an insurance system known as Weather Index Drought Insurance has been implemented for the last decade in several African countries including Niger [99], Ghana [100], Kenya [101] and Burkina Faso [102] as well as India [103]. The successful insurance organization called 'Kilimo Salama' which was initially established by Syngenta Foundation for Sustainable Agriculture (SFSA) and implemented in several East African countries has been recently transferred to the Agriculture and Climate Risk Enterprise Ltd. (ACRE) [104, 105].

5.5. Partnership in research and development

Collaborations among national and international institutions are required in both research and development, in order first to develop improved millet cultivars and later to disseminate them to the farming community. Among the institutions with a global mandate to improve millets, ICRISAT (International Crops Research Institute for Semi-Arid Tropics) has recently added tef to the list of its mandate crops [106]. With its headquarters in Patancheru, India and regional officers in Nairobi (Kenya) and Bamako (Mali), it has been focusing on the improvement of diverse millets. The centre is among the 15 international agricultural research centers that belong to the CGIAR (Consultative Group for International Agricultural Research), the global partnership that unites organizations engaged in research for food security. Hence, the research and development of tef, a vital crop in the Horn of Africa that feeds over 50 million people in Ethiopia alone, will receive a global partnership towards its improvement and development. In general, suggestions given to the improvement of understudied or orphan crops [107, 108] could also be applied to the research and development of millets.

6. Conclusions

Millets play a significant role in the livelihood of the population of developing world especially due to their enormous contribution to the food security of these countries. However, these crops have not been sufficiently studied and hence have been named orphan crops. Both conventional and modern improvement techniques have not yet been adequately implemented. It is believed that the changing climate will have significant effects on the types of crops cultivated in the next century. Currently, widely cultivated crops that provide the daily diet for many (such as wheat) might not be extensively cultivated in the future due to environmental stresses, especially the increase in global temperature. Millets might provide alternative climate-smart crops, as their adaptations to challenging environment are better than the current major crops of the world. Enhancing the productivity of millets requires concreted efforts of breeders, agronomists, policy makers and donors at both individual and institutional capacities.

7. Abbreviations

ABA; Abscisic acid

ACRE; Agriculture and Climate Risk Enterprise Ltd

AP or APX; Ascorbate peroxidase

CAT; Catalase

CGIAR; Consultative Group for International Agricultural Research

CSA; Central Statistical Agency (Ethiopia)

Ec-apx1; Ascorbate peroxidase

EIB; Ethiopian Institute of Biodiversity

FAO; Food and Agriculture Organization of the United Nations

FAOSTAT; Food and Agriculture Organization Statistics

GR; Glutathione reductase

ICRISAT; International Crops Research Institute for Semi-Arid Tropics

LEA; Late embryogenesis abundant

MDAR; Monodehydro-ascorbate reductase

Mt1D:; Mannitol dehydrogenase

*Pg*CA; β-carbonic anhydrase

POD; Peroxidase

PVC; Polyvinylchloride

ROS; Reactive oxygen species

SFSA; Syngenta Foundation for Sustainable Agriculture

SiARDP; *Setaria italica* ABA-responsive DREB-binding protein

SiLEA14; *Setaria italica* late embryogenesis abundant 14

SOD; Superoxide dismutase

UN; United Nations

WRSI; Water Requirement Satisfaction Index.

Acknowledgements

I would like to thank Syngenta Foundation for Sustainable Agriculture, SystemsX.ch and the University of Bern for supporting the Tef Improvement Project based at the Institute of Plant Sciences, University of Bern, Switzerland.

Author details

Zerihun Tadele[1,2*]

Address all correspondence to: zerihun.tadele@ips.unibe.ch

1 Institute of Plant Sciences, University of Bern, Bern, Switzerland

2 Addis Ababa University, Institute of Biotechnology, Addis Ababa, Ethiopia

References

[1] Seetharam, A., K.W. Riley, and G. Harinarayana, eds. *Small millets in global agriculture: Proceedings of the First International Small Millets Workshop, 29 October to 2 November 1986, Bangalore, India.* 1989, Oxford & IBH Publishing Co. Pvt. Ltd: New Delhi 413.

[2] ICRISAT-FAO, *The world sorghum and millet economies: facts, trends and outlook*. 1996: International Crops Research Institute for the Semi-Arid Tropics, India & Food and Agriculture Organization of the United Nations, Rome, Italy.

[3] Cannarozzi, G., et al., *Genome and transcriptome sequencing identifies breeding targets in the orphan crop tef (Eragrostis tef)*. BMC Genomics, 2014. 15: p. 581.

[4] Martel, E., et al., *Genome size variation and basic chromosome number in pearl millet and fourteen related Pennisetum species*. Journal of Heredity, 1997. 88(2): p. 139–143.

[5] Mann, D.G.J., et al., *Switchgrass (Panicum virgatum L.) polyubiquitin gene (PvUbi1 and PvUbi2) promoters for use in plant transformation*. BMC Biotechnology, 2011. 11: p. 74.

[6] Goron, T.L. and M.N. Raizada, *Genetic diversity and genomic resources available for the small millet crops to accelerate a New Green Revolution*. Frontiers in Plant Science, 2015. 6.

[7] FAOSTAT. *Crop production*. 2015 [cited 2015 September 7]; Available from: http://faostat3.fao.org/home/E.

[8] Bantilan, M.C.S. and U.K. Deb, *Impacts of genetic enhancement in pearl millet*, in *Crop Improvements and Its Effect on Productivity: The Impact of International Agricultural Research*, R.E. Evenson and D. Gollin, Editors. 2003, CABI International: UK. p. 215–240.

[9] CSA, *Agricultural Sample Survey for 2013/14*, in *Statistical Bulletin 532*. 2014: Addis Ababa, Ethiopia.

[10] Belton, P.S. and J.R.N. Taylor, *Sorghum and millets: protein sources for Africa*. Trends in Food Science & Technology, 2004. 15(2): p. 94–98.

[11] Baker, R.D., *Millet production. Guide A-414* 2003: Cooperative Extension Service, College of Agriculture and Home Economics, New Mexico University, Las Cruces, USA.

[12] Lu, H.Y., et al., *Earliest domestication of common millet (Panicum miliaceum) in East Asia extended to 10,000 years ago*. Proceedings of the National Academy of Sciences of the United States of America, 2009. 106(18): p. 7367–7372.

[13] Jones, M.K. and X. Liu, *Archaeology. Origins of agriculture in East Asia*. Science, 2009. 324(5928): p. 730–731.

[14] Brutnell, T.P., et al., *Setaria viridis: a model for C4 photosynthesis*. The Plant cell, 2010. 22(8): p. 2537–2544.

[15] Warner, D.A. and G.E. Edwards, *C-4 Photosynthesis and Leaf Anatomy in Diploid and Autotetraploid Pennisetum-Americanum (Pearl-Millet)*. Plant Science, 1988. 56(1): p. 85–92.

[16] Doust, A.N., et al., *Foxtail Millet: A Sequence-Driven Grass Model System*. Plant Physiology, 2009. 149(1): p. 137–141.

[17] Xu, W., et al., *A novel antifungal peptide from foxtail millet seeds.* Journal of the Science of Food and Agriculture, 2011. 91(9): p. 1630–1637.

[18] Nolde, S.B., et al., *Disulfide-stabilized Helical Hairpin structure and activity of a novel antifungal peptide EcAMP1 from seeds of barnyard grass (Echinochloa crus-galli).* Journal of Biological Chemistry, 2011. 286(28): p. 25145–25153.

[19] Saleh, A.S.M., et al., *Millet grains: nutritional quality, processing, and potential health benefits.* Comprehensive Reviews in Food Science and Food Safety, 2013. 12: p. 281–295.

[20] NAS, *Lost crops of Africa. Vol. I. Grains.* 1996, Washington, DC: National Academy of Science.

[21] Ravindran, G., *Studies on Millets – Proximate Composition, Mineral-Composition, and Phytate and Oxalate Contents.* Food Chemistry, 1991. 39(1): p. 99–107.

[22] Sripriya, G., U. Antony, and T.S. Chandra, *Changes in carbohydrate, free amino acids, organic acids, phytate and HCl extractability of minerals during germination and fermentation of finger millet (Eleusine coracana).* Food Chemistry, 1997. 58(4): p. 345–350.

[23] Kalinova, J. and J. Moudry, *Content and quality of protein in proso millet (Panicum miliaceum L.) varieties.* Plant Foods for Human Nutrition, 2006. 61(1): p. 45–49.

[24] Temple, V.J. and J.D. Bassa, *Proximate Chemical-Composition of Acha (Digitaria-Exilis) Grain.* Journal of the Science of Food and Agriculture, 1991. 56(4): p. 561–563.

[25] Vodouhè, R., *Promoting fonio production in West and Central Africa through germplasm management and improvement of post-harvest technology,* 2004: Benin. p. 18.

[26] Ragaee, S., E.M. Abdel-Aal, and M. Noaman, *Antioxidant activity and nutrient composition of selected cereals for food use.* Food Chemistry, 2006. 98(1): p. 32–38.

[27] FAO, *Sorghum and millets in human nutrition*. FAO Food and Nutrition Series No 27. 1995., Rome: FAO (Food and Agriculture Organization).

[28] Léder, I., *Sorghum and millets.* Füleky, G. Editors, Cultivated Plants Primarily as Food Sources. 2004, UNESCO, Eolss Publishers, Oxford, UK.

[29] Yami, A., *Tef straw: a valuable feed resource to improve animal production and productivity.*, in *Achievements and prospects of tef improvement.*, K. Assefa, S. Chanyalew, and Z. Tadele, Editors. 2013, EIAR-Uni. Bern: Bern. p. 233–251.

[30] Provost, C. and E. Jobson, *Move over quinoa, Ethiopia's teff poised to be next big super grain*, in *The Guardian*. 2014.

[31] NYDailyNews. *Ethiopia's teff grain set to be the world's next superfood. New York Daily News*. 2014 [cited 2015 September 14]; Available from: http://www.nydailynews.com/life-style/health/teff-world-grain-superfood-article-1.1716593.

[32] Jeffrey, J. *Will Ethiopia's teff be the next 'super grain'? BBC Business*. 2015 [cited 2015 September 14]; Available from: http://www.bbc.com/news/business-32128441.

[33] Spaenij-Dekking, L., Y. Kooy-Winkelaar, and F. Koning, *The Ethiopian cereal tef in celiac disease.* The New England journal of medicine, 2005. 353(16): p. 1748–1749.

[34] Chandrasekara, A. and F. Shahidi, *Antiproliferative potential and DNA scission inhibitory activity of phenolics from whole millet grains.* Journal of Functional Foods, 2011. 3(3): p. 159–170.

[35] Bhatt, D., et al., *Responses to drought induced oxidative stress in five finger millet varieties differing in their geographical distribution.* Physiology and molecular biology of plants: an international journal of functional plant biology, 2011. 17(4): p. 347–353.

[36] Rao, B.R., M.H. Nagasampige, and M. Ravikiran, *Evaluation of nutraceutical properties of selected small millets.* Journal of Pharmacy & Bioallied Sciences, 2011. 3(2): p. 277–279.

[37] Chandrashekar, A., *Finger Millet Eleusine coracana.* Advances in food and nutrition research, 2010. 59: p. 215–262.

[38] Itagi, S., R. Naik, and N. Yenag, *Versatile little millet therapeutic mix for diabetic and non-diabetics.* Asian Journal of Science and Technology, 2013. 4: p. 33–35.

[39] Pradeep, S.R. and M. Guha, *Effect of processing methods on the nutraceutical and antioxidant properties of little millet (Panicum sumatrense) extracts.* Food Chemistry, 2011. 126(4): p. 1643–1647.

[40] Ajithkumar, I.P. and R. Panneerselvam, *ROS Scavenging System, Osmotic Maintenance, Pigment and Growth Status of Panicum sumatrense Roth. Under Drought Stress.* Cell Biochemistry and Biophysics, 2014. 68(3): p. 587–595.

[41] Ramamasy, S. and S. Baas, *Climate variability and change: adaptation to drought in Bangladesh: A resource book and training guide.* 2007, Rome: Food and Agriculture Organization of the United Nations (FAO). 66.

[42] Webb, P. and J. Von Braun, *Famine and Food security in Ethiopia.* 1994: Wiley, New York.

[43] UN, *Global Drylands: A UN system-wide response. Prepared by the Environment Management Group.* 2011: United Nations.

[44] Wikipedia. *Sahel.* Available from: https://en.wikipedia.org/wiki/Sahel.

[45] Gebrekirstos, A., et al., *Climate-growth relationships of the dominant tree species from semi-arid savanna woodland in Ethiopia.* Trees-Structure and Function, 2008. 22(5): p. 631–641.

[46] Riché, B., et al., *Climate-related vulnerability and adaptive-capacity in Ethiopia's Borana and Somali communities. Final assessment report of CARE International and Save the Children UK.* 2009, International Institute for Sustainable Development (IISD). p. 82.

[47] Assefa, K., et al., *Breeding tef [Eragrostis tef (Zucc.) trotter]: conventional and molecular approaches*. Plant Breeding, 2011. 130(1): p. 1–9.

[48] Rosell, S. and B. Holmer, *Rainfall change and its implications for belg harvest in South Wollo, Ethiopia.* Geografiska Annaler Series a-Physical Geography, 2007. 89A(4): p. 287–299.

[49] Cheung, W.H., G.B. Senay, and A. Singh, *Trends and spatial distribution of annual and seasonal rainfall in Ethiopia*. International Journal of Climatology, 2008. 28(13): p. 1723–1734.

[50] ABCIC, *Effects of climate change on Eragrostis tef in Ethiopia: a call for action to avert food security crisis, in ABCIC Policy Brief No.1*. 2011.

[51] Matsuura, A., et al., *Effect of pre- and post-heading water deficit on growth and grain yield of four millets*. Plant Production Science, 2012. 15(4): p. 323–331.

[52] Winkel, T., J.F. Renno, and W.A. Payne, *Effect of the timing of water deficit on growth, phenology and yield of pearl millet [Pennisetum glaucum (L.) R Br] grown in Sahelian conditions*. Journal of Experimental Botany, 1997. 48(310): p. 1001–1009.

[53] Bidinger, F.R., V. Mahalakshmi, and G.D.P. Rao, *Assessment of Drought Resistance in Pearl-Millet [Pennisetum-Americanum (L) Leeke].2. Estimation of Genotype Response to Stress*. Australian Journal of Agricultural Research, 1987. 38(1): p. 49–59.

[54] Maqsood, M. and A.N.A. Ali, *Effects of drought on growth, development, radiation use efficiency and yield of finger millet (Eleucine coracana)*. Pakistan Journal of Botany, 2007. 39(1): p. 123–134.

[55] Takele, A., *Genotypic variability in dry matter production, partitioning and grain yield of tef [Eragrostis tef (Zucc.) Trotter] under moisture deficit.* SINET: Ethiopian Journal of Science, 1997. 20: p. 177–188.

[56] Fang, Y.J. and L.Z. Xiong, *General mechanisms of drought response and their application in drought resistance improvement in plants.* Cellular and Molecular Life Sciences, 2015. 72(4): p. 673–689.

[57] Kholová, J., *Understanding of terminal drought tolerance mechanisms in pearl millet [Pennisetum glaucum (L.) R. Br.]* in *Faculty of Science* 2010, Charles University in Prague Prague, p. 115.

[58] Kooyers, N.J., *The evolution of drought escape and avoidance in natural herbaceous populations*. Plant Science, 2015. 234: p. 155–162.

[59] Monneveux, P. and J.P. Ribaut, *Secondary traits for drought tolerance improvement in cereals*, in *Drought adaptation in cereals*, J.M. Ribaut, Editor. 2006, Food Products Press: New York. p. 97–143.

[60] Blum, A., *Drought resistance, water-use efficiency, and yield potential - are they compatible, dissonant, or mutually exclusive?* Australian Journal of Agricultural Research, 2005. 56(11): p. 1159–1168.

[61] Sivakumar, M.V.K., *Empirical-analysis of dry spells for agricultural applications in West Africa*. Journal of Climate, 1992. 5(5): p. 532–539.

[62] Agriinfo. *Plant Breeding for Drought Resistance*. 2015 [cited 2015 September 16]; Available from: http://www.agriinfo.in/default.aspx?page=topic&superid=3&topicid=2152.

[63] Vander Willigen, C., et al., *Mechanical stabilization of desiccated vegetative tissues of the resurrection grass Eragrostis nindensis: does a TIP 3;1 and/or compartmentalization of subcellular components and metabolites play a role?* Journal of Experimental Botany, 2004. 55(397): p. 651–661.

[64] Aparna, K., et al., *Seed number and 100-seed weight of pearl millet [Pennisetum glaucum L.) respond differently to low soil moisture in genotypes contrasting for drought tolerance.* Journal of Agronomy and Crop Science, 2014. 200(2): p. 119–131.

[65] Vadez, V., et al., *Small temporal differences in water uptake among varieties of pearl millet [Pennisetum glaucum (L.) R. Br.] are critical for grain yield under terminal drought.* Plant and Soil, 2013. 371(1-2): p. 447–462.

[66] Shao, H.B., et al., *Water-deficit stress-induced anatomical changes in higher plants.* C. R. Biologies, 2008. 331(3): p. 215–225.

[67] Biswal, A.K. and A. Kohli, *Cereal flag leaf adaptations for grain yield under drought: knowledge status and gaps.* Molecular Breeding, 2013. 31(4): p. 749–766.

[68] Balsamo, R.A., et al., *Drought tolerance of selected Eragrostis species correlates with leaf tensile properties.* Annals of Botany, 2006. 97(6): p. 985–991.

[69] Balsamo, R.A. and J.A.J. Orkwiszewski, *Leaf architecture, lignification, and tensile strength during vegetative phase change in Zea mays* Acta Societatis Botanicorum Poloniae, 2008. 77(3): p. 181–188.

[70] Sharma, P. and R.S. Dubey, *Drought induces oxidative stress and enhances the activities of antioxidant enzymes in growing rice seedlings.* Plant Growth Regulation, 2005. 46(3): p. 209–221.

[71] Smirnoff, N. and S.V. Colombe, *Drought influences the activity of enzymes of the chloroplast hydrogen-peroxide scavenging system.* Journal of Experimental Botany, 1988. 39(205): p. 1097–1108.

[72] Zhang, G., et al., *Genome sequence of foxtail millet (Setaria italica) provides insights into grass evolution and biofuel potential.* Nature Biotechnology, 2012. 30(6): p. 549–554.

[73] Bennetzen, J.L., et al., *Reference genome sequence of the model plant Setaria.* Nature Biotechnology, 2012. 30(6): p. 555–561.

[74] Qi, X., et al., *Genome-wide annotation of genes and noncoding RNAs of foxtail millet in response to simulated drought stress by deep sequencing*. Plant Molecular Biology, 2013. 83(4-5): p. 459–473.

[75] Wang, M.Z., et al., *SiLEA14, a novel atypical LEA protein, confers abiotic stress resistance in foxtail millet*. BMC Plant Biology, 2014. 14: p. 290.

[76] Parvathi, M.S., et al., *Expression analysis of stress responsive pathway genes linked to drought hardiness in an adapted crop, finger millet (Eleusine coracana)*. Journal of Plant Biochemistry and Biotechnology, 2013. 22(2): p. 193–201.

[77] Mariac, C., et al., *Genetic basis of pearl millet adaptation along an environmental gradient investigated by a combination of genome scan and association mapping*. Molecular Ecology, 2011. 20(1): p. 80–91.

[78] Yadav, R.S., et al., *Mapping and characterisation of QTL x E interactions for traits determining grain and stover yield in pearl millet*. Theoretical and Applied Genetics, 2003. 106(3): p. 512–520.

[79] Kaul, T., et al., *Biochemical and molecular characterization of stress-induced beta-carbonic anhydrase from a C(4) plant, Pennisetum glaucum*. Journal of Plant Physiology, 2011. 168(6): p. 601–610.

[80] Singh, R.K., et al., *Isolation and characterization of drought responsive EcDehydrin7 gene from finger millet [Eleusine coracana (L.) Gaertn.]*. Indian Journal of Genetics and Plant Breeding, 2014. 74(4): p. 456–462.

[81] Hema, R., et al., *Stable expression of mtlD gene imparts multiple stress tolerance in finger millet*. Plos One, 2014. 9(6): p. e99110.

[82] Bhatt, D., et al., *Cloning, expression and functional validation of drought inducible ascorbate peroxidase (Ec-apx1) from Eleusine coracana*. Molecular Biology Reports, 2013. 40(2): p. 1155–1165.

[83] Li, C., et al., *An ABA-responsive DRE-binding protein gene from Setaria italica, SiARDP, the target gene of SiAREB, plays a critical role under drought stress*. Journal of Experimental Botany, 2014. 65(18): p. 5415–5427.

[84] Wilson, P.B., et al., *The nucleotidase/phosphatase SAL1 is a negative regulator of drought tolerance in Arabidopsis*. The Plant Journal: For Cell and Molecular Biology, 2009. 58(2): p. 299–317.

[85] Wang, Y., et al., *Molecular tailoring of farnesylation for plant drought tolerance and yield protection*. The Plant Journal: For Cell and Molecular Biology, 2005. 43(3): p. 413–424.

[86] Tesema, A., *Genetic diversity of tef in Ethiopia*, in *Achievements and Prospects of Tef Improvement*, A. Assefa, S. Chanyalew, and A. Tadele, Editors. 2013, EIAR-University of Bern: Bern, Switzeralnd. p. 15–20.

[87] Saxena, N.P. and C. John, eds. *Field screening for drought tolerance in crop plants with emphasis on rice*. 2002, ICRISAT, Patancheru, India. 211.

[88] Ahloowalia, B.S., M. Maluszynski, and K. Nichterlein, *Global impact of mutation-derived varieties*. Euphytica, 2004. 135(2): p. 187–204.

[89] McCallum, C.M., et al., *Targeting induced local lesions IN genomes (TILLING) for plant functional genomics*. Plant Physiology, 2000. 123(2): p. 439–442.

[90] Comai, L., et al., *Efficient discovery of DNA polymorphisms in natural populations by Ecotilling*. Plant Journal, 2004. 37(5): p. 778–786.

[91] Shukla, V.K., et al., *Precise genome modification in the crop species Zea mays using zinc-finger nucleases*. Nature, 2009. 459(7245): p. 437–441.

[92] Townsend, J.A., et al., *High-frequency modification of plant genes using engineered zinc-finger nucleases*. Nature, 2009. 459(7245): p. 442–445.

[93] Cermak, T., et al., *Efficient design and assembly of custom TALEN and other TAL effector-based constructs for DNA targeting (vol 39, pg e82, 2011)*. Nucleic acids research, 2011. 39(17): p. 7879–7879.

[94] Miao, J., et al., *Targeted mutagenesis in rice using CRISPR-Cas system*. Cell Research, 2013. 23(10): p. 1233–1236.

[95] Jiang, W., et al., *Demonstration of CRISPR/Cas9/sgRNA-mediated targeted gene modification in Arabidopsis, tobacco, sorghum and rice*. Nucleic Acids Research, 2013. 41(20): p. e188.

[96] Gwata, E.T. and J. Mzezewa, *Optional crop technologies at a semi-arid ecotope in southern Africa*. Journal of Food Agriculture & Environment, 2013. 11(2): p. 291–295.

[97] Haussmann, B.I.G., et al., *Breeding strategies for adaptation of pearl millet and sorghum to climate variability and change in West Africa*. Journal of Agronomy and Crop Science, 2012. 198(5): p. 327–339.

[98] Mir, R.R., et al., *Integrated genomics, physiology and breeding approaches for improving drought tolerance in crops*. Theoretical and Applied Genetics, 2012. 125(4): p. 625–645.

[99] Leblois, A., et al., *Weather Index drought insurance: An ex ante evaluation for millet growers in Niger*. Environmental & Resource Economics, 2014. 57(4): p. 527–551.

[100] Molini, V., et al., *Safety nets and index-based insurance: historical assessment and semi-parametric simulation for Northern Ghana*. Economic Development and Cultural Change, 2010. 58(4): p. 671–712.

[101] Chantarat, S., et al., *Improving humanitarian response to slow-onset disasters using famine-indexed weather derivatives*. Agricultural Finance Review, 2008. 68(1): p. 169–195.

[102] Berg, A., P. Quirion, and B. Sultan, *Can weather index drought insurance benefit to Least Developed Countries' farmers? A case study on Burkina Faso*. Weather, Climate and Society, 2009. 1: p. 7184.

[103] Zant, W., *Hot stuff: Index insurance for Indian smallholder pepper growers*. World Development, 2008. 36(9): p. 1585–1606.

[104] SFSA. *Kilimo Salama is ACRE*. 2014; Available from: http://www.syngentafoundation.org/index.cfm?pageID=562.

[105] ACRE. *Acre Africa: Agriculture and Climate Risk Enterprise Ltd. (ACRE)*. Available from: http://acreafrica.com/.

[106] ICRISAT. *Tef: New superfood crop in ICRISAT's portfolio*. 2015 [cited 2015 September 7]; Available from: http://www.icrisat.org/newsroom/latest-news/happenings/happenings1689.htm.

[107] Tadele, Z. and K. Assefa, *Increasing Food Production in Africa by Boosting the Productivity of Understudied Crops*. Agronomy, 2012. 2(4): p. 240–283.

[108] Tadele, Z., *Role of crop research and development in food security of Africa*. International Journal of Plant Biology and Research, 2014. 2(3): p. 1019.

8

Antioxidant Enzyme Activities and Abiotic Stress Tolerance Relationship in Vegetable Crops

Sebnem Kusvuran, Sevinç Kiran and S. Sebnem Ellialtioglu

Abstract

Climatic changes can cause serious reductions in yield and crop quality. Under the threat of climatic changes, one of the precautions to cope is selection and development of resistant vegetable genotypes to abiotic stresses. Several physiological and biochemical reactions and different tolerance levels can occur according to plant species. When plants are subjected to environmental stresses such as salinity, drought, temperature extremes, herbicide treatment and mineral deficiency, the balance between the production of reactive oxygen species (ROS) and the quenching activity of antioxidants is upset, often resulting in oxidative damage. Since activated oxygen species can disrupt normal metabolism through oxidative damage to lipids, protein and nucleic acids, plants possess a number of antioxidant enzymes that protect them from these cytotoxic effects. To control the level of ROS and to protect cells under stress conditions, plant tissues contain several enzymes for scavenging ROS. The high levels of antioxidative enzyme activities were determined in the tolerant genotypes of tomatoes, eggplant, peppers, cucumbers, melons, squash, beans, okra, etc. to several abiotic stress factors. Both the whole plant and in vitro callus culture experiments gave similar results. Antioxidant enzymes can be useful for screening to determine the tolerant and sensitive plant genotypes against abiotic stresses.

Keywords: Drought, salinity, oxidative stress, ROS, chilling

1. Introduction

Stress has a strictly defined physical science definition describing the force per unit area acting upon a material, inducing strain and leading to dimensional change. Biologically, stress has also been defined as the overpowering pressure that affects the normal functions of individual life or the conditions in which plants are prevented from fully expressing their genetic potential for growth, development and reproduction. In the agricultural regard, stress has been described as a phenomenon that limits crop productivity or destroys biomass. It has become traditional to divide stresses experienced by plants into two major categories: biotic and abiotic stresses. Biotic stresses originate through interactions between organisms, while abiotic stresses are those that depend on the interaction between organisms and the physical environment. Abiotic stresses include potentially adverse effects of salinity, drought, flooding, chilling, metal toxicity, nutrient deficiency, UV exposure, air pollution, etc. [1]. The abiotic stresses represent the factors that most limit the agricultural productivity worldwide. These stresses not only have an impact on current crop species, but they are also significant barriers to the introduction of crop plants into areas that are not currently being used for agriculture [2].

When plants are subjected to environmental stresses such as salinity, drought, temperature extremes, herbicide treatment and mineral deficiency, the balance between the production of reactive oxygen species (ROS) and the quenching activity of antioxidants is upset, often resulting in oxidative damage.

In plants, there are a number of possible ROS sources. These include reactions such as photosynthesis and respiration found in the normal metabolism of plants. This is parallel with the well-known idea that ROS are certain to be one of the products output by aerobic respiration. Pathways that are embellished during abiotic stress also result in ROS production, such as during the photorespiration reaction, where glycolate oxidases in peroxisomes result in superoxidase production. Nonetheless, new sources of ROS have been found recently in plants, such as NADPH oxidases, cell wall-bound peroxidases and amine oxidases. They are involved in ROS production in such processes as cell death and are highly regulated [3]. The ROS are associated with several forms of cellular damage. Since activated oxygen species such as superoxide ($O_2^{\cdot-}$), hydrogen peroxide (H_2O_2) and the hydroxyl radical ($^{\cdot}OH$) can seriously disrupt normal metabolism through oxidative damage to lipids, protein and nucleic acids, plants possess a number of antioxidant enzymes that protect them from these potential cytotoxic effects [4–7].

Plant species and cultivars within a crop species differ greatly in their response to environmental stress. Plants with high levels of antioxidants, either constitutive or induced, have been reported to have greater resistance to oxidative damage [6–9]. Reports suggest that the extent of oxidative cellular damage in plants exposed to abiotic stress is controlled by the capacity of their antioxidant systems [10].

In general, two classes of nonenzymatic antioxidants are found. They are lipid-soluble membrane-associated antioxidants (e.g. α-tocopherol and β-carotene) and water-soluble reductants (e.g. glutathione, phenolics and ascorbate). Ascorbate peroxidase (APX), superoxide dismutase (SOD) and glutathione reductase (GR) compose enzymatic antioxidants and they are thought to search for H_2O_2 in chloroplast and mitochondria. Catalase (CAT) and peroxidase (POD) are the other enzymatic antioxidants and are able to remove H_2O_2 and can neutralise or scavenge oxyintermediates and free radicals [11]. Key enzymes involved in the detoxification of ROS are, namely, SOD, CAT, POD, APX and other enzymes implicated in the Halliwell and Asada cycle (ascorbate–glutathione pathway). Under stress conditions, these antioxidants enhance the activity of almost all of these enzymes [11]. Superoxide radicals that emerge as a result of stress in the plant tissues are transformed into hydrogen peroxide (H_2O_2)

by the SOD enzyme [12, 13]. The accumulation of H_2O_2, which results from the canalisation reaction of the SOD enzyme and is a powerful oxidant, is prevented by the ascorbate–glutathione cycle. The hydroxyl radical (OH), which is very reactive and the most toxic oxide, can react with all macromolecules without discrimination. SOD and CAT, by combining their actions, can prevent or decrease the formation of this oxide. Even though the particular scavengers are not fully known of the single radical oxygen or the hydroxyl radical, it is thought that SOD functions in removal via chemical reaction [11]. In the defence against intracellular antioxidants, SOD and GSH work together and SOD prevents the radical-mediated chain oxidation of GSH, thus helping GSH in its role as a free radical scavenger physiologically, without the accompaniment of oxidative stress [11, 14]. It was observed that with continued stress conditions, SOD enzyme activity, which acts by decreasing the oxidative oxygen species derived from stress, continued to increase. Even though the linearity of increased stress duration and the increase of SOD activity is concurrent, it was shown that genotypes with more tolerance are superior in this area. The CAT enzyme changes oxidative stress-induced reactive oxygen derivatives, like H_2O_2, into water and molecular oxygen [15]. CAT, found mostly in glyoxysomes of lipid-storing tissues in plants, contains a tetrameric haeme that catalyses the conversion of hydrogen peroxide, produced from the β-oxidation of fatty acids, into water and oxygen [11, 16]. The GR and APX enzymes, which are a part of the defence mechanism of tolerant genotypes against salt, drought and chilling stress, are generally effective in the reduction of hydrogen peroxide to water in chloroplasts and mitochondria, thereby detoxifying them [17, 18]. APX is one of the most important antioxidant enzymes of plants that detoxify H_2O_2 by using ascorbate for reduction. Different isoforms of APX are active in chloroplasts, cytosol and microsomes [11]. In the ascorbate–glutathione cycle, APX reduces H_2O_2 into water by oxidising ascorbate into monodehydroascorbate (MDHA), which is then converted into ascorbate via the MDHA reductase enzyme; thus, two MDHA molecules are changed into MDHA and dehydroascorbate (DHA) as a non-enzymatic side product in unequal amounts. Subsequently, the reduction of DHA occurs and ascorbate is produced by the action of dehydroascorbate reductase (DHAR) and GR. DHAR can then convert GSH into glutathione disulphide (GSSG) which then is reduced back into GSH by GR [18, 19]. Due to APX activity resulting in the need for regenerating ascorbic acid, it is thought that concurrently an increase in various other components of the antioxidative defence system is needed so that the protective mechanisms of plants can increase as necessary [11]. POD, CAT and APX appear to play an essential protective role in the scavenging process when coordinated with SOD activity. They scavenge H_2O_2 generated primarily through SOD action [11, 20].

The research indicates that APX, CAT, GR and SOD enzyme activities in large variation among cotton varieties [21], tomato [22–24] and melon genotypes [25] in their response to salinity have been observed.

The tolerance of plants to stress has been widely shown to vary with physiological growth stage, developmental phase and size of plants. There is also growing evidence of multiple tolerances to stress in plants with plants showing tolerance to more than one stress. Genetic variability within a species is a valuable tool for screening and breeding for higher stress tolerance.

2. Salinity stress

Salinity is one of the most important abiotic stresses that cause reduction in plant growth, development and productivity worldwide in arid and semi-arid regions, where soil salt content is naturally high and precipitation can be insufficient for leaching. The FAO estimates that 34 million hectares of irrigated land are salt-affected worldwide, and an additional 60–80 million hectares are affected by waterlogging and related salinity [26].

Salt stress changes the morphological, physiological and biochemical responses of plants. There is evidence that high salt concentrations cause an imbalance in cellular ions, resulting in ion toxicity and osmotic stress, leading to the generation of ROS, which cause damage to DNA, lipids and proteins. At the same time, ROS cause chlorophyll degradation and membrane lipid peroxidation, decreasing the membrane fluidity and selectivity. To prevent the negative effects of ROS, plants have developed various antioxidant enzyme systems including non-enzymatic antioxidants (e.g. ascorbic acid, glutathione and carotenoids) and antioxidative enzymes (e.g. GR, SOD and APX). While CAT and peroxidases detoxify the toxic hydrogen peroxide, superoxide is broken down into water and oxygen by catalyses from the SOD enzyme. APX reduces H_2O_2 using ascorbate as an electron donor in the ascorbate–glutathione cycle. Oxidised ascorbate is then reduced by GSH generated from GSSG catalysed by GR at the expense of NADPH. Previous studies showed that the level of antioxidative enzymes increases when plants are exposed to oxidative stress including salinity [27–29].

Plants with high levels of antioxidants, either constitutive or induced, have been reported to have greater resistance to oxidative damage. Genetic variability within a species is a valuable tool for screening and breeding for higher salt tolerance. Some authors have reported large variation among cotton varieties and tomato genotypes in their response to salinity. The response of plants to salt stress is variable and dependent on various factors, particularly plant genotype, extent of stress, age of plant and stage of plant growth when stress is experienced. Plants manifest themselves with various adaptive mechanisms (morphological, biochemical, enzymatic and physiological) to survive under stressed conditions. The extent of these mechanisms' adaptability is unique with every plant genotype [30].

In one recent study, the relationship between antioxidant enzymes and salt tolerance in the leaves of eggplant seedlings of two salt-tolerant varieties (Burdur Bucak and Mardin Kızıltepe) and two salt-sensitive genotypes (Giresun and Artvin Hopa) was examined. In salt-tolerant eggplants, APX, CAT, GR and SOD activities increased significantly when the seedlings were grown in the hydroponic system containing 150 mM NaCl [31]. It was observed, in the results of pumpkin salt tolerance studies, that in the salt-sensitive genotypes (CU-7 and A-24) less of the aforementioned enzymes showed activity as opposed to the increased activity seen in higher salinity levels and in the salt-tolerant genotypes (Iskenderun-4 and AB-44). It was also shown that these enzymes played a role in salinity tolerance of melons [32, 33], green bean [6] and soybean [8].

The salinity experiment in okra genotypes shows that NaCl-induced stress caused decreases in plant biomass, green pigments, photosynthetic activity, stomatal conduction, transpiration rate, number of stomata and stomatal size and resulted in alterations in enzymatic activities

(SOD, POD and CAT) and osmolyte accumulation (proline, glycine betaine, total free amino acids and total soluble sugars). The increase in Na and Cl and lipid peroxidation under saline conditions is the indication of ion toxicity and oxidative damage. However, the oxidative damage is controlled by a defensive system comprising various antioxidants, such as SOD, POD and CAT. The results depicted that salt-tolerant and salt-sensitive genotypes exposed to NaCl stress showed the highest activities of SOD, POD and CAT, both in root and leaf tissues of okra genotypes [30].

According to Yadav et al. [34], plants with high levels of antioxidants have been reported to have greater resistance to this oxidative damage and an increase in the activity of antioxidative enzymes in plants under salt stress. It was observed that increased levels of antioxidants in plants resulted in them having a stronger resistance against the oxidative damage, while plants under salt stress were shown to have the activity of antioxidative enzymes to become greater and that a correlation was seen between tolerance against salt levels and the amount of enzymes. It was also observed that in pea cultures, antioxidant enzyme activity increased when in saline environments, but SOD was unaffected in cucumber. The variations in these observations maybe due to the fact that the effects of salinity depend on a number of factors, for example, salt type, their concentration, plant genotype, growth stage and/or environmental conditions. The mechanism by which salinity affects the antioxidant responses is not yet clear. However, proposed that it might be via the change in membrane integrity caused by high Na^+ to Ca^{2+} ratio [34].

Salinity inhibition of plant growth is the result of low osmotic potential of soil (water stress), nutritional imbalance, specific ion effect (salt stress), or a combination of these factors. Grafted plants also exhibit phenotypic variations from scion and rootstock plants in terms of salinity tolerance, and grafting onto salt-tolerant rootstocks capable of inducing salt tolerance in the grafted shoots has been an effective method for improvement of salt tolerance in agricultural practices [35]. In parallel, *Solanum lycopersicum* L. ('Elazığ') grafted on *Nicotiana tabacum* L. ('Samsun') and *Nicotiana rustica* L. ('Hasankeyf'), namely "Tomacco" plant (patent no. TR-2008-05391-B), to 10-d high NaCl irrigation. Physical development, chlorophylls a and b, total chlorophyll, total carotenoid and anthocyanin levels were evaluated [36]. During an increase in osmotic stress levels, plants utilise various antioxidant enzymes and increase their activity; thus these enzymes play vital roles in ROS removal. SOD, which is an important scavenger, is used for producing water and oxygen by catalysing $O_2^{\cdot-}$; afterwards the ROS hydrogen peroxide is catalysed by POD and CAT to produce water and oxygen as well [37]. It was reported by Azevedo Neto et al. [38] that antioxidative enzymes (e.g. POD, SOD, GR, CAT and APX) in addition to low molecular mass antioxidants make up the complicated antioxidative defence system seen in plants. It was observed that grafted seedlings when under stress by over the normal amount of $Ca(NO_3)_2$, contents of hydrogen peroxide and malondialdehyde (MDA), had a much lower percentage of electrolyte leakage and O_2 production rate when compared to those of non-grafted seedlings from which can be deduced that there was less damage to the membrane of grafted seedlings than that of non-grafted seedlings when under excess $Ca(NO_3)_2$ stress. It was also observed that in grafted seedlings, free radical scavenging systems had a significant job when battling salinity stress, thus resulting in salinity tolerance [38]. In parallel with these results, Wei et al. [39] determined that due to the efficient scavenging system of free radicals in addition to the mechanisms that utilise antioxidative enzymes and polyamines for protection, eggplant seedlings that were grafted had a greater resistance against stress of over the normal amount of $Ca(NO_3)_2$ than when compared to non-grafted eggplant seedlings. For rootstock, they used the eggplant cultivar, Swartz cv. Torvum Vigor, which were tolerant to salinity [39]. Plant tissue culturing has been important in many fields including both agriculturally and commercially and has a significant role in the production of ornamental plants. It has also been invariably useful in manipulating plants for enhanced agronomic performances. In vitro culturing of plant cells provides a means for conducting many studies and improvement in scientific research, including helping to study plant physiology and genomics and their processes and enhancing genetic variability by providing the possibility to help in the breeding of improved cultivars. Thus, it has attracted a large amount of interest in later years. Regenerated plants are expected to have the same genotype as the donor plant; however, in some cases, somaclonal variants have been found among regenerated plants [40]. According to Kusvuran et al. [18], the responses to salt and drought stress of four pumpkin varieties (A-24, CU-7, Iskenderun-4 and AB-44) were investigated under in vitro culture conditions. In this study, it was observed that the tolerant genotypes had less damage on their cell walls (by lipid peroxidation) than that of the sensitive genotypes. While, it was observed that salinity stress had greater effect on the pumpkin genotype than stress by drought. The results of this study suggest that antioxidative defence mechanisms were effective in the pumpkin callus tissues during salt and drought stress. Thus, the increased activity of antioxidative enzyme activity as well as the lower increasing MDA content in the sạlt- and drought-adapted cells compared with the unadapted cells may contribute to salt and drought tolerance. The results obtained with the callus tissues are in agreement with those observed in studies using seedlings. The results demonstrate the selection of tolerant genotypes for oxidative stress such as drought; salinity could be used for in vitro methods [18]. Similarly, other research on squash, eggplant and melon indicated that SOD, CAT and APX enzyme activities in salt-tolerant genotypes are higher compared to salt-susceptible genotypes in both seedling and callus tissues [41–43].

Antioxidant enzymes such as SOD, POX and CAT are known to substantially reduce the levels of superoxide and hydrogen peroxide in plants. It is one of the most important enzymes used against oxidative stress in the plant defence system, and it occurs ubiquitously in every cell of all types of plants. The most common isoforms of SOD known in the literature are copper–zinc containing SOD (Cu/Zn-SOD), manganese containing SOD (Mn-SOD), iron containing SOD (Fe-SOD) and nickel containing SOD (Ni-SOD) [44]. Moharramnejad and Valizadeh [45] observed the three isoforms for SOD and POX and one isoform for CAT. Authors showed that in salt stress, the main activities of SOD, POX and CAT isozymes are significantly higher than normal conditions in red bean (*Phaseolus vulgaris* L.).

3. Drought stress

Drought stress is one of the most serious abiotic stresses that cause a reduction in plant growth, development and yield in many parts of the world [46–48]. However, plants have developed different morphological, physiological and biochemical mechanisms to withstand drought stress [49–51]. It is of considerable value to understand the reactions and responses of plants in drought environments as it is also a major part in making crops more tolerant towards stress. It is important to note that responses of plants towards water stress are seen to be considerably different at organisational levels according to the intensity and time they spend under the stressful environment as well as the period of growth they are in and their species [52]. When water becomes limited, the plant generally experiences stomatal closure in an effort to prevent further water loss, limiting the carbon dioxide available for fixation by photosynthesis and reducing NADP+ regeneration by the Calvin cycle [53]. These converse conditions increase ROS, such as hydrogen peroxide, superoxide, singlet oxygen and hydroxyl radicals [6, 54]. These ROS attack lipids, proteins and nucleic acids, causing lipid peroxidation, protein denaturing and DNA mutation. Plants possess several antioxidant enzyme systems that protect their cells from the negative effects of ROS. The role of antioxidant enzymes (APX, CAT, SOD and GR) as the components of the main tolerance mechanism is developed in response to different stress conditions. Many reports suggest that the extent of oxidative cellular damage in plants exposed to abiotic stress is controlled by the capacity of their antioxidant systems and the relationship between enhanced or constitutive antioxidant enzyme activities and an increased resistance to drought stress [55–57].

In one study [58], the effects of drought stress on plant growth, relative water content (RWC), ion concentration and activities of the antioxidant enzymes, APX and GR, in eight okra genotypes were investigated. Drought-resistant genotypes exhibit a better protection mechanism against oxidative damage by maintaining a higher inherited and induced activity of antioxidant enzymes than the sensitive genotypes. Previously, 31 different melon genotypes grown under salt and drought conditions were classified according to some growth parameters (i.e. shoot length, plant leaf area, leaf number, fresh and dry weight, leaf water content, ion accumulation and membrane injury index), as well as some antioxidant enzymes activities in vivo. At the end of the study, two salt- and drought-tolerant genotypes and two sensitive genotypes were selected according to the growth parameters measured. The aim of this study was to determine the activities of antioxidative stress enzymes in some salt- and drought-sensitive or salt- and drought-tolerant melon varieties grown in in vitro culture under salt and drought stress conditions. Another aim of this study was to determine whether in vitro callus culture can be used as a screening method for salt and drought stresses in a melon screening study. In contrast with the control, in all the different 8-day salinity and drought stress, it was discovered that growth prevention occurred in callus tissues within media containing 100 mM NaCl or 15% (w/v) PEG-6000 (polyethylene glycol). It was observed in the study from the MDA measurements, from which the amount of lipid peroxidation of the cell wall can be deduced, that the amount of damage to the cell wall (by lipid peroxidation) was more in the sensitive genotypes than that of the tolerant genotypes. The antioxidant enzyme (APX, CAT, SOD and GR) activities were investigated in the callus tissues of four melon genotypes under salt and drought stress. This study suggested that antioxidative defence mechanisms were effective in the melon callus tissues during salt and drought stress. The results demonstrate the selection of tolerant genotypes for oxidative stress such as drought; salinity could be used for in vitro methods [59]. Drought tolerance of tomato genotypes [60] was investigated and found biochemical changes (drought stress index, MDA content and antioxidant enzyme activities) that occur as a result of stress in plants were investigated. In salt-tolerant varieties, T-1 and T-2, the decrease of ions occurred at lower levels under drought conditions. APX, CAT, SOD and GR enzyme activities have increased in drought stress conditions. The four varieties showed an increase in MDA content under drought conditions, especially in the sensitive genotypes. The results indicate that the tomato seedlings respond with enzymatic defence systems against drought-induced oxidative stress.

Water stress tolerance is seen in all plant species but its extent varies from species to species. Effects of different PEG concentrations with drought stress on the activity of antioxidant enzymes, CAT and APX, were investigated in two melon genotypes. Drought tolerant (CU-196) and drought sensitive (CU-3) were grown in hydroponic conditions. Recently, PEG has being used as osmotic pressure inducer in drought physiology studies. In the study, 15, 30 and 45 mM PEG-6000 doses (–0.15 MPa, –0.52 MPa and –1.50 MPa, respectively) were compared to CAT and APX antioxidant enzymes in tolerant and sensitive melon genotypes. At the end of the study, CAT and APX enzyme activities significantly increased in CU-196 than CU-3 [61]. In the other study was conducted for determination of tolerance levels to drought of melon genotypes (Midyat, Şemame, Yuva and Ananas) that have determined the levels of tolerance to salt stress. In this study, three different irrigation methods have been applied to plants. (S0: control-plant-available water, 40% is consumed for irrigation, S1: plant-available water, 90% is consumed for irrigation, S2: during the period of 3–4 leaves of plants completely cut off from the irrigation). Morphological and biochemical changes that occur as a result of stress in plants were investigated. Drought stress applied to the visual scale evaluation of melon genotypes in terms of Midyat and Şemame melons had values close to controls. However, Yuva and Ananas genotypes were found to be more pronounced losses caused by drought. Under drought stress, for Midyat and Şemame genotypes that are tolerant to salt stress, plant fresh and dry weight, the values shown in chlorophyll were closer to control values. At the same time, SOD, CAT, GR and APX enzyme activities have increased in drought stress conditions. However, the susceptible varieties (Yuva and Ananas) compared to the control plants in terms of the parameters studied enzyme activities decreased to varying degrees. In general, it also drought-tolerant melon genotypes found to be tolerant groups, respectively. In particular, these melon varieties have enhanced levels of antioxidant enzyme activities by activating the tolerant concluded [62]. Kiran et al. [60] investigated that determination of tolerance levels to drought of tomato genotypes (TR-68516, Rio Grande, TR-63233, TR-63233 and H-2274). Authors indicated that SOD, CAT, GR and APX enzyme activities have increased in drought stress conditions. On the other hand, in the susceptible varieties (TR-63233 and H-2274) compared to the control plants in terms of the parameters studied enzyme activities decreased to varying degrees [60].

Smirnoff [63] indicated that oxidative damage is also manifest in effects on to proteins and nucleic acids, although these are rarely measured and can be affected by other factors. Oxidation of amino acid residues can be followed by the loss of catalytic activity and denaturation. The damaged proteins may be more susceptible to proteolytic degradation. DNA repairing enzymes may also be induced as a result of oxidative damage to DNA. The various repair and protection systems found in plants decrease the amount of open oxidative damage and consist of two groups. One of the groups includes CAT, ascorbate, PODs and SOD where in these systems react with oxygen forms that are active and keep them at a minimum level. The second group consists of GR, mono and DHARs, glutathione (GSH) and ascorbate, which are involved in the regeneration of antioxidants that are oxidised. For the purpose of maintaining the superoxide concentration at a minimum, and thus decreasing the production of hydroxyl radicals by the Haber–Weiss reaction of which the catalysis are done by metals, superoxide is converted into hydrogen peroxide by the SOD enzyme catalysis from the first group. Three types of SOD occur in plants: Cu/Zn-SOD, Mn-SOD and Fe-SOD. The latter two have similar amino acid sequences. SOD isoforms occur in most of the subcellular compartments (hydrogen peroxide is broken down to water by CAT which is located in peroxisomes and glyoxysomes). In the chloroplast, this function is fulfilled by ascorbate that also has a cytosolic isoform. In plants, a large amount of the PODs can have major roles other than antioxidants, while ascorbate, with both superoxide and singlet oxygen, has the ability to react non-enzymatically too. In the second group, GSH and ascorbate are the key players in the reactions involved in antioxidants becoming regenerated. In the reaction involving APX, the MDHA radical is the major product and goes on to react with NAD(P)H-dependant monodehydroascorbate reductase (MDHAR) to get reduced to ascorbate. Or, ascorbate and dehydroascorbate (DHA) can be produced by the non-enzymatic reaction between two molecules of MDHA, after which ascorbate is produced by the reduction of DHA by GSH and where the enzyme DHAR catalyses the reaction, with the second product as oxidised GSH (GSSG). An NADP-dependant GR can then reduce the oxidised GSH into GSH. In the chloroplast, these reactions, sometimes known as the Halliwell–Asada cycle, result in the catalysis of the light-dependant reduction reaction from hydrogen peroxide into water by the action of the reductant (NADPH) that is produced by photosynthesis. The key players of this cycle ascorbate and GSH, along with their isoforms (GR, DHAR and MDHAR), are found in large amounts in the chloroplast in addition to other subcellular compartments. The oxidation of GSH pools, which are also essential in keeping sulphydryl groups of enzymes in reduced forms, could result in enzymes that rely upon these reduced SH groups being inactivated [63].

Yasar et al. [64] investigated 38 genotypes of different pumpkin species for the relationship between the drought tolerance capacity and antioxidant enzyme activity. As a result, it was observed that the enzyme activities are extremely vital in the drought tolerance of the pumpkin genotypes, such as under dry conditions, the drought-tolerant pumpkin genotypes use antioxidative enzymes more actively compared to the drought susceptible genotypes. The genotypes exposed to drought stress had relatively inferior SOD enzyme activity compared to their controls. However, the CAT enzyme activities of these genotypes were found to be increased. Alternatively, the opposite situation was also observed; if the CAT enzyme activities were decreased compared to the controls, the SOD enzyme activities were observed to be increased compared to the genotypes in control group. However, such a relationship was not established for the APX enzyme activities [64].

Drought tolerance in black pepper is attained through osmotic adjustment and better ROS scavenging machinery, functioning through different antioxidant enzymes. The activities of antioxidant enzymes such as SOD and POD become higher during stress in tolerant variety [65].

4. Chilling stress

Low-temperature, or chilling, stress (damage caused by low, but above-freezing temperatures) has been recognised as a unique environmental impact on crop plant physiology [66]. The damage resulting from the symptoms of the chilling stress includes a decrease in growth and yield of the plant. These symptoms consist of the prevention of metabolic processes, rise in the permeability and seepage through the cell membrane due to alterations in the order of the molecules or in the physical form, wilting and chlorosis [67]. Prasad et al. [68] has suggested that mitochondria are critica1 organelles in the metabolic production of energy in the cell. The competence and the stability of mitochondria are very important for the seedlings to survive low-temperature stress, especially during early seedling growth. Low temperature induces oxidative stress in the cell [69]. Under aerobic conditions, superoxide radicals and H_2O_2 are found to be normal metabolites of plant cells [70, 71] as well as animal cells [72, 73] and are kept at low, steady-state levels by the action of antioxidant enzymes such as SOD, CAT, GSH POD and APX located in the organelles and cytosol [74–76].

Active oxygen species (AOS) has been proposed to be responsible for cold-induced injury because they are produced at higher concentration during cold stress and may initiate degradative reactions, causing lipid peroxidation, membrane deterioration, protein degradation and chlorophyll quenching. An efficient antioxidant activity is essential in order to maintain the concentration of AOS at relatively low levels [67]. On the other hand, the damage that occurs during chilling stress accompanying illumination was thought to be mediated by an oxygen radical. The defence mechanisms of the cell against oxidative stress involve antioxidants that can be found in many plant organs in large amounts to perform vital biological functions. These include the enzyme systems CAT, SOD and numerous PODs, e.g. APX and guaiacol peroxidases (POX) [69]. During photosynthesis, superoxide and hydrogen peroxide are produced as side products and need to be removed. This is achieved by the SOD enzyme and the enzymes GSH and ascorbate from the ascorbate–glutathione cycle. It was shown by Aroca et al. [77] that due to ROS being produced under chilled environments during light-induced photo-oxidation, the major damage from chilling stress occurs during this time. The reason for ROS production under these circumstances is because of the slowing down of the enzymes involved in the Calvin–Benson cycle, thus resulting in the limitation of the NADP + supplements receiving the electrons from the electron transport chain and inducing oxygen to absorb more energy than needed. To decrease photo-oxidation under chilling, there are three important mechanisms. The first one involves avoiding production of ROS by diminishing electron transport chain; the second one involves scattering surplus energy in the form of heat via violaxanthin de-epoxidation, and the third one involves scavenging ROS produced by antioxidant compounds and enzymes. Additionally, the water-water cycle in the chloroplast, where electrons flow in photosystem II from water to photosystem I to reduce oxygen without

O_2 levels having a net change, is said to be an active mechanism that can disperse energy from over excitation when there is an environmental stress. SOD, GR and APX are also some of the antioxidant enzymes involved in the function of this cycle [77]. The enzyme SOD is located in the cytoplasm, chloroplast, mitochondrion and peroxisome and acts as the first line of defence mechanism against ROS by dismutating O_2 into H_2O_2 [78]. Furthermore, the dismutation of superoxide radicals into H_2O_2 and oxygen is an important step in protecting the cell, and in that conversion, SOD is considered a key enzyme [69]. CAT also played a significant role in chilling tolerance and is especially important for removal of H_2O_2 in C3 plants. Exposure to low temperature may increase the amount of AOS not only in cold-sensitive but also in cold-tolerant plants. There was a correlation between the reduction in CAT activity and H_2O_2 accumulation [79].

Oxidative free radicals can be highly reactive towards cell components, and therefore, the ability of the cell to remove these undesirable species might be viewed as an important feature in improved resistance to chilling stress. The increases in the activities of CAT3 provide evidence for the increased production of superoxide and H_2O_2 in mitochondria of maize seedlings. Increases in superoxide and H_2O_2 can be expected in cases in which there is either high O_2 uptake or decreased ability of the electron transport pathway, which increases potential for higher electron leakage to O_2 for subsequent production of superoxide and H_2O_2 [68].

One major antioxidant that plays a role in the detoxification of ROS and plant protection against oxidative damage is glutathione. There are two versions in which glutathione can exist in which are the oxidised disulphide version (GSSG) and the reduced version (GSH). The function of glutathione as an antioxidant is mainly assigned to its reduced (GSH) version as this form is oxidised to form the oxidised (GSSG) version during its function as an antioxidant. Therefore, keeping the concentration of reduced glutathione, from the ratio GSH/GSSG, high is important for plants. The production of GSH can occur both in cytosol and the chloroplast in the leaves of the plant. Furthermore, in the ascorbate–glutathione cycle, GR catalyses GSSG reduction into GSH via donation of electrons from NADPH molecules. ROS detoxification in the chloroplast is known to be mostly carried out by the ascorbate–glutathione cycle, which is thus accepted as the main pathway in this process. In this cycle, ascorbate is also considered to be a major antioxidant in addition to GSH [80]. According to Prasad [81], rapidity with which GR enzyme was induced during the early stages of acclimation and remained induced during chilling and recovery clearly suggests that acclimation uniquely induces the antioxidant defence mechanism that is necessary for protecting the seedling from oxidative stress injury.

Chilling-sensitive pepper cultivars were investigated for SOD, CAT and POD enzyme activity under chilling stress condition. The results showed that the activity of CAT decreased, and both SOD and POD activities raised in two cultivars, However, permeability of plasma membrane was positively related to MDA content, SOD and POD activity and also negatively related to CAT activity variation. As low temperature treating was extended, permeability of plasma membrane, MDA content and POD activity increased and SOD and CAT activity decreased in two cultivars: Xiza No. 7 (less chilling-sensitive cv.) could maintain a higher protective enzyme activity, and permeability of plasma membrane and MDA content were low; Hajiao No. 1 (chilling-sensitive cv.) was quite the contrary [82].

Anderson et al. [83] indicated that the response of ascorbate and glutathione in mesocotyls to acclimation and chilling was generally the same as that in the coleoptile + leaf, although the increase in the total glutathione pools in response to acclimation was not as extensive. Induction of other antioxidants in the mesocotyl may reduce the need for GSH synthesis. In the roots, there was no effect of acclimation or chilling on any of the antioxidants tested. However, the fact that H_2O_2 levels in the roots were not greatly increased by chilling suggests that a modification of antioxidants was unnecessary to prevent oxidative stress [83]. Likewise, Lee and Lee [84] established that APX is also an important antioxidant enzyme in scavenging or utilising H_2O_2. Total APX activity increased when chilling stress occurred in the leaves of the cucumbers and seemed to be because of favoured induction of the isozymes APX-5 and APX-4. However, 24 h after the stress, the increase seen in APX activity was because of the favoured expression of the isoform APX-3 [84]. Chilling stress causes many physiological and biochemical changes. Kang and Saltveit [85] investigated that chilling tolerance in cucumber seedling radicals. Chilling seedlings with radicles 20-mm long for 48 h at 2.5°C inhibited subsequent growth by 36%, while it reduced the growth of 70-mm-long radicles by 63%. APX activity was higher in 20-mm-long radicals before chilling than in 70-mm-long radicles. It appears that higher APX, CAT and DPPH (the stable free radical 1,1-diphenyl-2-picryl-hydraz)-radical scavenging activities, and sustained APX activity during chilled and during subsequent growth at 25°C following chilling in 20-mm-long radicals corresponds with higher chilling tolerance. The activities of APX, CAT and DPPH appear to be positively correlated with chilling tolerance [85].

A study on the effects of chilling stress on two salt- and drought-tolerant and two sensitive pumpkin genotypes in callus culture [18] found that the tolerant genotypes showed lower increase in lipid peroxidation and a greater increase in APX, CAT and GR than the tolerant genotypes under stress conditions. It was observed that even though the increase of glucose, proline and fructose concentration went up with prolonging of the chilling effect in all the pumpkin genotypes, a much more significant increase was observed in the tolerant genotype than that of the sensitive genotype. Thus it can be concluded from these results that in pumpkin genotypes, chilling stress results in an increase in the peroxidation of lipids and in oxidative stress, due to reactive oxygen radical production. Song et al. [86] in their study observed that chilling stress, SOD and CAT activities decreased in some extent in both cultivars, in comparison to control in tomato cultivars. Compared to control, chilling stress resulted in significantly higher POD activity in cv. Mawa on day 6, whereas no significant changes of POD activity caused by chilling stress were observed in cv. Moneymaker at all time points tested. On the other hands, APX activities were increased in the two cultivars under chilling stress. GR activities increased in cv. Mawa after chilling stress, but almost no change was observed in cv. Moneymaker [86].

5. Nutrient deficiency and toxicity of heavy metal

The micronutrients essential for the normal growth and development of plants, as it is known to be required in several metabolic processes [87]. Deficiency of nutrients such as Zn, Mn, Cu,

Fe, Mg, B and K can modify the activities of several antioxidative enzymes [88]. Kosesakal and Unal [89] indicated that Zinc (Zn) is one of the essential micronutrients playing a significant role in many vital metabolic processes. Zinc deficiency is a major global problem hindering plant cultivation, and this problem is especially exacerbated in acidic calcareous soils, which is the most common soil type in arid and semi-arid regions of the world. It is known that magnesium deficiency results in the decrease of chlorophyll amounts in beans. It was observed by Welkie et al. [90] that in peppers, the amounts of chlorophyll and iron in leaves were directly proportional, while zinc was also shown to be beneficial in the synthesis of carotenoids and chlorophyll, thus being proved to be essential in the photosynthetic process in plants. Iron (Fe) is a cofactor of many antioxidant enzymes and could act as a pro-oxidant factor because free or loosely bound it catalyses free radical generation in the presence of reductants and peroxides through the Fenton reaction. The growth of sunflower plants under iron deficiency conditions affects POD isoforms differently, inducing a preferential reduction in activity of those isoforms involved in the detoxification processes [91].

Metal toxicities have received widespread attention as large amounts are released into the environment and affect living organisms. Heavy metal intoxication, especially by lead, cadmium, arsenic and mercury, constitutes serious threat to human health [92, 93]. Although information focussed on the relationship between heavy metals and oxidative stress in plants has been available in recent years, it is still difficult to draw a general conclusion about critical toxic metal concentrations in soils [94]. Heavy metals cause oxidative damage to plants, either directly or indirectly through AOS formations which are extremely toxic to living cells. Redox metals such as Cu or Fe appear to act directly on the production of AOS. Copper is among the major heavy metal contaminants in the environment with various anthropogenic and natural sources. Human health risk from heavy metal bioaccumulation in vegetables has been a subject of growing concern in recent years. Excess Cu inhibits plant growth and seed germination, induces chlorophyll degradation and interferes with photosystem activity. At the molecular level, Cu ions generate ROS. These reactive radicals cause oxidative damage of lipids, proteins and nucleic acids. Cu ions also are responsible for alterations of membrane integrity in plant cells. Cu-mediated membrane lipid peroxidation causes membrane damage, thus changing membrane permeability and leading to electrolyte leakage. Plants have evolved several antioxidant defence mechanisms to protect themselves from oxidative damage [95].

Zn is the second most abundant transition metal after iron (Fe) and is involved in various biological processes in organisms. Due to this, the results of the presence of Zinc were investigated, including zinc deficiency, hyperaccumulation and its protective role in plants. However, it is not clear what the implications of zinc stress are on antioxidant responses and the uptake of nutrition, though it is known that excess of zinc is not beneficial and can result in negative symptoms in plants. The symptoms that can be observed at the organism level include prevention of seed germination, of root development and of the growth of the plant, and chlorosis can be seen in the leaves. At the cellular level, excess Zn can significantly alter mitotic activity, affect membrane integrity and permeability and even kill cells. Investigates showed that Zn stress on the activity of many antioxidative enzymes (APX, SOD, POD and CAT) and antioxidant contents (ascorbate and GSH) in plants [96–105].

The availability of manganese (Mn) to plants is governed by redox processes, which depend on soil's Mn reserve, pH and the availability of electrons. However, excess Mn disturbs the metabolism of plants and inhibits the plant growth. Mn causes deficiency of Fe, Mg and Ca and induces inhibition of chlorophyll biosynthesis and a decline in the photosynthetic rate. The toxic effects of heavy metals, both essential and nonessential elements, have been linked to the production of ROS. To quickly get rid of ROS, which result in the disruption of cellular metabolism due to damage by oxidative stress to important molecules, numerous enzymatic and non-enzymatic ways have evolved in living organisms. As Mn^{2+} plays a role in numerous processes, it is thought that an excess of it results in oxidative stress [106]. Cadmium (Cd) is a non-redox metal unable to participate in Fenton-type reactions. Naturally occurring amounts of Cd are normally low, however, the concentration can be significantly increased by anthropogenic activities. The impact of the uptake of Cd by living cells has been shown to be drastic, inducing oxidative stress and normally leading to cell death depending on the metal dose and time length of exposure [107]. In general, Cd in plants reduces growth, both in roots and stems, due to suppression of the elongation growth rate of the cells [94]. According to Dinakar et al. [93], cadmium is easily translocated from plant roots to above-ground tissues and potentially threatens human health. Cadmium in plants interferes with physiological processes, resulting in declined productivity. Cadmium can harness photosynthetic activity, chlorophyll content, plant growth and induce oxidative stress. ROS are efficiently eliminated by non-enzymatic (glutathione, ascorbate, a-tocopherol and carotenoids) and enzymatic defence systems such as SOD, APX, POD and GR, which protect plants against oxidative damage. The detoxification of O_2 occurs due to the SOD enzyme, while H_2O_2 is detoxified by the enzymes CAT and PODs and thus OH radicals are not formed. In the detoxification of hydrogen peroxide from different compartments in the cell, glutathione reductase (GR) and APX are key players in the ascorbate–glutathione cycle. Glutathione is also the substrate for the biosynthesis. A constitutively high antioxidant capacity or increase in antioxidant level could prevent oxidative damage and improve tolerance to the oxidative stress established [108, 109]. Sandalio et al. [108] investigated effects of cadmium on antioxidative enzyme activity in pea. They said that the level of oxygen radicals in cells could be enhanced by a decrease of the enzymatic antioxidants involved in their detoxification, such as SOD.

It was recorded by Schützendübel and Polle [110] that antioxidative enzymes were prevented from functioning and that GSH was depleted for a short period by cadmium and other metals. It was also put forth that hydrogen peroxide accumulation resulted from the depletion in these antioxidants. These results were obtained by accessing models of antioxidative capacity. As more Cd tolerance was observed in plants when more GSH was synthesised, it can be deduced that the decrease in GSH levels is an important step for cadmium sensitivity [110]. Dong et al. [111] investigated that effect of Cd concentration in tomato seedling for antioxidative enzymes. From the results, it can be observed that POD and SOD activities significantly increased in plants that were given Cd with a concentration of 1–10 μM and that MDA levels also showed a significant increase, indicating that oxidative stress response was the result of Cd stress in tomato plants. Tanyolac et al. [112] reported that tolerance and protective mechanisms have evolved to scavenge free radicals such as superoxide, hydroxyl radicals and peroxides generated during various metabolic reactions. Antioxidative enzymes such as APX play a key role in controlling the cellular level of these radicals and peroxides. They found that APX activity was increased with Cu treatment [112]. Zhao et al. [113] investigated the different tolerance mechanisms to Cd stress between YSL189 and HZ903 at the seedling stage. When Cd concentration was >20 μM in the growing medium, the uptake rate of Cd was significantly higher in roots of YSL189 than in the roots of HZ903. When plants were supplied with 50- and 100-μM Cd in the growing medium, there were higher Cd concentration, higher biomass and plant height, shorter roots and higher expression levels of transporter genes natural resistance associated macrophage proteins (Nramp)2, Nramp3 and zinc and iron regulated transporter (ZIP) in roots of YSL 189 compared to HZ903. The high Cd accumulation in YSL189 was partly due to the higher Cd uptake rate and higher expression levels of Nramp2, Nramp3 and ZIP in its roots. At the same time, the degree of cell injury indicated by thiobarbituric acid reactive substance showed no significant differences in roots and stems between the two genotypes. The higher activities of SOD, POD and CAT in roots and stems of YSL189 were compared to HZ903 [113].

Xiong and Wang [95] indicated that Cu phytotoxicity in Brassica pekinensis. Cu treatments increased electrolyte leakage and POD activity, showing a significant correlation between Cu concentration in shoots with electrolyte leakage and POD activity. Oriental melon IVF09 was used as a scion, while the pumpkin Jinxinzhen No. 3 was used as stock to research the physiological characteristics of grafted melon (*Cucumis melo*) seedlings when copper stress was induced. It was observed from the results that copper stress resulted in the inhibition of the physical characteristics of the melon seeds. In the grafted seeds, as opposed to self-rooted seedlings, an increase was seen in the levels of glucose, photosynthetic pigments, fructose, the photosynthetic parameters, biomass, the phosphate and sucrose synthase activities, acid invertase and neutral invertase. When levels of Cu decreased and the levels of P, NA and K increased nutrients were taken up more easily. The concentration of CU in the leaves decreased by 31.3%, while a 15.2% decrease was seen in roots of the grafted seedlings when the levels of copper ion (Cu^{2+}) stress became 800 μM and it was shown that grafting resulted in better endogenous hormone balance in the seedlings. When compared to the control, it was observed that grafted seedlings had a higher concentration of IAA and that POD activity was increased, while concentrations of ABA and maleic dialdehyde and the CAT and SOD activities became less. Thus, it could be deduced that grafting of melon seedlings was beneficial to them when under copper stress and relieved the resulting physiological characteristics from the stress, showing that the resistance of the grafted seedlings to copper stress increased due to grafting [114].

Shi and Zhu [106] indicated that the accumulation of ROS significantly increased in cucumber leaves exposed to excess Mn. It was observed that cucumber leaves in the presence of excess Mn resulted in higher activity of SOD, DHAR, POD and GR while adding SA (salicylic acid) resulted in the inhibition of the activities of APX and CAT, thus showing that different antioxidant enzymes had different changes. When the cucumber leaves were treated with SA, in the presence of excess Mn, the concentrations of the essential glutathione and ascorbate antioxidants increased [106]. Human health risk from heavy metal bioaccumulation in vegetables has been a subject of growing concern in recent years. It was observed by Kiran et al. that when under abiotic stress, mainly heavy metal applications, the Burdur Merkez and Burdur Bucak genotypes that were salt-tolerant had a higher resistance as opposed to the sensitive genotypes. The results also showed that drought, heavy metal and salinity stress resistance was observed to have evolved in similar ways in plants [115]. Another heavy metal lead (Pb) exerts adverse effects on morphology, growth and photosynthetic processes of plants; causes inhibition of enzyme activities, water imbalance and alterations in membrane permeability; and disturbs mineral nutrition [116]. Wastewater, which is used in agriculture in order to provide growing water demand, might be included heavy metal and trace elements. Lead is one of the most hazardous heavy metals, and it causes an extensive pollution in the environment, and also it has adverse effect on the growing of plants. In the other study was conducted to evaluate the effects of Pb stress in on lettuce (*Lactuca sativa*). It was found that SOD and GR were increased with oxidative stress [115].

6. Conclusions

In conclusion, both the callus tissue and whole plant studies show a positive correlation between increased antioxidant activity and different abiotic tolerance. Antioxidative enzyme activities play an important role against stress. The tolerance level against salt, drought and chilling stress in callus culture can be utilised as an effective criterion in the plants with other physiological criteria. Therefore, it can be said that antioxidative defence mechanisms and effective working systems in the aspect of tolerance against stress conditions in the plants. The literature suggests that tolerant and sensitive genotypes show different responses under abiotic stress conditions, that antioxidative enzyme activities play a protective role against abiotic stress and that antioxidative defence mechanisms are effective in providing resistance to stress in plants. The results of the studies showed that the young plants of the tolerant genotypes may have better protection against stress by increasing the activity of antioxidant enzymes under different abiotic stresses.

Author details

Sebnem Kusvuran[1*], Sevinç Kiran[2] and S. Sebnem Ellialtioglu[3]

*Address all correspondence to: skusvuran@gmail.com

1 Cankiri Karatekin University, Kizilirmak Vocational High School, Cankiri, Turkey

2 Soil Fertilizer and Water Resources Central Research Institute, Ankara, Turkey

3 Department of Horticulture, Faculty of Agriculture, Ankara University, Ankara, Turkey

References

[1] Rehman S, Harris PJC, Ashraf M. Stress environments and their impact on crop production. Stress environments and their impact on crop production. Abiotic Stresses: Plant Resistance Through Breeding and Molecular Approaches. Haworth Press, New York. 2005: 3-18.

[2] Mittler R. Oxidative stress, antioxidants and stress tolerance. Trends in plant science. 2002; 7(9): 405–410.

[3] Gaspar T, Franck T, Bisbis B, Kevers C, Jouve L, Hausman JF, Dommes J. Concepts in plant stress physiology. Application to plant tissue cultures. Plant Growth Regulation. 2002; 37(3): 263–285.

[4] Li Y. Physiological responses of tomato seedlings (*Lycopersicon esculentum*) to salt stress. Modern Applied Science. 2009; 3(3): 171–176.

[5] Chookhampaeng S. The effect of salt stress on growth, chlorophyll content proline content and antioxidative enzymes of pepper (*Capsicum annuum* L.) seedling. European Journal of Scientific Research. 2011; 49: 103–109.

[6] Yasar F, Ellialtioglu S, Yildiz K. Effect of salt stress on antioxidant defence systems, lipid peroxidation, and chlorophyll content in green bean. Russian Journal of Plant Physiology. 2008; 55(6): 782–786.

[7] Dolatabadian A, Sanavy SAMM, Chashmi NA. The effects of foliar application of ascorbic acid (vitamin C) on antioxidant enzymes activities, lipid peroxidation and proline accumulation of canola (*Brassica napus* L.) under conditions of salt stress. Journal of Agronomy and Crop Science. 2008; 194: 206–213.

[8] Amirjani MR. Effects of salinity stress on growth, mineral composition, proline content, antioxidant enzymes of soybean. American Journal of Physiology. 2010; 5: 350–360.

[9] Siringam K, Juntawong N, Cha-Um S, Kirdmanee C. Salt stress induced ion accumulation, ion homeostasis, membrane injury and sugar contents in salt-sensitive rice (*Oryza sativa* L. spp. *indica*) roots under isoosmotic conditions. African Journal of Biotechnology. 2011; 10: 1340–1346.

[10] Silvana BD, Gallego SM, Benavides MP, Tomaro ML. Behaviour of antioxidant defense system in the adaptive response to salt stress in *Helianthus annuus* L. Cells. Plant Growth Regulation. 2003; 40: 81–88.

[11] Jaleel CA, Riadh K, Gopi R, Manivannan P, Inès J, Al-Juburi HJ, Panneerselvam R. Antioxidant defense responses: Physiological plasticity in higher plants under abiotic constraints. Acta Physiologiae Plantarum. 2009; 31(3): 427–436.

[12] Dixit V, Pandey V, Shyam R. Differential antioxidative responses to cadmium in roots and leaves of pea (*Pisum sativum* L. cv. Azad). Journal of Experimental Botany. 2001; 52: 1101–1109.

[13] Mittiova V, Tal M, Volokita M, Guy M. Salt stress induces up-regulation of an efficient chloroplast antioxidant system in the salt-tolerant wild tomato species *Lycopersicon pennellii* but not in the cultivated species. Physiologia Plantarum. 2002; 115(3): 393–400.

[14] Asada K. The water-water cycle in chloroplasts: Scavenging of active oxygens and dissipation of excess photons. Annual Review of Plant Physiology and Plant Molecular Biology. 1999; 50(1): 601–639.

[15] Dionisio-Sese ML, Tobita S. Antioxidant responses of rice seedling to salinity stress. Plant Science. 1998; 135: 1–9.

[16] Lopez-Huertas E, Charlton WL, Johnson B, Graham IA, Baker A. Stress induces peroxisome biogenesis genes. The EMBO Journal. 2000; 19(24): 6770–6777.

[17] Scandalios JG. Oxygen stress and superoxide dismutases. Plant Physiology. 1993; 101: 7–12.

[18] Kusvuran S, Ellialtioglu S, Polat Z. Antioxidative enzyme activity, lipid peroxidation, and proline accumulation in the callus tissues of salt and drought tolerant and sensitive pumpkin genotypes under chilling stress. Horticulture, Environment, and Biotechnology. 2013; 54(4): 319–325.

[19] Shalata A, Mittova V, Guy M, Tal M. Response of cultivated tomato and its wild salt-tolerant relative *Lycopersicon pennelli* to salt dependent oxidative stress: The root antioxidative system. Physiologia Plantarum. 2001; 112: 487–494.

[20] Chaitanya KV, Sundar D, Masilamani S, Reddy AR. Variation in heat stress-induced antioxidant enzyme activities among three mulberry cultivars. Plant Growth Regulation. 2002; 36(2): 175–180.

[21] Gossett DR, Millhollon EP, Lucas MC, Marney MM, Banks SW. The effects of NaCl on antioxidant enzyme activities in callus tissue of salt-tolerant and salt-sensitive cotton cultivars (*Gossypium hirsutum* L.). Plant Cell Reports. 1994b; 13: 498–503.

[22] Shannon MC, Gronwald J, Tal M. Effects of salinity on growth and accumulation of organic and inorganic ions in cultivated and wild tomato species. Journal of the American Society for Horticultural Science. 1987; 112: 416–423.

[23] Cuartero J, Fernandez-Munoz R. Tomato and salinity. Scientia Horticulturae. 1999; 78: 83–125.

[24] Dogan M. *In vivo* and *in vitro* investigation of the effect of salinity stress on some physiological parameters and antioxidant enzymes activities in the tomato (*Lycopersi-*

con sp.). PhD Thesis, Department of Biology, Institute of Natural and Applied Sciences, University of Hacettepe, Turkey, 2004; 182 p.

[25] Kusvuran S. Relationships between physiological mechanisms of tolerances to drought and salinity in melons. PhD Thesis, Department of Horticulture, Institute of Natural and Applied Sciences, University of Çukurova, Turkey, 2010; 356 p.

[26] Payen S, Basset-Mens C, Follain S, Grünberger O, Marlet S, Nunez M, Perret S. Pass the salt please! From a review to a theoretical framework for integrating salinization impacts in food LCA. In: Proceedings of the 9th International Conference on LCA in the Agri-Food Sector, 8–10 October – San Fransisco, 2014; 953–963.

[27] Sevengor S, Yasar F, Kusvuran S, Ellialtioglu S. The effect of salt stress on growth, chlorophyll content, lipid peroxidation and antioxidative enzymes of pumpkin seedling. African Journal of Agricultural Research. 2011; 6(21): 4920–4924.

[28] Zhu Z, Wei G, Li J, Qian Q, Yu J. Silicon alleviates salt stress and increases antioxidant enzymes activity in leaves of salt stresses cucumber (*Cucumis sativus* L.). Plant Science. 2004; 167: 527–533.

[29] Yasar F. Effects of salt stress on ion and lipid peroxidation content in green beans genotypes. Asian Journal of Biochemistry. 2007; 19(2): 1165–1169.

[30] Abbas T, Balal RM, Shahid MA, Pervez MA, Ayyub CM, Aqueel MA, Javaid MM. Silicon-induced alleviation of NaCl toxicity in okra (*Abelmoschus esculentus*) is associated with enhanced photosynthesis, osmoprotectants and antioxidant metabolism. Acta Physiologiae Plantarum. 2015; 37(2): 1–15.

[31] Yasar F, Talhouni M, Ellialtioglu S, Kusvuran S, Uzal O. SOD, CAT, GR and APX enzyme activities in callus tissues of susceptible and tolerant eggplant varieties under salt stress. Research Journal of Biotechnology. 2013; 8(11): 45–50.

[32] Kusvuran S, Ellialtioglu S, Yasar F, Abak K. Effects of salt stress on ion accumulations and some of the antioxidant enzymes activities in melon (*Cucumis melo* L.). Journal of Food, Agriculture and Environment. 2007; 2(5): 351–354.

[33] Yasar F, Kusvuran S, Ellialtioglu S. Determination of anti-oxidant activities in some melon (*Cucumis melo* L.) varieties and cultivars under salt stress. Journal of Horticultural Science and Biotechnology. 2006; 81: 627–630.

[34] Yadav S, Irfan M, Ahmad A, Hayat S. Causes of salinity and plant manifestations to salt stress: A review. Journal of Environmental Biology. 2011; 32(5): 667–685.

[35] Rivero Rosa M, Ruız Juan M, Romero L. Role of grafting in horticultural plants under stress conditions. Journal of Food, Agriculture and Environment. 2003; 1: 70–74.

[36] İşeri ÖD, Körpe DA, Sahin FI, Haberal M. High salt induced oxidative damage and antioxidant response differs in *Nicotiana tabacum* L. and *Nicotiana rustica* L. cultivars. Journal of Applied Biological Sciences. 2013; 7(1): 15–23.

[37] Zhang GW, Liu ZL, Zhou JG, Zhu YL. Effects of Ca $(NO_3)_2$ stress on oxidative damage, antioxidant enzymes activities and polyamine contents in roots of grafted and non-grafted tomato plants. Plant Growth Regulation. 2008; 56(1): 7–19.

[38] De Azevedo Neto AD, Prisco JT, Enéas-Filho J, de Abreu CEB, Gomes-Filho E. Effect of salt stress on antioxidative enzymes and lipid peroxidation in leaves and roots of salt-tolerant and salt-sensitive maize genotypes. Environmental and Experimental Botany. 2006; 56(1): 87–94.

[39] Wei GP, Yang LF, Zhu YL, Chen G. Changes in oxidative damage, antioxidant enzyme activities and polyamine contents in leaves of grafted and non-grafted eggplant seedlings under stress by excess of calcium nitrate. Scientia Horticulturae. 2009; 120(4): 443–451.

[40] Wani SH, Sofi PA, Gosal SS, Singh N. In vitro screening of rice (*Oryza sativa* L) callus for drought tolerance. Communications in Biometry and Crop Science. 2010; 5(2): 108–115.

[41] Sevengor S. Investigation of some antioxidant enzyme activities in pumpkin genotypes under salt stress in terms of in vitro and in vivo. PhD Thesis, Department of Horticulture, Institute of Natural and Applied Sciences, University of Ankara, Turkey, 2010; 165 p.

[42] Yasar F, Talhouni M, Ellialtioglu S, Kusvuran S, Uzal O. SOD, CAT, GR and APX enzyme activities in callus tissues of susceptible and tolerant eggplant varieties under salt stress. Research Journal of Biotechnology. 2013; 8(11): 45–50.

[43] Kusvuran S, Ellialtioglu S, Yasar F, Abak K. Antioxidative enzyme activities in the leaves and callus tissues of salt-tolerant and salt-susceptible melon varieties under salinity. African Journal of Biotechnology. 2012; 11(3): 635–641.

[44] Ashraf M. Biotechnological approach of improving plant salt tolerance using antioxidants as markers. Biotechnology Advances. 2009; 27(1): 84–93.

[45] Moharramnejad S, Valizadeh M. Assessment of oxidative stress tolerance in red bean (*Phaseolus vulgaris* L.) seedling under salinity. International Journal of Agronomy and Agricultural Research. 2014; 5(6): 49–56.

[46] Gong H, Zhu X, Chen K, Wang S, Chenglie Z. Silicon alleviates oxidative damage of wheat plants in pots under drought. Plant Science. 2005; 169: 313–321.

[47] Martinez JP, Silva H, Ledent JF, Pinto M. Effects of drought stress on the osmotic adjustment, cell wall elasticity and cell volume of six cultivars of common beans (*Phaseolus vulgaris* L.). European Journal of Agronomy. 2007; 26: 30–38.

[48] Kusvuran S, Dasgan HY, Abak K. Responses of different melon genotypes to drought stress. Yüzüncü Yıl University Journal of Agriculture Science. 2011; 21: 209–219.

[49] Li M, Wang GX, Lin JS. Application of external calcium in improving the PEG-induced water stress tolerance in liquorice cells. Botanical bulletin of Academia Sinica. 2003; 44: 275–284.

[50] Farooq M, Wahid A, Kobayashi N, Fujita D, Basra SMA. Plant drought stress: Effects, mechanisms and management. Agronomy for Sustainable Development. 2009a; 29: 185–212.

[51] Farooq M, Wahid A, Lee DJ, Ito O, Siddique KHM. Advances in drought resistance of rice. Critical Reviews in Plant Sciences. 2009b; 28: 199–217.

[52] Jaleel CA, Manivannan P, Wahid A, Farooq M, Somasundaram R, Panneerselvam R. Drought stress in plants: A review on morphological characteristics and pigments composition. International Journal of Agriculture and Biology. 2009; 11: 100–105.

[53] Kalefetoglu T, Ekmekci Y. The effects of drought on plants and tolerance mechanisms. Gazi University Journal of Science. 2005; 18: 723–740.

[54] Türkan I, Bor M, Özdemir F, Koca H. Differential responses of lipid peroxidation and antioxidants in the leaves of drought-tolerant P. *acutifolius* Gray and drought sensitive *P. vulgaris* L. subjected to polyethylene glycol mediates water stress. Plant Science. 2005; 168: 223–231.

[55] El-Tayeb MA. Differential response of two *Vicia faba* cultivars to drought: Growth, pigments, lipid peroxidation, organic solutes, catalase and peroxidase activity. Acta Agronomica Hungarica. 2006; 54: 25–37.

[56] Liu ZJ, Zhang XL, Bai JG, Suo BX, Xu PL, Wang L. Exogenous paraquat changes antioxidant enzyme activities and lipid peroxidation in drought-stressed cucumber leaves. Scientia Horticulturae. 2009; 121: 138–143.

[57] Basu S, Roychoudhury A, Paromita Saha P, Sengupta DN. Differential antioxidative responses of indica rice cultivars to drought stress. Plant Growth Regulation. 2010; 60: 51–59.

[58] Kusvuran S. Influence of drought stress on growth, ion accumulation and antioxidative enzymes in okra genotypes. International Journal of Agriculture & Biology. 2012; 14(3): 401–406.

[59] Kusvuran S, Ellialtioglu S, Talhouni M, Sonmez K, Kiran S. Effects of salt and drought stress on the growth, antioxidative enzyme activities and MDA content in callus tissues of four melon varieties. In: 6th Balkan Symposium on Vegetables and Potatoes, September 29–October 2, 2014, Zagreb, Croatia; 88.

[60] Kiran S, Kusvuran S, Talhouni M, Sonmez K, Ellialtioglu S, Ozkay F. The studies on some biochemical changes and ion regulation in the tomato genotypes exposed to drought stress. In: 6th Balkan Symposium on Vegetables and Potatoes, September 29–October 2, 2014, Zagreb, Croatia; 126.

[61] Kuşvuran S, Daşgan HY, Abak K. Effects of drought stress on antioxidant enzymes activities in melon. In:L VIII. National Vegetable Symposium, 23–25 June 2010, Van; 309–312.

[62] Kiran S, Kuşvuran S, Ellialtioglu S, Ozkay F. Studies on some morphological and biochemical changes in the melon genotypes exposed to drought stress. In: 1st Central Asia Congress on Modern Agricultural Techniques and Plant Nutrition, October 01–03, 2013, Bishkek, Kyrgyzstan, Soil Water Journal. 2013; 2(2): 1347–1354.

[63] Smirnoff N. The role of active oxygen in the response of plants to water deficit and desiccation. New Phytologist. 1993; 125(1): 27–58.

[64] Yasar F, Uzal O, Kose S, Yasar O, Ellialtioglu S. Enzyme activities of certain pumpkin (*Cucurbita* spp) species under drought stress. Fresenius Environmental Bulletin. 2014; 23(4): 1093–1099.

[65] Vijayakumari K, Puthur JT. Drought stress responses in tolerant and sensitive varieties of black pepper (*Piper nigrum* Linn.). Journal of Plantation Crops. 2014; 42: 78–85.

[66] Kratsch HA, Wise RR. The ultrastructure of chilling stress. Plant, Cell and Environment. 2002; 23: 337–350.

[67] Kang G, Wang C, Sun G, Wang Z. Salicylic acid changes activities of H_2O_2-metabolizing enzymes and increases the chilling tolerance of banana seedlings. Environmental and Experimental Botany. 2003; 50: 9–15.

[68] Prasad TK, Anderson MD, Stewart CR. Acclimation, hydrogen peroxide, and abscisic acid protect mitochondria against irreversible chilling injury in maize seedlings. Plant Physiology. 1994a; 105: 619–627.

[69] Prasad TK, Anderson MD, Martin BA, Stewart CR. Evidence for chilling-induced oxidative stress in maize seedlings and a regulatory role for hydrogen peroxide. Plant Cell. 1994b; 6: 65–74.

[70] Halliwell B. Oxidative damage, lipid peroxidation and antioxidant protection in chloroplasts. Chemistry and Physics of Lipids. 1987; 44: 327–340.

[71] Elstner EF. Mechanisms of oxygen activation in different compartments of plant cells. In: Pell EJ, Steffen KL, eds, Active Oxygen/Oxidative Stress and Plant Metabolism. Rockville, MD: American Society Plant Physiologists; 1994, pp. 13–25.

[72] Chance B, Sies H, Boveris A. Hydroperoxide metabolism in mammalian organs. Physiological Reviews. 1979; 59: 527–605.

[73] Halliwell B, Gutteridge JMC. Oxygen free radicals and iron in relation to biology and medicine: Some problems and concepts. Archives of Biochemistry and Biophysics. 1986; 246: 501–514.

[74] Jahnke LS, Hull MR, Long SP. Chilling stress; and oxygen metabolizing enzymes in *Zea mays* and *Zea diploperennis*. Plant Cell Environment. 1994; 14: 97–104.

[75] Cakmak I, Strbac D, Marschner H. Activities of hydrogen peroxide-scavenging enzymes in germinating wheat seeds. Journal of Experimental Botany. 1993; 44: 127–132.

[76] Walker MA, McKersie BD. Role of ascorbate-glutathione antioxidant system in chilling resistance of tomato. Journal of Plant Physiology. 1993; 141: 234–239.

[77] Aroca R, Irigoyen JJ, Sanchez-Dıaz M. Photosynthetic characteristics and protective mechanisms against oxidative stress during chilling and subsequent recovery in two maize varieties differing in chilling sensitivity. Plant Science. 2001; 161: 719–726.

[78] Li Q, Yu B, Gao Y, Dai AH, Bai JG. Cinnamic acid pretreatment mitigates chilling stress of cucumber leaves through altering antioxidant enzyme activity. Journal of Plant Physiology. 2011; 168: 927–934.

[79] Xu S, Li Y, Hu J, Guan Y, Ma W, Zheng Y, Zhu S. Responses of antioxidant enzymes to chilling stress in tobacco seedlings. Agricultural Sciences in China. 2010; 9(11): 1594–1601.

[80] Shu DF, Wang LY, Duan M, Deng YS, Meng QW. Antisense-mediated depletion of tomato chloroplast glutathione reductase enhances susceptibility to chilling stress. Plant Physiology and Biochemistry. 2011; 49(10): 1228–1237.

[81] Prasad TK. Mechanisms of chilling-induced oxidative stress injury and tolerance in developing maize seedlings: Changes in antioxidant system, oxidation of proteins and lipids, and protease activities. The Plant Journal. 1996; 10(6): 1017–1026.

[82] Zhirong Z, Guoyi L. The effect of chilling stress on membrane lipid peroxidation and protective enzyme in pepper seedlings [J]. Acta Agriculturae Boreali-Occidentalis Sinica. 1994; 1994-03.

[83] Anderson MD, Prasad TK, Stewart CR. Changes in isozyme profiles of catalase, peroxidase, and glutathione reductase during acclimation to chilling in mesocotyls of maize seedlings. Plant Physiology. 1995; 109(4): 1247–1257.

[84] Lee DH, Lee CB. Chilling stress-induced changes of antioxidant enzymes in the leaves of cucumber: in gel enzyme activity assays. Plant Science. 2000; 159(1): 75–85.

[85] Kang HM, Saltveit ME. Reduced chilling tolerance in elongating cucumber seedling radicles is related to their reduced antioxidant enzyme and DPPH-radical scavenging activity. Physiologia Plantarum. 2002; 115(2): 244–250.

[86] Song Y, Diao Q, Qi H. Putrescine enhances chilling tolerance of tomato (*Lycopersicon esculentum* Mill.) through modulating antioxidant systems. Acta Physiologiae Plantarum. 2014; 36(11), 3013–3027.

[87] Candan N, Tarhan L. Changes in chlorophyll-carotenoid contents, antioxidant enzyme activities and lipid peroxidation levels in Zn-Stressed *Mentha pulegium.* Turkish Journal of Chemistry. 2002; 27: 21–30.

[88] Yu Q, Osborne L, Rengel Z. Micronutrient deficiency changes activities of superoxide dismutase and ascorbate peroxidase in tobacco plants. Journal of Plant Nutrition. 1998; 21(7): 1427–1437.

[89] Kosesakal T, Unal M. Role of zinc deficiency in photosynthetic pigments and peroxidase activity of tomato seedlings. IUFS Journal of Biology. 2009; 68(2): 113–120.

[90] Welkie GW, Hekmat-Shoar H, Miller GW. Responses of pepper (*Capsium annuum*) plants to iron deficiency: Solution pH and riboflavin. Plant Nutrition—Physiology and Applications. 1990; 41: 207–211.

[91] Ranieri A, Castagna A, Baldan B, Soldatini F. Iron deficiency differently affects peroxidase isoforms in sunflower. Journal of Experimental Botany. 2001; 52(354): 25–35.

[92] Wennberg PO, Cohen RC, Stimpfle RM, Koplow JP, Anderson JG, Salawitch RJ, Wofsy SC. Removal of stratospheric O_3 by radicals: In situ measurements of OH, HO_2, NO, NO_2, ClO, and BrO. Science. 1994; 266(5184): 398–404.

[93] Dinakar N, Nagajyothi PC, Suresh S, Udaykiran Y, Damodharam T. Phytotoxicity of cadmium on protein, proline and antioxidant enzyme activities in growing *Arachis hypogaea* L. Seedlings. Journal of Environmental Sciences. 2007; 20: 199–206.

[94] Gomes-Junior RA, Moldes CA, Delite FS, Pompeu GB, Grata PL, Mazzafera P, Lea PJ, Azevedo RA. Antioxidant metabolism of coffee cell suspension cultures in response to cadmium. Chemosphere. 2006; 65: 1330–1337.

[95] Xiong ZT, Wang H. Copper toxicity and bioaccumulation in chinese cabbage (*Brassica pekinensis* Rupr.). Environmental Toxicology. 2005; 20(2): 188–194.

[96] Mrozek JE, Funicelli NA. Effect of zinc and lead on germination of *Spartina alterniflora* Loisel seeds at various salinities. Environmental and Experimental Botany. 1982; 22: 23–32.

[97] Ebbs S, Uchil S. Cadmium and zinc induced chlorosis in Indian mustard [*Brassica juncea* (L.) Czern] involves preferential loss of chlorophyll b. Photosynthetica. 2008; 46: 49–55.

[98] Cakmak I. Possible roles of zinc in protecting plant cells from damage by reactive oxygen species. New Phytologist. 2000; 146 (2): 185–205.

[99] Auld DS. Zinc coordination sphere in biochemical zinc sites. Biometals. 2001; 14: 271–313.

[100] Stoyanova Z, Doncheva S. The effect of zinc supply and succinate treatment on plant growth and mineral uptake in pea plant. Brazilian Journal of Plant Physiology. 2002; 14: 111–116.

[101] Rout GR, Das P. Effect of metal toxicity on plant growth and metabolism: I. Zinc. Agronomie. 2003; 23: 3–11.

[102] Hacisalihoglu G, Kochian LV. How do some plants tolerate low levels of soil zinc? Mechanisms of zinc efficiency in crop plants. New Phytologist. 2003; 159: 341–350.

[103] Broadley MR, White PJ, Hammond JP, Zelko I, Lux A. Zinc in plants. New Phytologist. 2007; 173: 677–702.

[104] Lingua G, Franchin C, Todeschini V, Castiglione S, Biondi S, Burlando B, Parravicini V, Torrigiani P, Berta G. Arbuscular mycorrhizal fungi differentially affect the response to high zinc concentrations of two registered poplar clones. Environmental Pollution. 2008; 153: 137–147.

[105] Wang C, Zhang SH, Wang PF, Hou J, Zhang WJ, Li W, Lin ZP. The effect of excess Zn on mineral nutrition and antioxidative response in rapeseed seedlings. Chemosphere. 2009; 75: 1468–1476.

[106] Shi Q, Zhu Z. Effects of exogenous salicylic acid on manganese toxicity, element contents and antioxidative system in cucumber. Environmental and Experimental Botany. 2008; 63: 317–326.

[107] Benavides MP, Susana MG, María LT. Cadmium toxicity in plants. Brazilian Journal of Plant Physiology. 2005; 17(1): 21–34.

[108] Sandalio LM, Dalurzo HC, Gomez M, Romero-Puertas MC, Del-Rio LA. Cadmium-induced changes in the growth and oxidative metabolism of pea plants. Journal of Experimental Botany. 2001; 52(364): 2115–2126.

[109] Ekmekci Y, Tanyolac D, Ayhan B. Effects of cadmium on antioxidant enzyme and photosynthetic activities in leaves of two maize cultivars. Journal of Plant Physiology. 2008; 165: 600–611.

[110] Schützendübel A, Polle A. Plant responses to abiotic stresses: Heavy metal-induced oxidative stress and protection by mycorrhization. Journal of Experimental Botany. 2002; 53(372): 1351–1365.

[111] Dong J, Wu F, Zhang G. Influence of cadmium on antioxidant capacity and four microelement concentrations in tomato seedlings (*Lycopersicon esculentum*). Chemosphere. 2006; 64: 1659–1666.

[112] Tanyolac¸ D, Ekmekci Y, Unalan S. Changes in photochemical and antioxidant enzyme activities in maize (*Zea mays* L.) leaves exposed to excess copper. Chemosphere. 2007; 67: 89–98.

[113] Zhao S, Zhang Y, Ye X, Zhang Q, Xiao W. Responses to cadmium stress in two tomato genotypes differing in heavy metal accumulation. Turkish Journal of Botany. 2015; 39.

[114] Tan M, Zhang XY, Fu QS, He ZQ, Wang HS. Effects of grafting on physiological characteristics of melon (*Cucumis melo*) seedlings under copper stress. The Journal of Applied Ecology. 2014; 25(12): 3563–3572.

[115] Kıran S, Özkay F, Kuşvuran Ş, Ellialtıoğlu Ş. The effect of humic acid applied to the plants of lettuce (*Lactuca sativa* var. *crispa*) irrigated with water with high content of lead on some characteristics. Research Journal of Biological Sciences. 2014; 7(1): 14–19.

[116] Sharma P, Dubey RS. Lead toxicity in plants. Brazilian Journal of Plant Physiology. 2005; 17: 35–52.

9

Abiotic Stress Alleviation with Brassinosteroids in Plant Roots

Sevgi Marakli and Nermin Gozukirmizi

Abstract

This chapter covers the advances in establishment and optimization of brassinosteroids (BRs) in the alleviation of abiotic stresses such as water, salinity, temperature, and heavy metals in plant system, especially roots. Plant roots regulate their developmental and physiological processes in response to various internal and external stimuli. Studies are in progress to improve plant root adaptations to stress factors. BRs are a group of steroidal hormones that play important roles in a wide range of developmental phenomena, and recently they became an alleviation agent for stress tolerance in plants. This review is expected to provide a resource for researchers interested in abiotic stress alleviation with BRs.

Keywords: Water stress, salt stress, temperature stress, heavy metal stress

1. Introduction

Abiotic stress responses in plants occur at various organ levels among which the root-specific processes are of particular importance. Under normal growth condition, root absorbs water and nutrients from the soil and supplies them throughout the plant body, thereby playing pivotal roles in maintaining cellular homeostasis. However, this balanced system is altered during the stress period when roots are forced to adopt several structural and functional modifications. Examples of these modifications include molecular, cellular, and phenotypic changes such as alteration of metabolism and membrane characteristics, hardening of cell wall, and reduction of root length [1, 2]. The root system has the crucial role of extracting nutrients and water through a complex interplay with soil biogeochemical properties and of maintaining these functions under a wide range of stress scenarios to ensure plant survival and reproduction [3].

Water stress is characterized by a reduction of water content and leaf water potential, closure of stomata, and decreased growth. Severe water stress may result in the arrest of photosynthesis, disturbance of metabolism, and finally the death of plant [4]. This water loss causes a loss of turgor pressure that may be accompanied by a decrease in cell volume depending on the hardness of the cell wall [5]. The cells of the root must activate processes to limit water loss and mitigate its harmful effects.

Salinity also affects plant growth, activity of major cytosolic enzymes by disturbing intracellular potassium homeostasis, causing oxidative stress and programmed cell death, reducing nutrient uptake, genetic and epigenetic effects, metabolic toxicity, inhibition of photosynthesis, decreasing CO_2 assimilation, and reducing root respiration [6, 7, 8, 9]. Salt stress affects the root in all developmental zones. Cell division decreases in the meristematic zone and cell expansion attenuates in the elongation zone, resulting in reduced overall growth [10]. Cells also expand radially in the elongation zone [11], and root hair outgrowth suppresses in the differentiation zone [12]. Salt stress additionally results in agravitropic growth [13] as well as reduced lateral root number under high-salt conditions and enhanced lateral root number under moderate-salt conditions [14, 15]. Salt stress developes from excessive concentrations of salt, especially sodium chloride (NaCl) in soil. Root is the primary organ of exposure and hence responds rapidly [16]. Salt stress is known to increase Na^+/K^+ ratio in the root that leads to cell dehydration and ion imbalance [17, 18, 19].

High temperature increases the permeability of plasma membrane [20], and also reduces water availability [21]. Moreover, low temperature (chilling and frost stress) is also a major limiting factor for productivity of plant indigenous to tropical and subtropical climates [22]. Chilling stress has a direct impact on the photosynthetic apparatus, essentially by disrupting the thylakoid electron transport, carbon reduction cycle, and stomatal control of CO_2 supply, together with an increased accumulation of sugars, peroxidation of lipids, and disturbance of water balance [23].

Heavy metal contamination in soil could result in inhibition of plant growth and yield reduction and even poses a great threat to human health via food chain [24]. Among heavy metals, Cadmium (Cd) in particular causes increasingly international concern [25]. Cd-contaminated soil results in considerable accumulation of Cd in edible parts of crops, and then it enters the food chain through the translocation and accumulation by plants [26, 27]. Another metal, chromium (Cr III or VI), is not required by plants for their normal plant metabolic activities. On the contrary, excess of Cr (III or VI) in agricultural soils causes oxidative stress for many crops. Reactive oxygen species (ROS), like hydrogen peroxide (H_2O_2), hydroxyl radical ($OH^{\cdot}$), and superoxide radical (O_2^{-}) generated under Cr-stress, are highly reactive and cause oxidative damages to DNA, RNA, proteins, and pigments [28, 29]. Nickel (Ni) is one of the most abundant heavy metal contaminants of the environment due to its release from mining and smelting practices. It is classified as an essential element for plant growth [30]. However, at higher concentrations, nickel is an important environmental pollutant. Ni^{2+} ions bind to proteins and lipids such as specific subsequences of histones [31] and induce oxidative damage. Copper (Cu) is also an essential micronutrient for most biological organisms. It is a cofactor for a large array of proteins involved in diverse physiological processes, such as photosynthesis, electron transport chain, respiration, cell wall metabolism, and hormone signaling [32, 33]. Cu has emerged as a major environmental pollutant in the past few decades because of its excessive use in manufacturing and agricultural industries [34]. Zinc (Zn) is one of the other essential microelement, the second most abundant transition metal, and plays roles in many metabolic reactions in plants [35, 36]. However, high concentrations of Zn are toxic, induce structural disorders, and cause functional impairment in plants. At organism level, Zn stress reduces rooting capacity, stunted growth, chlorosis, and at cellular level alters mitotic activity [37, 38].

The key to find out abiotic stress tolerance resides in understanding the plant's capacity to accelerate/maintain or repress growth. Most plant hormones play a role in development and have been implicated in abiotic stress responses. One of these hormones, BRs, are a group of steroidal hormones that play significant roles in a wide range of developmental phenomena including cell division and cell elongation in stems and roots, photo-morphogenesis, reproductive development, leaf senescence, and also in stress responses [39]. Mitchell et al. [40] discovered BRs which were later extracted from the pollen of *Brassica napus* by Grove et al. [41]. To date, more than 70 BR-related phytosteroids have been identified in plants [42].

BRs increase adaptation to various abiotic stresses such as light [43, 44], low or high temperature [45], drought [46, 47, 48], salt stress [9], and heavy metal stress [49, 50]. BRs may be applied/supplied to plants at different stages of their life cycle such as meiosis stage [51], anthesis stage [52], and root application [9, 53].

In this chapter, the potential role of BRs in alleviating the adverse effects of water, salt, low/high temperature stresses, and heavy metals on plants, especially roots, were discussed.

2. Water stress

Water shortage is predicted as one of the most important environmental problem for the 21st century that limits crop production [54]. Although drought stress inhibites the plant–water relations, exogenous application of BRs maintaines tissue–water status [55] by stimulating the proton pumping [56], activating nucleic acid and protein synthesis [57] and regulation of genes expressions [58]. It has been shown that 24-epibrassinolide (24-epiBL)-treated *Arabidopsis* and *B. napus* seedlings had a higher survival rate when subjected to drought [59], and in another study BR-treated sorghum (*Sorghum vulgare*) showed increased germination and seedling growth under osmotic stress [60].

Root nodulation is a fundamental developmental event in leguminous crops, and is sensitive to water shortage [61, 62, 63]. As endogenous hormones play an important role in the organogenesis and initial growth of nodules in roots, attempts have been made to increase root nodulation by growth regulator treatments [64, 65]. The potential of BRs in the improvement of root nodulation and yield have been reported in groundnut [66]. Upreti and Murti [67] also studied the effects of two BRs, epibrassinolide (EBL) and homobrassinolide (HBL), on root nodulation and yield in *Phaseolus vulgaris* L. cv. Arka Suvidha under water stress. They concluded that water stress negatively influenced nodulated root, but BRs increased tolerance to water stress and EBL was relatively more effective than HBL.

Several researchers have found that increased proline levels can protect plants from water stress. BR treatment increased the contents of proline and protein under water stress [68]. Zhang et al. [69] also indicated that BR treatment promoted the accumulation of osmoprotectants, such as soluble sugars and proline. It may be due to the fact that BRs activated the enzymes of proline biosynthesis, which caused an additive effect on the proline content [70].

Drought stress causes increment in H_2O_2 due to decrease in antioxidative enzyme activities [71]. Plants have improved various defense mechanisms to respond and adapt to water stress [72]. Vardhini et al. [73] studied with sorghum seedlings grown under PEG-imposed water stress and investigated the effects of HBL and 24-epiBL on the activities of four oxidizing enzymes: superoxide dismutase (SOD), glutathione reductase (GR), IAA oxidase, and polyphenol oxidase (PPO). They found that supplementation of both the BRs resulted in enhanced SOD and GR but lowered IAA oxidase and PPO. Li and Feng [68] also reported that treatment of brassinolide significantly increased peroxidase (POD), catalase (CAT), and ascorbate peroxidase (APX) activities of seedlings under normal water and mild water stress. Therefore, increment in enzyme activities provided tolerance of *Xanthoceras sorbifolia* seedlings to drought stress. It has been found that BRs can induce the expression of some antioxidant genes and enhance the activities of antioxidant enzymes such as SOD, POD, CAT, and APX [74, 75].

3. Salt stress

Salinity stress is one of the most serious abiotic stress factors. It causes morphological, biochemical, cytogenetic, and molecular changes in plants [9, 76, 77, 78]. Root lengths, shoot lengths, and root numbers decrease in plants exposed to salt stress [7]. Moreover, salinity also induces oxidative stress in plants due to production ROS [79, 80]. These ROS are produced in the cell and interacted with a number of vital cellular molecules and metabolites, thereby leading to a number of destructive processes causing cellular damage [81].

BRs reduce impacts of salt stress on ROS, gene expression, mitotic index, nutrient uptake, and growth [9, 82, 83–88]. There are lots of studies to analyse alleviation of salt stress by using BRs. In these studies, different parameters have been investigated to understand the mechanism of BRs on salt stress (Table 1).

References	[9]	[84]	[85]	[86]	[87]	[88]
Effects on gene expression	---	---	---	---	Increased *Cu/Zn-SOD*, *APX*, *CAT*, *GR* and *OsBRI* expressions	---

					but reduced *Fe-SOD* and *Mn-SOD, OsDWF4* and *SalT*	
Effects on protein content	Increased	---	Increased	---	Increased	---
Effects on enzyme activities	Increased SOD and CAT activities	Increased CAT, GR, POX and SOD activities	Increased POX and SOD activities	Increased SOD and POD activities	Showed varying results depending on 24-epiBL concentration for SOD, APX, CAT, GR	Increased CAT, POX and SOD activities
Effects on growth and/or cell division	Increased	Increased	Increased	Increased	Increased	Increased
Methods of salt and BRs applications	Seeds were grown under both 150–250 mM salt concentrations and 0.5 and 1 μM HBR at 48 h and 72 h.	Plants received 100 mM NaCl as well as 0.01 μM of HBL during 18 days after sowing	25, 50, 100, and 150 mM NaCl were applied and then sprayed twice with 0.05 ppm brassinolide during 25 days. At 45 days from sowing, the plants were collected	Seedlings were exposed to 90 mM NaCl with 0, 0.025, 0.05, 0.10, and 0.20 mg dm^{-3} 24-epiBL for 10 days	Seeds were soaked for 8 h in different concentration s of 24-epiBL (10^{-11}, 10^{-9} and 10^{-7} M). After 24-epiBL application, The seeds were sown in autoclaved sand moistened with different concentration s of NaCl (0, 75, 100, 125 mM) during 12 days	The 15-day-old plants were exposed to 100 mM NaCl and they were subsequently treated by exogenous 24-epiBL (10^{-8} M). The plants were harvested after 30 days of growth
Plant Species	*Hordeum vulgare* L.	*Vigna radiata* L.	*Vigna sinensis* L.	*Solanum melongena* L.	*Oryza sativa* L. var. Pusa Basmati-1 cv. indica	*Cucumis sativus L.*

Table 1. Effects of BRs on plants subjected to salt stress Dashes indicate that there are no results in study.

4. Temperature stress

4.1. High temperature

In general, a transient elevation in temperature (usually 10–15°C above environment) causes heat shock or heat stress [89]. High-temperature effects can be seen at the biochemical and molecular level in plant organs (especially leaves). Heat stress induces decrease in duration of developmental phases, leading to fewer organs, smaller organs, reduce light perception over the shortened life cycle, and finally play an important role in losing the product [90, 91, 92].

High-temperature stress often induces the overproduction of ROS [93] which can cause membrane lipid peroxidation, protein denaturation, and nucleic acid damage [94, 95]. Many studies have demonstrated that ROS scavenging mechanisms play an important role in protecting plants from high-temperature stress [96, 97]. BRs applications decrease ROS levels and increase antioxidant enzyme activities to provide thermotolerance to elevated temperatures [98].

4.2. Low temperature

Chilling and frost stresses affect growth, development, survival, and crop productivity in plants [99, 100, 101]. However, BRs treatments enhance seedling tolerance to chilling stress [101] and increase the height, root length, root biomass, and total biomass of rice under low-temperature conditions [102, 103]. In another study, Krishna [104] reported the same results in maize. They postulated that treatments with BRs promoted growth recovery of maize seedlings following chilling treatment (0–3°C).

Chilling stress increases the proline, betaine, soluble protein, soluble sugar contents of plants [79, 105]. Studies showed that BRs treatment enhanced proline content and therefore increased plant chilling resistance and cell membrane stability [99, 100, 106, 107].

Chilling stress could trigger the production of antioxidant enzymes in plants to prevent the chilling injury [108]. In the previous investigations, it was reported that treatment with BRs further increased the activities of antioxidant enzymes under chilling stress as well [99, 100, 107, 109]. The enhanced activities of the antioxidative enzymes as a result of BRs applications may occur with increasing de novo synthesis or activation of the enzymes, which is mediated through transcription and/or translation of specific genes to gain tolerance [57].

5. Heavy metal stresses

5.1. Cd stress

Cd toxicity has emerged as one of the major agricultural problems in many soils around the world [110]. It has been shown to interfere with the uptake, transport, and utilization of essential nutrients and water, change enzyme activities, cause symptoms (chlorosis, necrosis), decrease in fresh and dry mass of root and shoot and also their lengths [110, 111, 112].

There are lots of studies to investigate the effects of BRs on Cd stress in plant species [110, 113, 114]. In these studies, results showed that BRs change different parameters such as germination, plant dry biomass, protein content, and antioxidant enzyme activities (Table 2). It is proposed that the changes induced by BRs are mediated through the repression and/or derepression of specific genes [58]. Microarray experiments evaluating gene expression changes in *Arabidopsis* roots and shoots under Cd stress were performed [115]. Moreover, studies showed that gene expression in response to Cd mimics a BR increase, and Cd exposure most probably induces an activation of the BR signaling pathway in *Arabidopsis* [116].

5.2. Cr stress

Cr (III or VI) is not required by plants for their normal plant metabolic activities [117]. The entry of Cr into a plant system occurs through roots via using the specialized uptake systems of essential metal ions required for normal plant metabolism [118]. On the contrary, excess of Cr (III or VI) in agricultural soils causes oxidative stress to many crops. Reduced seed germination, disturbed nutrient balance, wilting, and plasmolysis in root cells and thus effects on root growth of plants have been documented in plants under Cr stress [118, 119].

Choudhary et al. [120] reported that EBL treatment improved seedlings growth under Cr (VI) stress. Ability of EBL to increase seedling growth under this metal stress could be attributed to the capacity of BRs to regulate cell elongation and divisional activities, by enhancing the activity of cell wall loosening enzymes (xyloglucan transferase/hydrolase, XTH) [121]. Studies also indicated that increment in antioxidant activities as a result of BRs application (Table 2) provide plant tolerance to grow under Cr stress.

5.3. Ni stress

The heavy metals that affect (either positively or negatively) plants include Fe, Cu, Zn, Mn, Co, Ni, Pb, Cd, and Cr, but out of them, nickel has recently been defined as an essential micronutrient, because of its involvement in urease activity in legumes [122]. Excess Ni causes different problems. These symptoms include the inhibition in root elongation, photosynthesis and respiration, and interveinal chlorosis [123]. Moreover, the toxic concentration of Ni also inhibits enzyme activities and protein metabolism [124]. This metal also accelerates the activities of antioxidative enzymes [125, 126].

BRs effect on Ni stress in plants has been studied to understand the relationship between BRs and this stress (Table 2). One of these studies was carried out by Yusuf et al. [49]. They showed that seed germination and seedling growth were significantly reduced by Ni treatment, but HBL treatment enhanced germination percentage as well as shoot and root lengths in Ni-stressed seedlings. BRs confer tolerance against heavy metals either by reducing their uptake or by stimulating the antioxidative enzymes in *B. juncea* [127, 128]. The exogenous application of BRs in nickel-stressed *R. sativus* L., and *Triticum aestivum* L. plants enhanced the pool of antioxidant enzyme activity, thus alleviating the toxic effects of this stress [129].

5.4. Cu stress

Among the pollutants of agricultural soils, Cu has become increasingly hazardous due to its involvement in fungicides, fertilizers, and pesticides [130]. In addition, Cu present in excess has been known to decrease root biomass and alter plant metabolism [131, 132]. Sharma and Bhardwaj [127] demonstrated decrease in growth parameters of *Brassica juncea* grown under Cu stress. The reduction in growth parameters due to the Cu stress occurred as a result of decreasing mitotic activity and cell elongation [133, 134]. Moreover, Chen et al. [130] suggested a different opinion. They concluded that Cu-induced inhibition in root growth of rice seedlings was due to the stiffening of the cell wall. Moreover, excess of Cu ion leads to the generation of harmful ROS via the formation of free radicals [135].

Effects of exogenous application of BRs were studied on *Raphanus sativus* seedlings under Cu stress. It was found that 24-epiBL promoted the shoot and root growth by overcoming the Cu toxicity [136]. The growth-promoting effects of BRs on seedlings under Cu stress may be linked to the general ability of BRs to promote cell elongation and cell cycle progression [137, 138] as well as the stimulation of genes encoding xyloglucanses and expansins [139]. BRs applications also increase antioxidant enzyme activities [140, 141]. Increasing all parameters as a result of BRs application improves plant tolerance against Cu stress, and finally plant development (Table 2).

5.5. Zn stress

Zn is an essential microelement, the second most abundant transition metal after iron (Fe), and has a role in many metabolic reactions in plants [35, 36]. However, high concentrations of Zn are toxic, induce structural disorders, and cause functional problems in plants. At organism level, Zn stress causes reduced rooting capacity, growth, and at cellular level alters mitotic activity [37, 38]. It induces oxidative stress by promoting ROS production as a result of indirect consequence of Zn toxicity [142].

Application of BRs on plants alleviates Zn stress via increasing protein content and antioxidant enzyme activities (Table 2). Çağ et al. [143] reported that EBL application effectively enhanced the protein content in *Brassica oleraceae* cotyledons. Sharma et al. [144] also reported that pre-sowing treatments of HBL lowered the uptake of metal and enhanced the activities of antioxidative enzymes and protein concentration of *B. juncea* seedlings under Zn stress. Moreover, Ramakrishna and Rao [145] also reported that the application of 24-epiBL significantly alleviated the Zn-induced oxidative stress.

References	[114]	[146]	[30]	[147]	[141]	[145]
Effects on gene expression	---	---	---	---	---	---
Effects on protein content	---	Increased	Decreased protein content	---	Showed varying results	---

	Cd Stress	Cr Stress	Ni Stress	Cu Stress		Zn Stress
Effects on enzyme activities	Increased CAT, POX and SOD activities	Decreased CAT level	Decreased SOD but increased POD activities	Increased SOD, CAT and POX activities	Showed varying results	Increased
Effects on growth and/or cell division	Alleviated	Improved	Improved	Improved	Increased	Increased
Methods of metal and BRs applications	Soil amended with 50 μM Cd and foliage sprayed with 10^{-8} M HBL at 20 days after sowing. The plants were sampled at 30 days after sowing	Seeds were treated with eEBL (10^{-9} M) and 1.2 mM Cr(VI) (K_2CrO_4) solution at 7 days	Seeds were soaked for 8 h in different concentrations of 24-epiBL (0, 10^{-7}, 10^{-9} and 10^{-11} M). Then, seeds were grown under different (0, 0.5, 1.0, 1.5, and 2.0 mM) Ni concentrations	Seeds were grown in different levels of Cu^{2+} (50 or 100 mg kg^{-1} of soil. At 15 and 20 days stage, 10^{-5} mM HBL was applied. At 45 days, plants were collected	Seeds were grown under both 30 μM – 40 μM Cu and 2 μM HBR at 48 h and 72 h	EBR (0.5, 1, and 2 μM) and 5 mM of Zn were applied to seeds. Seven day old seedlings were collected
Plant Species	*Triticum aestivum* L.	*Raphanus sativus* L.	*B. juncea* L.	*Vigna radiata* L.	*Helianthus annuus* L.	*Raphanus sativus* L.

Table 2. Effects of BRs on plants subjected to Cd, Cr, Ni, Cu and Zn stresses. Dashes indicate that there are no results in study

6. Conclusion

Roots are very important plant organs whose architecture is determined by endogenous and environmental conditions to adjust water and nutrient uptake from soil [148, 149]. BRs, one of the plant hormones, have both positive and negative effects on root growth related to hormone concentrations [150]. Experimental condition is one of the most important factors for analysing BRs effects on root development. The procedures using BRs to alleviate abiotic systems generally are easy, time saving, and one of the most reliable systems [53]. Therefore, BRs open up new approaches for plant tolerance against hazardous environmental conditions [151]. Morphological, biochemical, and molecular analyses have been performed to analyse the effects of BRs. However, detailed analyses should be performed to investigate the relationship between abiotic stresses and BRs, especially gene expression studies will provide knowledge about interaction at molecular level in plants [152]. We tried to cite as many papers as possible. Yet we apologize to authors whose works are gone unmentioned in this chapter.

Acknowledgements

We are grateful to the Research Fund of Istanbul University for financial support (Projects 39824, 43403, and UDP54481).

Author details

Sevgi Marakli* and Nermin Gozukirmizi

*Address all correspondence to: sevgi.marakli@istanbul.edu.tr

Istanbul University, Faculty of Science, Department of Molecular Biology and Genetics, Vezneciler, Istanbul, Turkey

References

[1] Gowda VRP, Henry A, Yamauchi A, Shashidhar HE, Serraj R. Root biology and genetic improvement for drought avoidance in rice. Field Crops Res 2011;122:1–13. DOI: 10.1016/j.fcr.2011.03.001

[2] Atkinson NJ, Urwin PE. The interaction of plant biotic and abiotic stresses: from genes to the field. J Exp Bot 2012; 63:3523–3543. DOI: 10.1093/jxb/ers100

[3] Ahmadi N, Audebert A, Bennett MJ. The roots of future rice harvests. Rice 2014;7:29. DOI: 10.1186/s12284-014-0029-y

[4] Jaleel CA, Manivannan P, Lakshmanan GMA, Gomathinayagam M, Panneerselvam R. Alterations in morphological parameters and photosynthetic pigment responses of *Catharanthus roseus* under soil water deficits. Colloids Surfaces B: Biointerf 2008;61:298–303. DOI: 10.1016/j.colsurfb.2007.09.008

[5] Verslues PE, Agarwal M, Katiyar-Agarwal S, Zhu J, Zhu, JK. Methods and concepts in quantifying resistance to drought, salt and freezing, abiotic stresses that affect plant water status. Plant J 2006;45:523–39. DOI: 10.1111/j.1365-313X.2005.02593.x

[6] Abogadallah GM. Antioxidative defense under salt Stress. Plant Signal Behavior 2010;5:369-374. DOI: 10.4161/psb.5.4.10873

[7] Demirkiran A, Marakli S, Temel A, Gozukirmizi N. Genetic and epigenetic effects of salinity on *in vitro* growth of barley. Genet Molecul Biol 2013;36(4):566–70. DOI: 10.1590/S1415-47572013000400016

[8] Liu J, Gao H, Wang X, Zheng Q, Wang C, Wang X, Wang Q. Effects of 24-epibrassinolide on plant growth, osmotic regulation and ion homeostasis of salt stressed canola. Plant Biol 2014;16(2):440–50. DOI: 10.1111/plb.12052

[9] Marakli S, Temel A, Gozukirmizi N. Salt stress and homobrassinosteroid interactions during germination in barley roots. Not Bot Horti Agrobo 2014;42:446–52. DOI: 10.15835/nbha4229461

[10] West G, Inze D, Beemster GTS. Cell cycle modulation in the response of the primary root of *Arabidopsis* to salt stress. Plant Physiol 2004;135:1050–8. DOI: 10.1104/pp.104.040022

[11] Burssens S, Himanen K, van de Cotte B, Beeckman T, Van Montagu M, Inze D, Verbruggen N. Expression of cell cycle regulatory genes and morphological alterations in response to salt stress in *Arabidopsis thaliana*. Planta 2000;211:632–40. DOI: 10.1007/s004250000334

[12] Halperin SJ, Gilroy S, Lynch JP. Sodium chloride reduces growth and cytosolic calcium, but does not affect cytosolic pH, in root hairs of *Arabidopsis thaliana* L. J Exp Bot 2003;54:1269–80. DOI: 10.1093/jxb/erg134

[13] Sun F, Zhang W, Hu H, Li B, Wang Y, Zhao Y, Li K, Liu M, Li X. Salt modulates gravity signaling pathway to regulate growth direction of primary roots in *Arabidopsis*. Plant Physiol 2008;146:178–188. DOI: 10.1104/pp.107.109413

[14] Wang Y, Li K, Li X. Auxin redistribution modulates plastic development of root system architecture under salt stress in *Arabidopsis thaliana*. J Plant Physiol 2009;166:1637–45. DOI: doi:10.1016/j.jplph.2009.04.009

[15] Zolla G, Heimer YM, Barak S. Mild salinity stimulates a stress-induced morphogenic response in *Arabidopsis thaliana* roots. J Exp Bot 2010;61:211–224. DOI: 10.1093/jxb/erp290

[16] Ghosh D, Xu J. Abiotic stress responses in plant roots: a proteomics perspective, Front Plant Sci 2014;5:6. DOI: 10.3389/fpls.2014.00006

[17] Tester M, Davenport R. Na^+ tolerance and Na^+ transport in higher plants. Annal Bot 2003;91:503–27. DOI: 10.1093/aob/mcg058

[18] Cavalcanti FR, Santos Lima JPM, Ferreira-Silva SL, Viegas RA, Silveira JAG, Roots and leaves display contrasting oxidative response during salt stress and recovery in cowpea. J Plant Physiol 2007;164:591–600. DOI: 10.1016/j.jplph.2006.03.004

[19] Munns R, Tester M. Mechanisms of salinity tolerance. Ann Rev Plant Biol 2008;59:651–81. DOI: 10.1146/annurev.arplant.59.032607.092911

[20] Zhang JH, Huang WD, Liu YP, Pan QH. Effects of temperature acclimation pretreatment on the ultrastructure of mesophyll cells in young grape plants (*Vitis vinifera* L. cv. Jingxiu) under cross-temperature stresses. J Integrat Plant Biol 2005;47:959–70. DOI: 10.1111/j.1744-7909.2005.00109.x

[21] Simoes-Araujo JL, Rumjanek NG, Margis-Pinheiro M. Small heat shock proteins genes are differentially expressed in distinct varieties of common bean. Brazil J Plant Physiol 2003;15:33–41. DOI: 10.1590/S1677-04202003000100005

[22] Salveit ME. Chilling injury is reduced in cucumber and rice seedlings in tomato pericarp discs by heat-shocks applied after chilling. Postharvest Biol Technol 2001;21:169–77. DOI: doi:10.1016/S0925-5214(00)00132-0

[23] Allen DJ, Ort DR. Impact of chilling temperatures on photosynthesis in warm-climate plants. Trends Plant Sci 2001;6:36–42. DOI: doi:10.1016/S1360-1385(00)01808-2

[24] Lux A, Martinka M, Vaculik M, White PJ. Root responses to cadmium in the rhizosphere: a review. J Exp Bot 2011;62:21–37. DOI: 10.1093/jxb/erq281

[25] Mulligan CN, Yong RN, Gibbs BF. Remediation technologies for metal-contaminated soils and groundwater: an evaluation. Eng Geol 2001;60:193–207. DOI: doi:10.1016/S0013-7952(00)00101-0

[26] Uraguchi S, Mori S, Kuramata M, Kawasaki A, Arao T, Ishikawa S. Root-to-shoot cd translocation via the xylem is the major process determining shoot and grain cadmium accumulation in rice. J Exp Bot 2009;60:2677–88. DOI: 10.1093/jxb/erp119

[27] Cai Y, Cao F, Wei K, Zhang G, Wu F. Genotypic dependent effect of exogenous glutathione on cd-induced changes in proteins, ultrastructure and antioxidant defense enzymes in rice seedlings. J Hazard Mater 2011;192:1056–66. DOI: 10.1016/j.jhazmat.2011.06.011

[28] Ali S, Bai P, Zeng F, Cai S, Shamsi IH, Qui B, Wu F, Zhang G. The ecotoxicological and interactive effects of chromium and aluminum on growth, oxidative damage and antioxidant enzymes on two barley genotypes differing in Al tolerance. Environ Exp Bot 2011;70:185–91. DOI: 10.1016/j.envexpbot.2010.09.002

[29] Sharma I, Pati PK, Bhardwaj R. Effect of 28-homobrassinolide on antioxidant defence system in *Raphanus sativus* L. under chromium toxicity. Ecotoxicol 2011;20:862–74. DOI: 10.1007/s10646-011-0650-0

[30] Kanwar MK, Bhardwaj R, Chowdhary SP, Arora P, Sharma P, Kumar S. Isolation and characterization of 24-Epibrassinolide from *Brassica juncea* L. and its effects on growth, Ni ion uptake, antioxidant defense of *Brassica* plants and in vitro cytotoxicity. Acta Physiol Plant 2013;35:1351–62. DOI: 10.1007/s11738-012-1175-8

[31] Bal W, Kasprzak KS. Induction of oxidative damage by carcinogenic metal. Toxicol Lett 2002;127:55–62. DOI: doi:10.1016/S0378-4274(01)00483-0

[32] Bhakuni G, Dube B.K, Sinha P, Chatterjee C. Copper stress affects metabolism and reproductive yield of chickpea. J Plant Nutri 2009;32:703–11. DOI: 10.1080/01904160902743258

[33] Andre CM, Larondelle Y, Evers D. Dietary antioxidants and oxidative stress from a human and plant perspective: a review. Curr Nutri Food Sci 2010;6:2–12. DOI: 10.2174/157340110790909563

[34] Bouazizi H, Jouili H, Geitmann A, El Ferjani E. Cell wall accumulation of Cu ions and modulation of lignifying enzymes in primary leaves of bean seedlings exposed

to excess copper. Biol Trace Element Res 2011;139:97–107. DOI: 10.1007/s12011-010-8642-0

[35] Gayor A, Srivastava PS, Iqbal M. Morphogenic and biochemical responses of *Bacopa monniera* cultures to zinc toxicity. Plant Sci 1999;143:187–93. DOI: 10.1016/S0168-9452(99)00032-1

[36] Vaillant N, Monnet F, Hitmi A, Sallanon H, Coudret A. Comparative study of responses in four datura species to zinc stress. Chemosphere 2005;59:1005–13. DOI: 10.1016/j.chemosphere.2004.11.030

[37] Castiglione S, Franchin C, Fossat T, Lingua G, Torrigiani P, Biondi S. High zinc concentrations reduce rooting capacity and alter metallothionein gene expression in white poplar (*Populus alba* L. cv. Villafranca). Chemosphere 2007;67:1117–26. DOI: 10.1016/j.chemosphere.2006.11.039

[38] Tewari RK, Kumar P, Sharma PN. Morphology and physiology of zinc-stressed mulberry plants. J Plant Nutri Soil Sci 2008;171:286–94. DOI: 10.1002/jpln.200700222

[39] Choudhary SP, Yu JQ, Yamaguchi-Shinozaki K, Shinozaki K, Phan Tran L-S. Benefits of brassinosteroid crosstalk. Trends Plant Sci 2012;17:594–605. DOI: 10.1016/j.tplants.2012.05.012

[40] Mitchell JW, Mandava NB, Worley JE, Plimmer JR, Smith MV. Brassins: a family of plant hormones from rape pollen. Nature 1970;225:1065–66. DOI: 10.1038/2251065a0

[41] Grove MD, Spencer GF, Rohwededer WK, Mandava N, Worley JF, Warthen JR JD, Steffens GL, Flippen-Anderson JL, Cook JR JC. Brassinolide, a plant promoting steroid isolated from *Brassica napus* pollen. Nature 1979;281:216–7. DOI: 10.1038/281216a0

[42] Zhao B, Li J. Regulation of brassinosteroid biosynthesis and inactivation. J Integrat Plant Biol 2012;54:746–59. DOI: 10.1111/j.1744-7909.2012.01168.x.

[43] Wang M, Jiang WJ, Yu HJ. Effects of exogenous epibrassinolide on photosynthetic characteristics in tomato (*Lycopersicon esculentum* Mill) seedlings under weak light stress. J Agricult Food Chem 2010;58:3642–5. DOI: 10.1021/jf9033893

[44] Kurepin LV, Joo SH, Kim S-K, Pharis RP, Back TG. (2012). Interaction of brassinosteroids with light quality and plant hormones in regulating shoot growth of young sunflower and *Arabidopsis* seedlings. J Plant Growth Regulat 2012;31:156–64. DOI: 10.1007/s00344-011-9227-7

[45] Wang Q, Ding T, Gao L, Pang J, Yang N. Effect of brassinolide on chilling injury of green bell pepper in storage. Sci Horticult 2012;144:195–200. DOI: 10.1016/j.scienta.2012.07.018

[46] Anjum SA, Wang LC, Farooq M, Hussain M, Xue LL, Zou CM. Brassinolide application improves the drought tolerance in maize through modulation of enzymatic antioxidants and leaf gas exchange. J Agronomy Crop Sci 2011;197:177–85.

[47] Li YH, Liu YJ, Xu XL, Jin M, An LZ, Zhang H, Effect of 24-epibrassinolide on drought stress-induced changes in *Chorispora bungeana*. Biol Plant 2012;56:192–6. DOI: 10.1007/s10535-012-0041-2

[48] Mahesh B, Parshavaneni B, Ramakrishna B, Rao SSR. Effect of brassinosteroids on germination and seedling growth of radish (*Raphanus sativus* L.) under PEG-6000 induced water stress. Am J Plant Sci 2013;4:2305–13. DOI: 10.4236/ajps.2013.412285

[49] Yusuf M, Fariduddin Q, Hayat S, Hasan SA, Ahmad A. Protective responses of 28-Homobrssinolide in cultivars of *Triticum aestivum* with different levels of nickel. Arch Environ Contam Toxicol 2011;60:68–76. DOI: 10.1007/s00244-010-9535-0

[50] Yusuf M, Fariduddin Q, Ahmad A. 24-Epibrassinolide modulates growth, nodulation, antioxidant system, and osmolyte in tolerant and sensitive varieties of *Vigna radiate* under different levels of nickel: a shotgun approach. Plant Physiol Biochem 2012;57:143–53. DOI: 10.1016/j.plaphy.2012.05.004

[51] Saka H, Fujii S, Imakawa A.M, Kato N, Watanabe S, Nishizawa T, Yonekawa S. Effect of brassinolide applied at the meiosis and flowering stages on the levels of endogenous plant hormones during grain filling in rice plant (*Oryza sativa* L.). Plant Prod Sci 2003;6:36–42. DOI: 10.1626/pps.6.36

[52] Liu H, Guo T, Zhu Y, Wang C, Kang G. Effects of Epibrassinolide (epi-BR) application at anthesis on starch accumulation and activities of key enzymes in wheat grains. Acta Agronom Sinica 2006;32:924–30.

[53] Kartal G, Temel A, Arican E, Gozukirmizi N. Effects of brassinosteroids on barley root growth, antioxidant system and cell division. Plant Growth Regulat 2009;58:261–7. DOI: 10.1007/s10725-009-9374-z

[54] Yuan G-F, Jia C-G, Li Z, Sun B, Zhang L-P, Liu N, Wang Q-M. Effect of brassinosteroids on drought resistance and abscisic acid concentration in tomato under water stress. Sci Horticult 2010;126:103–8. DOI: 10.1016/j.scienta.2010.06.014

[55] Farooq M, Wahid A, Basra SMA, Islam-ud-Din. Improving water relations and gas exchange with brassinosteroids in rice under drought stress. J Agron Crop Sci 2009;195:262–9. DOI: 10.1111/j.1439-037X.2009.00368.x

[56] Khripach VA, Zhabinski VN, Khripach NB. New practical aspects of brassinosteroids and results of their ten year agricultural use in Russia and blakanes. In: Hayat S, Ahmad A, editors. Brassinosteroids; Bioactivity and Crop Productivity. Kluwer Academic Publisher, Dordrecht; 2003. pp. 189–230.

[57] Bajguz A. Effect of brassinosteroids on nucleic acid and protein content in cultured cells of *Chlorella vulgaris*. Plant Physiol Biochem 2000;38:209–15. DOI: 10.1016/S0981-9428(00)00733-6

[58] Felner M. Recent progress in brassinosteroid research: hormone perception and signal transduction. In: Hayat S, Ahmad A, editors. Brassinosteroids: Bioactivity and Crop Productivity. Kluwer Academic Publishers, Dordrecht; 2003. pp. 69–86.

[59] Kagale S, Divi UK, Krochko JE, Keller WA, Krishna P. Brassinosteroid confers tolerance in *Arabidopsis thaliana* and *Brassica napus* to a range of abiotic stresses. Planta 2007;225:353–64. DOI: 10.1007/s00425-006-0361-6

[60] Vardhini BV, Rao SSR. Amelioration of osmotic stress by brassinosteroids on seed germination and seedling growth of three varieties of sorghum. Plant Growth Regulat 2003;41:25–31. DOI: 10.1023/A:1027303518467

[61] Bordeleau LM, Prevost D. Nodulation and nitrogen fixation in extreme environments. Plant Soil 1994;161:115–25. DOI: 10.1007/BF02183092

[62] Ramos MLG, Gordon AJ, Minchin FR, Sprent JI, Parsons R. Effect of water stress on nodule physiology and biochemistry of a drought tolerant cultivar of common bean (*Phaseolus vulgaris* L.). Annals Bot 1999;83:57–63. DOI: 10.1006/anbo.1998.0792

[63] Serraj R, Sinclair TR, Purcell LC. Symbiotic N_2 fixation response to drought. J Exp Bot 1999;50:143–55. DOI: 10.1093/jxb/50.331.143

[64] Singh T, Kumar V. Nodulation and plant growth as influenced by growth regulators in some legumes. Acta Bot Ind 1989;17:177–81.

[65] Fedorova EE, Zhiznevskaya GY, Al'zhapparova ZK, Izmailov SF. Phytohormones in nitrogen-fixing nodules of leguminous plants. Fiziologiya I Biokhimiya Kul'Turnykh Rastenii. 1991;23:426–38.

[66] Vardhini BV, Rao SSR. Effect of brassinosteroids on nodulation and nitrogenase activity in groundnut (*Arachis hypogaea* L.). Plant Growth Regulat 1999;28:165–7. DOI: 10.1023/A:1006227417688

[67] Upreti KK, Murti GSR. Effects of brassinosteroids on growth, nodulation, phytohormone content and nitrogenase activity in French bean under water stress. Biologia Plant 2004;48:407–11. DOI: 10.1023/B:BIOP.0000041094.13342.1b

[68] Li KR, Feng CH. Effects of brassinolide on drought resistance of *Xanthoceras sorbifolia* seedlings under water stress. Acta Physiol Plant 2011;33:1293–300. DOI: 10.1007/s11738-010-0661-0

[69] Zhang M, Zhai Z, Tian X, Duan L, Li Z. Brassinolide alleviated the adverse effect of water deficits on photosynthesis and the antioxidant of soybean (*Glycine max* L.). Plant Growth Regulat 2008;56:257–64. DOI: 10.1007/s10725-008-9305-4

[70] Fariduddin Q, Khanam S, Hasan SA, Ali B, Hayat S, Ahmad A. Effect of 28-homobrassinolide on the drought stress-induced changes in photosynthesis and antioxi-

dant system of *Brassica juncea* L. Acta Physiol Plant 2009;31:889–97. DOI: 10.1007/s11738-009-0302-7

[71] Reddy AR, Chaitanya KV, Vivekanandan M. Drought induced responses of photosynthesis and antioxidant metabolism in higher plants. J Plant Physiol 2004;161:1189-1202. DOI:10.1016/j.jplph.2004.01.013

[72] Xiong L, Schumaker KS, Zhu JK. Cell signaling during cold, drought, and salt stress. Plant Cell 2002;14:S165–83. DOI: 10.1105/tpc.000596

[73] Vardhini BV, Sujatha E, Rao SSR. Brassinosteroids: alleviation of water stress in certain enzymes of sorghum seedlings. J Phytol 2011;3(10):38–43.

[74] Mazorra LM, Nunez M, Hechavarria M, Coll F, Sanchez-Blanco MJ. Influence of brassinosteroids on antioxidant enzymes activity in tomato under different temperatures. Biol Plant 2002;45:593–6. DOI: 10.1023/A:1022390917656

[75] Cao S, Xu Q, Cao Y, Qian K, An K, Zhu Y, Binzeng, H, Zhao H, Kuai B. Loss of function mutation in *DET2* gene lead to an enhanced resistance to oxidative stress in *Arabidopsis*. Physiol Plant 2005;123:57–66. DOI: 10.1111/j.1399-3054.2004.00432.x

[76] Eraslan F, Inal A, Gunes A, Alpaslan A. Impact of exogenous salicylic acid on the growth, antioxidant activity and physiology of carrot plants subjected to combined salinity and boron toxicity. Sci Horticult 2007;113:120–8. DOI: 10.1016/j.scienta.2007.03.012

[77] Tuteja N. Mechanisms of high salinity tolerance in plants. Methods Enzymol 2007;428:419–38. DOI: 10.1016/S0076-6879(07)28024-3

[78] Munns R, Tester M. Mechanisms of salinity tolerance. Ann Rev Plant Biol 2008;59:651–81. DOI: 10.1146/annurev.arplant.59.032607.092911

[79] Ashraf M, Foolad MR. Improving plant abiotic-stress resistance by exogenous application of osmoprotectants glycinebetaine and proline. Environ Exp Bot 2007;59:206–16.

[80] Daneshmand F, Arvin MJ, Kalantari KM. Acetylsalicylic acid ameliorates negative effects of NaCl or osmotic stress in *Solanum stoloniferum* in vitro. Biol Plant 2010;54:781–4. DOI: 10.1007/s10535-010-0142-8

[81] Ashraf M. Biotechnological approach of improving plant salt tolerance using antioxidants as markers. Biotechnol Adv. 2009;27:84–93. DOI: 10.1016/j.biotechadv.2008.09.003

[82] Talaat NB, Shawky BT. 24-Epibrassinolide alleviates salt-induced inhibition of productivity by increasing nutrients and compatible solutes accumulation and enhancing antioxidant system in wheat (*Triticum aestivum* L.). Acta Physiol Plant 2013;35:729–40. DOI: 10.1007/s11738-012-1113-9

[83] Divi UK, Rahman T, Krishna P. Gene expression and functional analyses in brassinosteroid-mediated stress tolerance. Plant Biotechnol J 2015, DOI: 10.1111/pbi.12396.

[84] Hayat S, Hasan SA, Yusuf M, Hayat Q, Ahmad A. Effect of 28-homobrassinolide on photosynthesis, fluorescence and antioxidant system in the presence or absence of salinity and temperature in *Vigna radiata*. Environ Exp Bot 2010;69:105–12. DOI: 10.1016/j.envexpbot.2010.03.004

[85] El-Mashad AAA, Mohamed HI. Brassinolide alleviates salt stress and increases antioxidant activity of cowpea plants (*Vigna sinensis*). Protoplasma 2012;249:625–35. DOI: 10.1007/s00709-011-0300-7

[86] Ding H-D, Zhu X-H, Zhu Z-W, Yang S-J, Zha D-S, Wu X-X. Amelioration of salt-induced oxidative stress in eggplant by application of 24-epibrassinolide. Biol Plant 2012;56(4):767–70. DOI: 10.1007/s10535-012-0108-0

[87] Sharma I, Ching E, Saini S, Bhardwaj R, Pati PK. Exogenous application of brassinosteroid offers tolerance to salinity by altering stress responses in rice variety Pusa Basmati-1. Plant Physiol Biochem 2013;69:17–26. DOI: 10.1016/j.plaphy.2013.04.013

[88] Fariduddin Q, Mir BA, Yusuf M, Ahmad A. 24-epibrassinolide and/or putrescine trigger physiological and biochemical responses for the salt stress mitigation in *Cucumis sativus* L. Photosynthetica 2014;52(3):464–74. DOI: 10.1007/s11099-014-0052-7

[89] Bajguz A. Brassinosteroid enhanced the level of abscisic acid in *Chlorella vulgaris* subjected to short-term heat stress. J Plant Physiol 2009;166:882–6. DOI: 10.1016/j.jplph.2008.10.004

[90] Stone P. The effects of heat stress on cereal yield and quality. In: Basra AS. (Ed.) Crop Responses and Adaptations to Temperature Stress. Binghamton, NY: Food Products Press; 2001. p. 302.

[91] Rane J, Chauhan H. Rate of grain growth in advanced wheat (Triticum aestivum) accession under late-sown environment. Ind J Agricult Sci 2002;72:581–5.

[92] Hussain SS, Mudasser M. Prospects for wheat production under changing climate in mountain areas of Pakistan: an econometric analysis. Agricult Sys 2007;94:494–501. DOI: 10.1016/j.agsy.2006.12.001

[93] Janeczko A, Oklestkova J, Pociecha E, Koscielniak J, Mirek M. Physiological effects and transport of 24-epibrassinolide in heat-stressed barley. Acta Physiol Plant 2011;33:1249–59. DOI: 10.1007/s11738-010-0655-y

[94] Yin H, Chen QM, Yi MF. Effects of short-term heat stress on oxidative damage and responses of antioxidant system in *Lilium longiflorum*. Plant Growth Regulat 2008;54:45–54. DOI: 10.1007/s10725-007-9227-6

[95] Bartwal A, Mall R, Lohani P, Guru SK, Arora S. Role of secondary metabolites and brassinosteroids in plant defense against environmental stresses. J Plant Growth Regulat 2012;32:216–32. DOI: 10.1007/s00344-012-9272-x

[96] Hu WH, Xiao YA, Zeng JJ, Hu XH. Photosynthesis, respiration and antioxidant enzymes in pepper leaves under drought and heat stresses. Biol Plant 2010;54:761–5. DOI: 10.1007/s10535-010-0137-5

[97] Asthir B, Koundal A, Bains NS. Putrescine modulates antioxidant defense response in wheat under high temperature stress. Biol Plant 2012;56:757–61. DOI: 10.1007/s10535-012-0209-1.

[98] Wu X, Yao X, Chen J, Zhu Z, Zhang H, Zha D. Brassinosteroids protect photosynthesis and antioxidant system of eggplant seedlings from high-temperature stress. Acta Physiol Plant 2014;36:251–61. DOI: 10.1007/s11738-013-1406-7

[99] Liu Y, Zhao Z, Si J, Di C, Han J, An L. Brassinosteroids alleviate chilling-induced oxidative damage by enhancing antioxidant defense system in suspension cultured cells of *Chorispora bungeana*. Plant Growth Regulat 2009;59:207–14. DOI: 10.1007/s10725-009-9405-9

[100] Liu YJ, Jiang HF, Zhao ZG, An LZ. Abscisic acid is involved in brassinosteroids-induced chilling tolerance in the suspension cultured cells from *Chorispora bungeana*. J Plant Physiol 2011;168:853–62. DOI: 10.1016/j.jplph.2010.09.020

[101] Wang B, Zeng G. Effect of epibrassinolide on the resistance of rice seedlings to chilling injury. J Plant Physiol Molecul Biol 1993;19:38–42.

[102] Kim KS, Sa JG. Effects of plant growth regulator, brassinolide, on seedling growth in rice (*Oryza sativa* L). Res Rep Rural Develop Admin Rice 1989;31(1):49–53.

[103] Hirai K, Fujii S, Honjo K. The effect of brassinolide on ripening of rice plants under low temperature condition. Japan J Crop Sci 1991;60(1):29–35.

[104] Krishna P. Brassinosteroid-mediated stress responses. J Plant Growth Regulat 2003;22:289–97. DOI: 10.1007/s00344-003-0058-z

[105] Burbulis N, Jonytiene V, Kupriene R, Blinstrubiene A. Changes in proline and soluble sugars content during cold acclimation of winter rapeseed shoots in vitro. J Food, Agricult Environ 2011;9:371–4.

[106] Hu WH, Wu Y, Zeng JZ, He L, Zeng QM. Chill-induced inhibition of photosynthesis was alleviated by 24-epibrassinolide pretreatment in cucumber during chilling and subsequent recovery. Photosynthetica 2010;48:537–44. DOI: 10.1007/s11099-010-0071-y

[107] Fariduddin Q, Yusuf M, Chalkoo S, Hayat S, Ahmad A. 28-homobrassinolide improves growth and photosynthesis in *Cucumis sativus* L. through an enhanced antiox-

idant system in the presence of chilling stress. Photosynthetica 2011;49(1):55–64. DOI: 10.1007/s11099-011-0022-2

[108] Oidaira H, Sano S, Koshiba T, Ushimaru T. Enhancement of antioxidative enzyme activities in chilled rice seedlings. J Plant Physiol 2000;156:811–3. DOI: 10.1016/S0176-1617(00)80254-0

[109] Ahammed GJ, Yuan HL, Ogweno JO, Zhou YH, Xia XJ, Mao WH, Shi K, Yu JQ. Brassinosteroid alleviates phenanthrene and pyrene phytotoxicity by increasing detoxification activity and photosynthesis in tomato. Chemosphere 2012;86:546–55. DOI: 10.1016/j.chemosphere.2011.10.038

[110] Hasan SA, Hayat S, Ahmad A. Brassinosteroids protect photosynthetic machinery against the cadmium induced oxidative stress in two tomato cultivars. Chemosphere 2011;84:1446–51. DOI: 10.1016/j.chemosphere.2011.04.047

[111] López-Millán AF, Sagardoy R, Solanas M, Abadía A, Abadía J. Cadmium toxicity in tomato (*Lycopersicon esculentum*) plants grown in hydroponics. Environ Exp Bot 2009;65:376–85. DOI: 10.1016/j.envexpbot.2008.11.010

[112] Hayat S, Alyemeni MN, Hasan SA. Foliar spray of brassinosteroid enhances yield and quality of *Solanum lycopersicum* under cadmium stress. Saudi J Biol Sci 2012;19:325–35. DOI: 10.1016/j.sjbs.2012.03.005

[113] Cao F, Liu L, Ibrahim W, Cai Y, Wu F. Alleviating effects of exogenous glutathione, glycinebetaine, brassinosteroids and salicylic acid on cadmium toxicity in rice seedlings (*Oryza sativa*). Agrotechnology 2013;2:107. DOI:10.4172/2168-9881.1000107

[114] Hayat S, Khalique G, Wani AS, Alyemeni MN, Ahmad A. Protection of growth in response to 28-homobrassinolide under the stress of cadmium and salinity in wheat. Int J Biol Macromolecul 2014;64:130–6. DOI: 10.1016/j.ijbiomac.2013.11.021

[115] Herbette S, Taconnat L, Hugouvieux V, Piette L, Magniette M-LM, Cuine S, Auroy P, Richaud P, Forestier C, Bourguignon J, Renou J-P, Vavasseur A, Leonhardt N. Genome-wide transcriptome profiling of the early cadmium response of *Arabidopsis* roots and shoots. Biochimie 2006;88:1751–65. DOI: 10.1016/j.biochi.2006.04.018

[116] Villiers F, Jourdain A, Bastien O, Leonhardt N, Fujioka S, Tichtincky G, Parcy F, Bourguignon J, Hugouvieux V. Evidence for functional interaction between brassinosteroids and cadmium response in *Arabidopsis thaliana.* J Exp Bot 2012;63:1185–200. DOI: 10.1093/jxb/err335

[117] Gill A, Sagoo MIS. Mutagenic potential and nutritive quality of turnip plants raised over chromium amended soils. Int J Bot 2010;6:127–31.

[118] Shanker AK, Cervantes TC, Loza-Tavera H, Avudainayagam S. Chromium toxicity in plants. Environ Int 2005;31:739–53. DOI: 10.1016/j.envint.2005.02.003

[119] Panda SK, Choudhury S. Chromium stress in plants. Brazil J Plant Physiol 2005;17(1): 95–102. DOI: 10.1590/S1677-04202005000100008

[120] Choudhary SP, Kanwar M, Bhardwaj R, Gupta BD, Gupta RK. Epibrassinolide ameliorates Cr (VI) stress via influencing the levels of indole-3-acetic acid, abscisic acid, polyamines and antioxidant system of radish seedlings. Chemosphere 2011;84:592–600. DOI: 10.1016/j.chemosphere.2011.03.056

[121] Sun Y, Veerabomma S, Abdel-Mageed HA, Fokar M, Asami T, Yoshida S, Allen RD. Brassinosteroid regulates fiber development on cultured cotton ovules. Plant Cell Physiol 2005;46:1384–91. DOI: 10.1093/pcp/pci150

[122] Welch RM. Micronutrient nutrition of plants. Crit Rev Plant Sci 1995;14:49–82. DOI: 10.1080/07352689509701922

[123] Marschner H. Mineral Nutrition of Higher Plants. 2nd Ed. Academic Press, London 1995.

[124] Kevrešan S, Petrovič N, Popovič M, Kandrač M. Effect of heavy metals on nitrate and protein metabolism in sugar beet. Biol Plant 1998;41:235–40. DOI: 10.1023/A:1001818714922

[125] Schickler H, Caspi H. Response of antioxidative enzymes to nickel and cadmium stresss in hyperaccumulator plants of the genus *Alyssum*. Physiol Plant 1999;105:39–44. DOI: 10.1034/j.1399-3054.1999.105107.x

[126] Prasad SM, Dwivedi R, Zeeshan M. Growth, photosynthetic electron transport, and antioxidant responses of young soybean seedlings to simultaneous exposure of nickel and UV-B stress. Photosynthetica 2005;43:177–85. DOI: 10.1007/s11099-005-0031-0

[127] Sharma P, Bhardwaj R. Effects of 24-epibrassinolide on growth and metal uptake in *Brassica juncea* L. under copper metal stress. Acta Physiol Plant 2007;29:259–63. DOI: 10.1007/s11738-007-0032-7

[128] Kanwar MK, Bhardwaj R, Arora P, Choudhary SP, Sharma P, Kumar S. Plant steroid hormones produced under Ni stress are involved in the regulation of metal uptake and oxidative stress in *Brassica juncea* L. Chemosphere 2012;86:41–9. DOI: 10.1016/j.chemosphere.2011.08.048

[129] Sharma I, Pati PK, Bhardwaj R. Effect of 24-epibrassinolide on oxidative stress markers induced by nickel-ion in *Raphanus sativus* L. Acta Physiol Plant 2011;33:1723–35. DOI: 10.1007/s11738-010-0709-1

[130] Chen LM, Lin CC, Kao CH. Copper toxicity in rice seedlings: changes in antioxidative enzyme activities, H_2O_2 level and cell wall peroxidase activity in roots. Bot Bull Acad Sin 2000;41:99–103.

[131] Sheldon AR, Menzies NW. The effect of copper toxicity on the growth and root morphology of Rhode grass (*Chloris gayana* Knuth.) in resin buffered solution culture. Plant Soil 2005;278:341–9. DOI: 10.1007/s11104-005-8815-3

[132] Chatterjee C, Sinha P, Dube BK, Gopal R. Excess copper induced oxidative damages and changes in radish physiology. Commun Soil Sci Plant Anal 2006;37:2069–76. DOI: 10.1080/00103620600770425

[133] Alaoui-Sossé B, Genet P, Vinit-Dunand F, Toussaint ML, Epron D, Badot PM. Effect of copper on growth in cucumber plants (*Cucumis sativus*) and its relationships with carbohydrate accumulation and changes in ion contents. Plant Sci 2004;166:1213–8. DOI: 10.1016/j.plantsci.2003.12.032

[134] Maksymiec W, Krupa Z. Effects of methyl jasmonate and excess copper on root and leaf growth. Biol Plant 2007;51:322–6. DOI: 10.1007/s10535-007-0062-4

[135] Ducic T, Polle A. Transport and detoxification of manganese and copper in plants. Brazil J Plant Physiol 2005;17:103–12. DOI: 10.1590/S1677-04202005000100009

[136] Choudhary SP, Bhardwaj R, Gupta BD, Dutt P, Gupta RK, Kanwar M, Biondi S. Enhancing effects of 24-epibrassinolide and Putrescine on the antioxidant capacity and free radical scavenging activity of *Raphanus sativus* seedlings under Cu ion stress. Acta Physiol Plant 2011;33:1319–33. DOI: 10.1007/s11738-010-0665-9

[137] Guo H, Li L, Ye H, Yu X, Algreen A, Yin Y. Three related receptor-like kinases are required for optimal cell elongation in *Arabidopsis thaliana*. Proc Nat Acad Sci USA 2009;106:7648–53. DOI: 10.1073/pnas.0812346106

[138] Gonzalez-Garcia MP, Vilarrasa-Blasi J, Zhiponova M, Divol F, Mora-Garcia S, Russinova E, Cano-Delgado AI. Brassinosteroids control meristem size by promoting cell cycle progression in *Arabidopsis* roots. Development 2011;138:849–59. DOI: 10.1242/dev.057331

[139] Gudesblat GE, Russinova E. Plants grow on brassinosteroids. Curr Opin Plant Biol 2011;14:530–7. DOI: 10.1016/j.pbi.2011.05.004

[140] Filová A, Sytar O, Krivosudská E. Effects of brassinosteroid on the induction of physiological changes in *Helianthus annuus* L. under copper stress. Acta Universitatis Agriculturae et Silviculturae Mendelianae Brunensis 2013;61:623–9. DOI: 10.11118/actaun201361030623

[141] Betul Burun K, Marakli S, Gozukirmizi N. Alleviation of copper stress with brassinosteroid in germinating sunflower roots. J Animal Plant Sci 2015; in press.

[142] Vázquez MN, Guerrero YR, González LM, de la Noval WT. Brassinosteroids and plant responses to heavy metal stress. An overview. Open J Metal 2013;3:34–41. DOI: 10.4236/ojmetal.2013.32A1005.

[143] Çağ S, Gören-Sağlam N, Çıngıl-Barış Ç, Kaplan E. The effect of different concentration of epibrassinolide on chlorophyll, protein and anthocyanin content and peroxidase activity in excised red cabbage (*Brassica oleraceae* L.) cotyledons. Biotechnol Biotechnol Equip 2007;21:422–5. DOI: 10.1080/13102818.2007.10817487

[144] Sharma P, Bhardwaj R, Arora N, Arora HK. Effect of 28- homobrassinolide on growth, Zn metal uptake and antioxidative enzyme activities in *Brassica juncea* L.

seedlings. Brazil J Plant Physiol 2007;19(3):203–10. DOI: 10.1590/S1677-04202007000300004

[145] Ramakrishna B, Rao SSR. 24-Epibrassinolide alleviated zinc-induced oxidative stress in radish (*Raphanus sativus* L.) seedlings by enhancing antioxidative system. Plant Growth Regulat 2012;68:249–59. DOI: 10.1007/s10725-012-9713-3

[146] Choudhary SP, Kanwar M, Bhardwaj R, Yu J-Q, Pan Tran L-S. Chromium stress mitigation by polyamine-brassinosteroid application involves phytohormonal and physiological strategies in *Raphanus sativus* L. PLoS ONE 2012;7(3):e33210. DOI: 10.1371/journal.pone.0033210

[147] Fariduddin Q, Khan TA, Yusuf M. Hydrogen peroxide mediated tolerance to copper stress in the presence of 28-homobrassinolide in *Vigna radiata*. Acta Physiol Plant 2014;36:2767–78. DOI: 10.1007/s11738-014-1647-0

[148] Malamy JE. Intrinsic and environmental response pathways that regulate root system architecture. Plant Cell Environ 2005;28:67–77.

[149] Vriet C, Russinova E, Reuzeau C. Boosting crop yields with plant steroids. Plant Cell 2012;24:842–57. DOI: 10.1105/tpc.111.094912

[150] Müssig C, Shin G-H, Altmann T. Brassinosteroids promote root growth in *Arabidopsis*. Plant Physiol 2003;133:1261–71. DOI: 10.1104/pp.103.

[151] Bajguz A, Hayat S. Effects of brassinosteroids on the plant responses to environmental stresses. Plant Physiol Biochem 2009;47:1–8. DOI:10.1016/j.plaphy.2008.10.002

[152] Shu H, Ni W, Guo S, Gong Y, Shen X, Zhang X, Xu P, Guo Q. Root-applied brassinolide can alleviate the NaCl injuries on cotton. Acta Physiol Plant 2015;37:75. DOI 10.1007/s11738-015-1823-x

10

Stress-Induced Proteins in Recalcitrant Seeds during Deep Dormancy and Early Germination

Marina I. Azarkovich

Abstract

The role and functions of dehydrins and heat shock proteins in seeds (especially in desiccation-sensitive recalcitrant seeds) are discussed.

During the periods of dehydration, a wide variety of plants can express dehydration proteins (dehydrins), which are also members of the plant late embryogenesis abundant (LEA) protein family. Dehydrins have been most extensively studied in relation to drought and cold stresses. Dehydrins are synthesized in orthodox seeds, and their development at the final stage is associated with genetically determined seeds drying. Dehydrins amount can reach 4% of total cell proteins. At the same time, dehydrins are found in desiccation-sensitive recalcitrant seeds. The following are the functions of dehydrins with experimental evidence: binding to water and ions, binding to phospholipids, radical scavenging, phosphorylation, binding to calcium, protection of enzymes, binding to cytoskeletons, and binding to nucleic acids.

It seems evident that, in the embryo cells, heat shock induced changes in gene expression and HSP synthesis but did not result in translational discrimination of mRNAs for non-heat shock proteins. Such specific feature has been observed earlier for orthodox seeds during their development and early stages of germination. It is suggested that such response to HS is characteristic just of embryo tissues; it could be considered an additional molecular mechanism improving embryo tolerance to unfavorable environmental conditions.

Keywords: Recalcitrant seeds, temperature stress, dehydrins, heat shock proteins

1. Introduction

Plant seeds are a unique object for studying the mechanisms of tolerance and adaptation to abiotic stresses. The seeds could not escape unfavorable environmental conditions but must adapt to overcome them, to retain a capability of germination and to fulfill their physiological destination, that is, species preservation and distribution. Furthermore, in seeds the developmental program of the individual plant is switched over from embryogenesis to germination; in the periods of seed development and germination, seed embryos, being subjected to the action of unfavorable conditions, must change cell activity on the level of gene expression and induce the synthesis of anti-stress proteins to protect themselves and overcome stress effects. On the other hand, embryos have to provide expression of genes for proteins required for further development, that is, germination *per se*.

Different environmental stresses to a plant may result in similar responses at the cellular and molecular level. This is due to the fact that the impacts of the stressors trigger similar strains and downstream signal transduction chains. Stressors such as drought (lack of environmental water), salinity (high osmolarity), and cold, especially frost (lack of liquid water), induce water deficiency [1]. All three forms of abiotic stress affect the water relations of a plant on the cellular as well as whole plant level, causing specific as well as unspecific reactions, damages and adaptation reactions. The stabilizing effect of liquid water on the membrane can be supported by compatible solutes and special proteins. At the metabolic level, osmotic adjustment by synthesis of low-molecular osmolytes (carbohydrates, betains, proline) can counteract cellular dehydration and turgor loss [1].

During the periods of dehydration, a wide variety of plants can express dehydration proteins (dehydrins), which are also members of the plant late embryogenesis abundant (LEA) protein family. For the first time, LEA proteins were characterized in ripening seeds [2]. But they are widely investigated now [3–5]. However, their precise role has not been clear yet. These proteins are supposed to protect cells from water loss; they can behave as molecular chaperones [2]. Now the expression of LEA proteins has been shown not only in ripen seeds but also in other plant (or animal) tissues. LEA proteins are induced by osmotic or cold stress, by exogenous ABA (abscisic acid) [6]. It is believed that LEA proteins are non-catalytic proteins. Practically, all LEA proteins are water-soluble hydrophilic heat-stable and unfolded proteins. LEA proteins are synthesized in orthodox seeds, and the development of which at the final stage is associated with genetically determined drying. Their amount can reach 4% of total cell proteins [7]. At the same time, LEA proteins are found in desiccation-sensitive recalcitrant seeds. That is why the study of recalcitrant seeds is of great interest. On the one hand, these seeds produce LEA proteins, and on the other hand, they are sensitive to desiccation. Therefore, a characterization of LEA proteins in recalcitrant seeds can help understanding their role, for example when plant cells are challenged by abiotic stress as cold or salt stress.

LEA proteins are classified into more than seven distinct groups [8]. Among the induced LEA proteins, dehydrins (group II of late embryogenesis abundant proteins) have been most commonly studied, but our knowledge of their fundamental role in the cell is incomplete.

2. Dehydrins—stress-induced proteins

The accumulation of dehydrin transcripts and proteins during dehydration and a correlation between the level of drought tolerance and the amount of dehydrin present strongly suggest that they are involved in protecting the plant from the negative effects of dehydration [9–11]. In recent years, dehydrins are studied very intensively, which resulted in the appearance of numerous reviews and research papers [12–18].

The dehydrins are very hydrophilic proteins and exhibit an unusually low level of recognizable structure [10, 19]. The main characteristic of the dehydrins is the presence of one or more lysine-rich stretches of 15 amino acids (the sequence EKKGIMKIKEKLPG), called the K motifs (or K-segments), that are predicted to form class A amphipathic alpha-helices [20, 21]. All dehydrins investigated have K-segment; it is usually used for preparing anti-dehydrin antibodies [12, 13]. The K-segment occurs 1–12 times, with 1 or 2 repeats being the most common. Since the K-segments can form amphipathic alfa-helices, they may stabilize membranes against dehydration. Dehydrins can also contain two other motifs: an N-terminal Y segment (consensus V/TDE/QYGNP) and a serine-rich S segment [11]. The Y-segment is similar to the nucleotide-binding domain found in bacteria. Typically, 1–3 Y-segments are present at the N-terminus of a dehydrin [11]. The S-segment contains a tract of Ser residues and is present in one or no copies in a dehydrin. Dehydrins extracted from drought-stressed plants are phosphorylated on these serines [22]. The role of phosphorylation is not clear but may be correlated with translocation of dehydrins to the nucleus [23] or the increased negative charge could enhance the ability of the protein to bind divalent cations such as zinc. It has been proposed that the short amphipathic K segments of dehydrin polypeptides interact with solvent-exposed hydrophobic patches on proteins undergoing partial denaturation and thereby interfere with protein aggregate formation [10]. Amphipathic K helixes could also be involved in binding membrane lipids and thus could play a more specific role in protecting lipoproteins, proteins located in membranes, and/or the membrane structure itself [10, 24].

According to the presence of the K-, S-, and Y-segments, dehydrins can be divided into five structural subgroups: Kn, SKn, KnS, YnKn, and YnSKn [10, 11]. Although not specifically included in the YSK naming system, dehydrins also contain Φ-segments, which are rich in Gly, Thr, and many other polar amino acids. This poorly conserved segment tends to be located between the Y-, S-, and K-segments.

Dehydrins are evolutionarily conserved among photosynthetic organisms including angiosperms, gymnosperms, ferns, mosses, liverworts, algae and cyanobacteria, as well as in some non-photosynthetic organisms such as yeast [17, 25]. Dehydrins seem to be very ancient proteins—a 40-kD protein was observed in Calothrix sp. strain PCC 7601, and in *Nostoc* sp. strain Mac-R2, an osmotic-induced doublet at 39 and 40 kD was observed. It appears that cyanobacteria produce a dehydrin-like protein under osmotic stress [25].

The expression of many dehydrins increases by the phytohormone abscisic acid (ABA), they are also referred as RAB proteins (Responsive to ABA) [26–28]. On the other hand, protein level of some dehydrins is regulated by low temperature only [22]. In particular, studies of stress-induced accumulation of five dehydrins in *Arabidopsis* revealed that two of them (LTI30 and COR47) accumulated primarily in response to low temperature. The level of another two proteins (ERD14 and LTI29) was upregulated by ABA and low temperatures, whereas RAB18 was only found in ABA-treated plants [22]. Borovsky and coworkers [15] have found that not only cold but also drought, freezing, and exogenous ABA treatment also result in accumulation of heat-stable dehydrin-like proteins in plant mitochondria. The most tolerant winter wheat and rye accumulate more of the heat-stable dehydrins than maize. Cold-induced accumulation of the heat-stable mitochondrial dehydrin-like proteins in all species studied was accompanied by increasing of plant cryotolerance [15].

Dehydrins have been most extensively studied in relation to drought and cold stresses [29–35]. Some experimental studies provide evidence that dehydrins contribute to freezing stress tolerance in plants and suggest that this could be partly due to their protective effect on membranes [36]. Dehydrins stabilize plant plasma and organellar membranes in conditions of stress, and further zinc may be an important co-factor in stabilization [37, 38].

The hypothesis that dehydrins have detergent and chaperone-like properties and may interact with compatible solutes to serve as structural stabilizers of macromolecules under conditions of water deficit [10] is now experimentally evident [17, 39].

Numerous *in vitro* functions have been described and proposed for dehydrins, including cryoprotection of lactate dehydrogenase (LDH), cryoprotection of purified protoplasts and chloroplasts, prevention of water loss, binding of excess ions, binding of nucleic acids, prevention of protein aggregation at elevated temperatures, and prevention of ice crystal growth. RAB18 (Y_2SK_2) accumulates in response to the phytohormone abscisic acid (ABA), drought, and low temperature [22]. LTI29 and COR47 (SK_3) accumulate primarily in response to low temperature but also to ABA and salt stress [22]. LTI30 (K_6) accumulates mainly under cold stress [22]. ERD14 (SK_2) is present in non-stressed plants although the protein level is upregulated by stress, particularly drought stress [22]. Dehydrin XERO (YSK_2) mRNA has been found to be constitutively expressed [39].

Of these functions, the most extensively studied has been the cryoprotection of LDH, where it has been shown that dehydrins are more effective than small molecules such as sucrose at protecting LDH activity from freeze-thaw damage [39].

The following are the functions of dehydrins with experimental evidence: binding to water and ions, binding to phospholipids, radical scavenging, phosphorylation, binding to calcium, protection of enzymes, binding to cytoskeletons, and binding to nucleic acids [17].

DHN genes are also expressed significantly in seeds toward the end of maturation, a period when the seed undergoes a developmentally programmed reduction in water content [22, 40]. The LEA/dehydrin proteins have been estimated to comprise up to 4% of the total seed protein [7].

3. Dehydrins in orthodox and recalcitrant seeds

The categories "orthodox" and "recalcitrant" seeds are used to describe the storage behavior of seeds. Orthodox seeds undergo maturation drying and are shed from the parent plant at low moisture contents. During maturation, they acquire desiccation tolerance, allowing them to be dried without irreversible damage. Because of this ability, seeds can be stored for long periods in cold and dry vaults. On the other hand, recalcitrant seeds do not undergo maturation drying and are shed at relatively high moisture contents.

It is believed that, in orthodox seeds, dehydrins favor the development of tolerance to osmotic stress at seed dehydration during their maturation [2, 6].

The lack of resistance of recalcitrant seeds to drying was thought to be the result of the absence of dehydrins [9, 41]. Subsequent studies, however, demonstrated that dehydrins are present in the fraction of heat-stable proteins in recalcitrant seeds of many woody species of the temperate climatic zone [41–43], including those of horse chestnut, but they were not found in species inhabiting humid tropics. This raises the question of the function performed by dehydrins in recalcitrant seeds.

In this connection, the investigation of dehydrin functions, properties, and distribution in recalcitrant seeds becomes actual. The data available so far indicate that dehydrins are present in some but not all recalcitrant seed species. They appear in response to low-temperature stress, an increase in the ABA content, and natural or artificial limited dehydration [42, 44–46].

Two tropical recalcitrant species exhibited a differential capacity to produce dehydrin-related proteins during seed maturation [43]. Dehydrins were present in axes and cotyledons of *Castanospermum australe* seeds during mid-maturation and at maturity. However, in *Trichilia dregeana*, no dehydrin-related polypeptides were detected in the mature seed. During the development of *C. australe* seeds, the nature of the dehydrin-related polypeptides accumulated in the cotyledons and axis changed and new polypeptides were detected in the mature seeds. The dehydrins present in cotyledons of mature seeds (31, 37, and 40 kDa) were still detectable after germination (i.e., in untreated seedlings) [43].

Kalemba and Pukacka [47] have compared mature and dried seeds from three species of the Acer genus, which differed in desiccation tolerance. Seeds of three Acer species—*Acer platanoides* L. (Norway maple, orthodox), *Acer pseudoplatanus* L. (sycamore, recalcitrant), and *Acer saccharinum* (silver maple, recalcitrant)—harvested during various cropping years were compared and analyzed to determine whether a genetic or an environmental influence dominated the regulation of dehydrin protein expression. The authors compared the appearance of dehydrins and small heat shock proteins in seedlots originating from cropping years that differed in weather conditions, which were monitored in detail during seed development. The experiments showed that three main dehydrins with approximate molecular weights of 46, 35, and 23 kDa were characteristic of all examined Acer species seeds. The three proteins were present in two seedlots of the orthodox Norway maple seeds and were noted either individually or together in all seedlots of recalcitrant Acer seeds. The modulation of dehydrin expression by environmental factors such as developmental heat sum and rainfall is supposed [47].

The presence of dehydrins alone in recalcitrant seeds is not sufficient to prevent desiccation injury [42, 9].

In two papers published recently [48, 49], the authors suggest an interesting point of view and "paradigm change": LEA proteins are synthesized as response to drought stress, which takes place at the end of maturation (of orthodox seeds). But it is known that dehydrins/LEA proteins are synthesized before maturation drying [6]. And maturation drying is genetically determined in orthodox seeds. During embryogenesis and maturation drying of orthodox seeds, dehydrins are synthesized and seeds may survive dry storage. "Typical" recalcitrant seeds [48, 49] such as *Avicennia* and *Brunguiera* have no dehydrins and high water content and they do not survive water loss during storage. Intermediate seeds (*Coffea, Barringtinia*) have no dehydrins during embryogenesis but have genes of dehydrins and may synthesize dehydrins after partial water loss in storage, so these seeds may survive [48, 49]. And "atypical" recalcitrant seeds (*Camellia, Castanea, Euterpe, Quercus*) accumulate some dehydrins at the end of embryogenesis and undergo a weak maturation drying and can be stored for a limited time [48, 49]. So, the role of dehydrins in recalcitrant seeds is not clear, and further comprehensive studies are required.

4. Dehydrins in horse chestnut seeds during dormancy and germination

The mature horse chestnut (*Aesculus hippocastanum* L.) seeds are not tolerant to dehydration (i.e., belong to the recalcitrant seed type), but they are resistant to long cold stress, for instance, during winter in the central Russia when seeds are under the snow cover. The second biological peculiarity of horse chestnut seeds is that they are in the state of deep dormancy, that is, they are incapable of germinating under favorable conditions without prior prolonged incubation at low above-zero temperatures and high water content, known as stratification.

The analysis of horse chestnut seed proteins made it possible to reveal a number of unique characteristics of their proteome, which distinguish recalcitrant horse chestnut seeds from the majority of orthodox seeds and is evidently related to the specific features of their physiological behavior. These characteristics include extremely low content of globulins, predominance of water-soluble proteins located in the cytosol, and the high level of non-compartmentalized heat-stable proteins [50, 52].

The presence of heat-stable proteins (i.e., proteins resistant to high-temperature denaturation) in horse chestnut seeds is the most interesting feature of the object of our study. The results obtained demonstrated that heat-stable proteins accumulate during maturation of horse chestnut seeds and are present in freshly picked seeds in considerable amounts. These proteins account for more than 30% of soluble cytosolic proteins of the axial organs and most of soluble proteins of cotyledons (more than 80%). It is possible that in some seeds certain heat-stable proteins may function as storage deposits, being a source of nitrogen for the seedling. The involvement of heat-stable proteins in some other specific functions related to their extreme temperature resistance, characteristic amino acid composition, and high cellular content cannot be ruled out.

In order to identify and characterize stress-induced dehydrin-like polypeptides in mature recalcitrant horse chestnut seeds, we analyzed the fraction of cytosolic heat-stable proteins isolated in the period of seed dormancy and germination [52]. In our experiments, in tissues of dormant seeds, dehydrin was identified by immunoblotting as a single bright band of the polypeptide with a mol wt of about 50 kD. During radicle emergence, not only the fraction of heat-stable proteins was reduced but also the proportion of dehydrins in it decreased. Apparently, recalcitrant seed germination is accompanied by dehydrin disappearance, like this occurs during orthodox seed germination.

Since horse chestnut seeds contain along with heat-stable proteins numerous heat-sensitive proteins, it was of importance to elucidate whether dehydrin-like proteins are present among the latter. Nobody analyzed this protein fraction for the presence of dehydrin-like proteins. In the fraction of heat-sensitive proteins of horse chestnut seeds, we detected a component with a mol wt of 80 kD, which cross-reacted with anti-dehydrin antibody, that is, it was immunologically revealed as dehydrin-like protein. The analysis of different subcellular protein fractions of axial organs, cotyledons, and cotyledon petioles showed that both 50-kD dehydrin and 80-kD dehydrin-like protein could be detected in the total homogenate protein. It seems important that one of the axis heat-sensitive polypeptides cross-reacted with the anti-dehydrin antibody but differed from heat-stable dehydrin by a higher molecular weight (about 80 kD). This is the first indication on the possible presence of dehydrin-like proteins among heat-sensitive polypeptides of horse chestnut seeds [52].

During stratification, there were no substantial changes in the content of 50-kD dehydrin. However, this protein disappeared rapidly during seed germination. Since a small heat shock protein ubiquitin plays an important role in cell protein degradation, it was of interest to elucidate whether there is any connection between dehydrin, which should disappear during seed germination, and ubiquitin, which marks specifically proteins destined for degradation. It turned out that 50-kD dehydrin cross-reacted with anti-ubiquitin antibody. This means that dehydrin ubiquitination might provide for dehydrin rapid disappearance after radicle protrusion. This fact seems very interesting because just 50-kD dehydrin disappeared firstly during horse chestnut seed germination.

Earlier we have established that embryo axes of dormant seeds are not in the dormant state and could germinate *in vitro* (72 h on water at 28 °C) in each period of stratification of dormant seeds [53]. Like during seed germination *in vivo*, 50-kD dehydrin was not detected in such axes germinated *in vitro*. As we have demonstrated [53], treatment of excised axes with ABA (10^{-5} M), cycloheximide, or α-amanitin suppressed their germination. Under these conditions of suppressed growth, 50-kD dehydrin remained in the axes and was easily detected by immunoblotting. The mechanism of ABA inhibitory action on excised axis germination remains unknown. It is also unknown whether ABA induces 50-kD dehydrin synthesis in axes or simply prevents its degradation, resulting in its presence in axes on the level detected before germination. However, at the comparison of ABA-treated axes with those treated with cycloheximide, or α-amanitin, the substantially stronger signal may be noted in the case of ABA treatment. This may indicate indirectly on the induction of dehydrin synthesis at germination inhibition by ABA [52].

The molecular weight of the single dehydrin we detected in the horse chestnut seeds was slightly above 50 kD. At the same time, other researchers reported other values [41, 42], for example, 12, 14, and 18 kD or 30–55 kD. So far it is difficult to explain such differences in dehydrin sizes. It is not excluded that this is related to some specific feature of horse chestnut plants or their populations in different countries. It seems more likely that these differences are related to the influence of different growth conditions. Thus, seeds collected in different years differed in the size of dehydrins: 12, 14, and 18 kD in seeds collected in 1992; 14 kD in seeds collected in 1993, and 23, 30, and 35–55 kD in seeds collected in 1994 [41, 42].

However, dehydrins comprise only a small part of heat-stable proteins of recalcitrant seeds. The functions of other heat-stable proteins accumulating in horse chestnut seeds in great amounts during dormancy and germination, the reasons for their extreme heat resistance, and their relation to the low-temperature action during stratification remain unclear. These questions are of great interest and require further study.

We believe that the presence of the proteome of horse chestnut seeds of hydrophilic proteins capable of holding moisture may be related to the recalcitrant character of these seeds. Heat-stable proteins also promote the resistance of highly watered seed to cold stress under the conditions of stratification, thereby keeping the embryo viable. From our point of view, the mere presence of a large amount of heat-stable proteins in dormant recalcitrant horse chestnut seeds is of great interest.

5. Heat shock proteins

Another interesting family of stress-induced proteins is the heat shock family [54, 55].

It is known that heat stress may inhibit growth and development of plants. This may be due to heat stress itself or due to formation of ROS and other oxidants induced by heat stress. Heating also changes the structure of proteins, up to complete denaturation, and alters the activity of many enzymes. Disturbances in membranes structure, in their permeability and fluidity may cause partial or total disintegration of cells. Heat shock, that is, a short-term increase in temperature by 8–10 °C above the optimum one, is well known to induce rapid transient and reversible changes in gene expression in all living organisms. These changes result in the synthesis of specific group of polypeptides called heat shock proteins and suppression (complete or partial) of the synthesis of "normal" cell proteins synthesized by the cells before heat shock. This general biological phenomenon was qualified as a response to heat shock [55, 56]. The universality and conserved character of this response indicate its importance in cell physiology. Since heat shock proteins accumulation at heat shock in the cells of plant vegetative organs and seedlings was correlated with the development of plant tolerance to subsequent action of lethal temperatures, it was suggested that the response to heat shock is a manifestation of molecular mechanisms providing cell heat tolerance [57–59].

The induction of transcription of heat shock proteins is a common phenomenon in all living organisms. These proteins are grouped in plants into five classes according to their approximate molecular weight: HSP100, HSP90, HSP70 (chaperones), HSP60 (chaperonins), small heat shock proteins (sHSPs), and ubiquitin (8.5 kD). Higher plants have at least 20 sHSPs and there might be 40 kinds of these sHSPs in one plant species. All of the major HSPs (that is, those expressed in very high amounts in response to heat and other stresses) have related functions: they ameliorate problems caused by protein misfolding and aggregation. However, each major

HSP family has a unique mechanism of action. Some promote the degradation of misfolded proteins (ubiquitin, and various ubiquitin-conjugating enzymes); others bind to different types of folding intermediates and prevent them from aggregating (HSP70 and HSP60); and still another (HSP100) promotes the reactivation of proteins that have already aggregated [55].

During two recent decades occurred after heat shock proteins discovery, the notions concerning their properties and role gradually widened and became much more complex. Now, we know that many heat shock proteins are molecular chaperons and facilitate protein–protein interactions in the cell, that heat shock proteins can be present in normal cells not subjected to stress, that they can be expressed at some developmental stages in the absence of heat shock, and that their synthesis can be induced by other stress types [57, 60–62]. Changes in gene expression resulting in heat shock proteins accumulation in the cells are evidently play an important physiological role and somehow protect cell structures and separate protein components against injuries induced by various stressors and increase cell tolerance and their adaptation to unfavorable environmental conditions [63]. Nevertheless, so far we did not decipher completely heat shock proteins functions and molecular mechanisms of their action. The possible heat shock effects on the synthesis of normal non-heat-shock proteins in various plant tissues are still less studied; it is not clear whether these effects are universal to the same degree as those of heat shock proteins gene expression. In many cases, heat shock suppressed the total protein synthesis and especially that of non-heat-shock proteins. This effect was evidently controlled on the level of translation because the normal pattern of protein synthesis was rapidly restored after the change in the temperature and transcription suppression with α-amanitin. It is known that heat shock inhibited total protein synthesis in the vegetative organs of seedlings and plant cell cultures. This was related not to the inhibition of non-heat-shock mRNA synthesis but to incapability of these mRNAs to be translated under heat shock conditions. However, this specific response to heat shock was not evidently universal because it was not observed in seed embryos during seed development and germination. Thus, it has been shown for soybean and common bean seeds that the synthesis of storage proteins and many other non-heat shock proteins and their mRNAs was not reduced and even was activated under heat shock and occurred along with the synthesis of heat shock proteins [64, 65]. In embryos of wheat [66], sorghum [67], maize [68], and pea [69], heat shock markedly activated protein synthesis during early stages of germination, and heat shock proteins synthesis was induced simultaneously with the synthesis of the bulk of proteins produced by embryo tissues before heat shock. On the basis of these facts, it was concluded that such a specific response of protein synthesis to heat shock observed in seed embryos of many grasses during seed development and germination could have a definite physiological significance; this could be a manifestation of additional molecular mechanisms improving embryo tolerance to unfavorable environmental conditions and, as a consequence, their viability [69].

Small heat shock proteins function as intracellular chaperones for other proteins. They play an important role in protein–protein interactions such as folding. They assist in the establishment of proper protein conformation and prevent unwanted protein aggregation. By helping to stabilize partially unfolded proteins, HSPs aid in transporting proteins across membranes within the cell [57]. HSPs recognize and bind to other proteins when these other proteins are in non-native conformations. The non-native conformations of these proteins could be due to protein-denaturing stress or due to immature peptides folded, assembled, or localized to an appropriate cellular compartment. In the presence of ATP at normal physiological concentrations, sHSPs change their conformation and releases denatured protein, allowing other molecular chaperones such as HSP70 to renature the protein and renew its biological activity. In the absence of ATP, sHSPs such as α-crystallin are more efficient than HSP70 in preventing stress-induced protein aggregation. *In vitro*, sHSPs selectively bind and stabilize proteins and prevent their aggregation at elevated temperatures in an ATP-independent way and protect enzymes against heat-induced inactivation.

In recalcitrant of chestnut (*Castanea sativa*) seeds, a 20-kD protein has been purified from cotyledons, where it accumulates at levels comparable to those of major seed storage proteins [58]. This protein, termed Cs sHSP 1, forms homododecameric complexes under non-denaturing conditions and appears to be homologous to cytosolic class I small heat shock proteins (sHSPs) from plant sources. *In vitro* evidence has been obtained that the isolated protein can function as a molecular chaperone; it not only increases, at stoichiometric levels, the renaturation yields of chemically denatured citrate synthase but also prevents the irreversible thermal inactivation of this enzyme. Although a role in desiccation tolerance has been hypothesized for seed sHSPs, this does not seem to be the case for Cs sHSP 1. The presence of immunologically related proteins in orthodox and recalcitrant seeds of 13 woody species has been investigated [58]. The results indicate that the presence of Cs sHSP 1- like proteins, even at high levels, is not enough to confer desiccation tolerance and that the amount of these proteins does not furnish a reliable criterion to identify desiccation-sensitive seeds. Additional proteins or mechanisms appear necessary to keep the viability of orthodox seeds upon shedding. The synthesis of small HSPs may be an important part of seed developmental program [61]. As detected in pea [70], and in other species, the class I sHSPs accumulate in Arabidopsis seeds at midmaturation and decline during germination [61]. The similarity of sHSP regulation in such diverse species supports the conclusion that there is a selective advantage to this pattern of sHSP accumulation. The correlation of sHSP expression with the development of desiccation tolerance and dormancy suggests a possible role for the sHSPs in either or both of these processes, as has been previously hypothesized [62, 71].

It was found that one major small heat shock protein existed with a molecular mass of 22 kDa and was detectable at high concentrations in seeds of three Acer species (orthodox and recalcitrant). After the seeds were dried, the content of this protein significantly increased. The largest content of this protein was observed in the oldest seeds, especially in embryonic axes. The proteins identified may play a protective role during water deficit and storage [47].

6. Heat shock proteins in recalcitrant horse chestnut seeds

In the work [72], first data are obtained about functioning of molecular mechanisms providing for perception and transduction of heat signal and inducing heat shock proteins synthesis in the cells of embryo axes of dormant recalcitrant seeds, which are in metabolically active state but could not germinate. Embryo axes, cotyledon pieces, and cotyledon petioles were excised from recalcitrant horse chestnut (*Aesculus hippocastanum* L.) seeds in different times after the start of cold stratification and incubated at 28 or 40 °C on the medium containing ^{35}S-methionine and 50 μg/ml chloramphenicol for 4 h. The total rate of protein synthesis, the pattern of synthesized polypeptides, their distribution between subcellular fractions, and their relation to heat denaturing (5°min at 75 °C) were assessed. It was established that in all embryo parts, especially in axes, heat shock markedly activated protein synthesis in the beginning of stratification and to a lesser degree after ten weeks of stratification; heat shock suppressed protein synthesis at radicle emersion and especially during axial organ growth. Independently of the duration of stratification, which gradually released seed deep dormancy, isolated axes, cotyledons, and cotyledon petioles synthesized *in vivo* numerous diverse polypeptides at both 28 and 40°C. Newly synthesized polypeptides were present in the fractions of cell structures and cytosol; they differed in molecular weights, the intensity of labeling, and tolerance to heat denaturing. None of the dominating polypeptides present initially in all embryo parts and belonging mainly to heat-stable proteins was synthesized either at 28°C or 40°C. Some proteins synthesized at 40°C could be considered heat shock proteins because they were not synthesized at 28°C or their synthesis was markedly activated by heat shock. No less than 10 proteins behaved as obvious heat shock proteins; they were predominantly heat-sensitive soluble cytosolic proteins. All heat shock proteins, except those with mol wts of 220 and 34 kD, were highly labeled proteins. Some of them were characteristic of cell structures (220, 90, 20, and 18 kD); others were detected only in the cytosol (100, 80, and 34 kD). The synthesis of heat shock proteins did not depend on transcription and occurred on pre-existing mRNAs. An embryo capability of responding to heat shock did not depend on the seed physiological state and their germinability; it was similarly manifested in stratified and non-stratified seeds [72].

The analysis of heat shock action on gene expression in embryo tissues in dormant horse chestnut seeds demonstrated translation activation leading to the induction of a wide set of heat shock protein synthesis at the maintenance of the bulk of normal (non-shock) cellular protein synthesis. Due to this specific embryo response to heat shock, at early stages of germination and even under unfavorable conditions, embryo cells retain a capability of continuation or supporting on the sufficient level of the synthesis of proteins required for cell activity switching over to new developmental program, from embryogenesis to germination, and thus increase the reliability of germination. We believe that the absence of discrimination of non-heat shock mRNA translation during heat shock is specific to embryo tissues and could be considered an additional mechanism facilitating seed adaptation to unfavorable environmental conditions and successful germination.

Horse chestnut seeds survive successfully (and even require for deep dormancy release) the period of long cooling (18–22 weeks), retaining the high water content in their cells, and thus they are well adapted to overcome or correct damages arising under these conditions. However, according to current knowledge, heat shock proteins just fulfill this protective function. Therefore, it might be that, in mature dormant horse chestnut seeds, some amounts of required heat shock proteins are already present. These heat shock proteins could be synthesized and accumulated under the influence of elevated temperatures in the embryo cells during seed development or after their falling, that is, in response to heat shock, and they were preserved in the cells after seed entry into deep dormancy in the metabolically active state;

they could improve embryo tolerance to unfavorable environmental conditions during stratification and thus increase seed viability.

We believe that the absence of discrimination of non-heat shock mRNA translation during heat shock is specific to embryo tissues and could be considered an additional mechanism facilitating seed adaptation to unfavorable environmental conditions and successful germination.

Induction of heat shock proteins synthesis is a universal feature of the response to heat shock. According to our data, all tissues isolated from the embryos of dormant recalcitrant horse chestnut seeds responded to heat shock not only by continuation of non-heat shock protein synthesis but also by induction of similar sets of heat shock proteins. Two observations are of interest. We did not observe any dramatic changes in the set of heat shock proteins synthesized by isolated axes in the response to heat shock in the course of stratification, which evidently facilitate seed deep dormancy release. Moreover, heat shock proteins synthesis was readily detected in axes excised from non-stratified seeds, that is, it did not depend on seed capability of germination. This indicates independence of heat shock proteins synthesis at heat shock in the course of stratification of the embryo physiological state and its capability of germination. At the same time, some of our data indicate that stratification still somehow affected embryo physiological state. Thus, in the course of stratification, sensitivity of isolated axis growth to abscisic acid and indol-3-acethyc acid decreased [53], some characteristics of the proteome changed [50], and sensitivity of isolated axis translation to heat shock changed as well. However, this did not affect embryo tissue capacity to respond to heat shock. It is likely that signals providing for dormancy state, its release, and seed germination do not interact with signals leading to heat shock proteins synthesis induction. Furthermore, heat shock proteins' gene expression in isolated axes of dormant recalcitrant horse chestnut seeds was not dependent on transcription and was controlled predominantly on the level of translation. This means that all components required for the complex molecular mechanism of heat shock proteins gene expression were present in axis cells of mature seeds and were evidently produced still during seed development, may be under the influence of elevated temperatures. After mature seed falling, this mechanism is retained in the cells in the functionally active state and is capable of a rapid initiation of heat shock proteins synthesis in response to heat shock or another stress. However, the realization of this mechanism of heat shock proteins accumulation under natural conditions of stratification seems not very probable because the rate of protein synthesis under low temperature is low and heat signal is absent. Nevertheless, horse chestnut seeds survive successfully the period of long cooling (18–22 weeks), retaining the high water content in their cells and thus they are well adapted to overcome or correct damages arising under these conditions. However, according to current knowledge, heat shock proteins just fulfill this protective function. Therefore, it might be that, in mature dormant horse chestnut seeds, some amounts of required heat shock proteins are already present. These heat shock proteins could be synthesized and accumulated under the influence of elevated temperatures in the embryo cells during seed development or after their falling, that is, in response to heat shock, and they were preserved in the cells after seed entry into deep dormancy in metabolically active state; they could improve embryo tolerance to unfavorable environmental conditions during stratification and thus increase seed viability. This suggestion is supported by our observation that one of heat shock proteins, ubiquitin, was present in dormant horse chestnut seeds in functionally active state (i.e., in association with dehydrins) [52].

7. Conclusions

Plants cannot avoid the exposure to different abiotic factors but adapt morphologically and physiologically by some other mechanisms. Almost all stresses induce the production of groups of proteins called dehydrins and heat shock proteins (HSPs), which comprise several evolutionarily conserved protein families. Accumulation of dehydrins can be induced not only by drought but also by cold, salinity, and treatment with abscisic acid.

Since HSPs accumulation at heat shock in the cells of plant vegetative organs and seedlings was correlated with the development of plant tolerance to subsequent action of lethal temperatures, it was suggested that the response to heat shock is a manifestation of molecular mechanisms providing cell heat tolerance.

Heat stress induces the known genes for HSPs and chaperones: *hspA, groES, groEL1, groEL2, dnaJ, htpG, dnaK2, clpB1*, and *htrA* for protease, *sigB* for the σ-factor of RNA polymerase, *hik34* for sensory histidine kinase, *sodB* for superoxide dismutase, and some other genes. Using DNA microarrays, it has been shown that none of the aforementioned genes is induced by heat stress specifically [73, 74]. Expression of these genes is induced by high osmolarity, NaCl, oxidative stress (H_2O_2), high light, and UV-B. This phenomenon has been observed earlier. However, before the application of DNA microarrays, the information on HSPs that respond to various stresses has been fragmented and limited to studies of individual genes. Now it is clear that the genes whose transcription is specifically induced by high temperatures are classified as unknown. The entire list of genes of the genuine HSPs is limited to following titles: *sll0441, sll0688, sll1106, sll1884, slr0852, slr0095*, and *slr1597* [75]. The remaining genes that are induced by heat shock belong rather to a group of the general-stress-responsive (GSR) genes [76], and HSP may be renamed as general stress protein (GSP) [76, 77].

Acknowledgements

The work was supported by Presidium of RAS (program Molecular and Cell Biology).

Author details

Marina I. Azarkovich*

Address all correspondence to: m-azarkovich@ippras.ru

Timiryazev Institute of Plant Physiology, Russian Academy of Sciences, Moscow, Russia

References

[1] Beck EH, Fettig S, Knake C, Hartig K, and Bhattarai T: Specific and unspecific responses of plants to cold and drought stress. J. Biosci. 2007; 32:501–510.

[2] Dure L. III, Greenway SC, and Galau GA. Developmental biochemistry of cottonseed embryogenesis and germination – changing messenger ribonucleic acid population as shown by *in vitro* and *in vivo* protein synthesis. Biochemistry.1981; 20:4162–4168.

[3] Wise MJ and Tunnacliffe A. POPP the question: What do LEA proteins do? Trends Plant Sci. 2004; 9:13–17.

[4] Goyal K, Walton LJ and Tunnacliffe A. LEA proteins prevent protein aggregation due to water stress. Biochem. J. 2005; 388:151–157.

[5] Battaglia M, Olvera-Carrillo Y, Garciarrubio A, Campos F and Covarrubias AA. The enigmatic LEA proteins and other hydrophillins. Plant Physiol. 2008; 148:6–24, doi: 10.1104/pp.108.120725.

[6] Galau GH, Huges DW and Dure L. III. Abscisic acid induction of cloned cotton late embryogenesis-abundant (LEA) mRNAs. Plant Mol. Biol. 1986; 7:150–170.

[7] Roberts JK, DeSimone NA, Lingle WL and Dure L. III. Cellular concentrations and uniformity of cell-type accumulation of two LEA proteins in cotton embryos. Plant Cell. 1993; 5:769–780.

[8] Hundermark M and Hincha DK. LEA (late embryogenesis abundant) proteins and their encoding genes in *Arabidopsis thaliana*. BMC Genomics. 2008; 9:118–124.

[9] Bradford KJ and Chandler PM. Expression of "dehydrin-like" proteins in embryos and seedlings of *Zizaniapalustris* and *Oryza sativa* during dehydration. Plant Physiol. 1992; 93:488–494.

[10] Close TJ. Dehydrins: emergence of a biochemical role of a family of plant dehydration proteins. Physiol. Plantarum. 1996; 97:795–803.

[11] Close TJ. Dehydrins: a commonality in the response of plants to dehydration and low temperature. Physiol. Plant. 1997; 100:291–296, doi: 10.1111/j.1399-3054.1997.tb. 04785.x

[12] Allagulova CR, Gimalov FR, Shakirova FM and Vakhitov VA. The plant dehydrins: structure and putative functions. Biochemistry. 2003; 68:945–951.

[13] Rorat T. Plant dehydrins – tissue location, structure and function. Cell Mol. Biol. Lett. 2006; 11:536–556, doi: 10.2478/s11658-006-0044-0

[14] Shinozaki K and Yamaguchi-Shinozaki K. Gene networks involved in drought stress response and tolerance. Journal of Experimental Botany. 2007; 58:221–227, doi: 10.1093/jxb/erl164

[15] Borovskii GB, Stupnikova IV, Antipina AI, Vladimirova SV and Voinikov VK. Accumulation of dehydrin-like proteins in the mitochondria of cereals in response to cold, freezing, drought and ABA treatment. BMC Plant Biol. 2002; 2:5–11.

[16] Kosova K, Vitamavas P and Prasil IT. The role of dehydrins in plant response to cold. Biol. Plant. 2007; 51:601–617.

[17] Hara M. The multifunctionality of dehydrins. Plant Signaling and Behavior. 2010; 5:503–508.

[18] Hanin M, Brini F, Ebel C, Toda Y, Takeda S and Masmoudi K. Plant dehydrins and stress tolerance. Plant Signaling and Behavior.2011; 6:1503–1509, doi:10.4161/psb.6.10.17088

[19] Kovacs D, Kalmar E, Torok Z and Tompa P. Chaperon activity of ERD10 and ERD14, two disordered stress-related plant proteins. Plant Physiology. 2008; 147:381–390, doi:10.1104/pp.108.118208

[20] Close TJ, Fenton RD and Moonan F. A view of plant dehydrins using antibodies specific to the carboxy terminal peptide. Plant Mol. Biol.1993; 23:279–286.

[21] Dure III LS. A repeating 11-mer amino acid motif and plant desiccation. The Plant Journal. 1993;3: 363–9.

[22] Nylander M, Svensson J, Palva ET, Welin B. Stress-induced accumulation and tissue specific localization of dehydrins in *Arabidopsis thaliana*. Plant Mol. Biol. 2001; 45:263–279.

[23] Goday A, Jensen AB, Culiáñez-Macià FA, Albà MM, Figueras M, Serratosa J, Torrent M, Pagès M. The maize abscisic acid-responsive protein Rab17 is located in the nucleus and interacts with nuclear localization signals. Plant Cell. 1994; 6:351–60.

[24] Koag MCh, Fenton RD, Wilkens S, Close TJ. The binding of maize DHN1 to lipid vesicles. Gain of structure and lipid specificity. Plant Physiol. 2003; 131:309–316, doi: 10.1104/pp.011171.

[25] Close TJ, Lammers PJ. An osmotic stress protein of cyanobacteria is immunologically related to plant dehydrins. Plant Physiol. 1993; 101:773–779.

[26] Close TJ, Kortt AA, Chandler PM. A cDNA-based comparison of dehydration-induced proteins (dehydrins) in barley and corn. Plant Molecular Biology. 1989; 13:95–108.

[27] Sharma AD, Vasudeva R, Kaur R. Expression of a boiling-stable protein (BsCyp) in response to heat shock, drought and ABA treatments in *Sorghum bicolor*. Plant Growth Regulation. 2006; 50:249–254.

[28] Chung S, Parish RW. Combinatorial interactions of multiple cis-elements regulating the induction of the Arabidopsis XERO2 dehydrin gene by abscisic acid and cold. The Plant Journal. 2008; 54:15–29.

[29] Bravo LA, Close TJ, Corcuera LJ, Guy CL. Characterization of an 80-kDa dehydrin-like protein in barley responsive to cold acclimation. Physiol. Plant. 1999; 106:177–183.

[30] Ismail AM, Hall AE, Close TJ. Purification and partial characterization of a dehydrin involved in chilling tolerance during seedling emergence of cowpea. Plant Physiol. 1999; 120:237–244.

[31] Marian CO, Krebs SL, Arora R. Dehydrin variability among rhododendron species: a 25-kDa dehydrin is conserved and associated with cold acclimation across diverse species. New Phytologist. 2003; 161:773–780, doi: 10.1111/j.1469-8137.2003.01001.x

[32] Kosová K, Vitámvás P, Prášil IT. Wheat and barley dehydrins under cold, drought and salinity – what can LEA-II proteins tell us about plant stress response? Frontiers in Plant Science. 2014; 5:Article 343:1–6, doi: 10.3389/fpls.2014.00343

[33] Wisniewski M, Webb R, Balsamo R, Close TJ, Yu XM, Griffith M. Purification, immunolocalization, cryoprotective, and antifreeze activity of PCA60: a dehydrin from peach (*Prunus persica*). Physiol. Plantarum. 1999; 105:600–608, doi: 10.1034/j. 1399-3054.105402.x

[34] Puhakainen T, Hess MW, Mäkelä P, Svensson J, Heino P, Palva ET. Overexpression of multiple dehydrin genes enhances tolerance to freezing stress in *Arabidopsis*. Plant Molecular Biology. 2004; 54:743–753.

[35] Wellin BV, Olson A, Palva ET. Structure and organization of two closely-related low-temperature-induced *Dhn/Lea/Rab*-like genes in *Arabidopsis thaliana* (L.)Heynh. Plant Mol. Biol. 1995; 29: 391–395.

[36] Chakrabortee S, Boschetti C, Walton LJ, Sarkar S, Rubinsztein DC, Tunnacliffe A. Hydrophilic protein associated with desiccation tolerance exhibits broad protein stabilization function. PNAS. 2007; 104:18073–18078.DOI:10.1073/pnas.0706964104

[37] Hara M, ShinodaY, TanakaY, Kuboi T. DNA binding of citrus dehydrin promoted by zinc ion. Plant, Cell and Environment. 2009; 32: 532–541, doi: 10.1111/j. 1365-3040.2009.01947.x

[38] Rahman LN, Bamm VV, Voyer JAM, Smith GST, Chen L, Yaish MW, Moffatt BA, John R. Dutcher JR, Harauz G. Zinc induces disorder-to-order transitions in free and membrane-associated *Thellungiella salsuginea* dehydrins TsDHN-1 and TsDHN-2: a solution CD and solid-state ATR-FTIR study. Amino Acids. 2011; 40:1485–1502, doi: 10.1007/s00726-010-0759-0

[39] Huges S, Graether S. Cryoprotective mechanism of a small intrinsically disordered dehydrin protein. Protein Science. 2011; 20:42–50.

[40] Blackman SA, Obendorf RL, Leopold AC. Maturation proteins in desiccation tolerance of developing soybean seeds. Plant Physiol. 1992; 100:225–230.

[41] Finch-Savage WE, Pramanik SK, Bewly JD. The expression of dehydrin proteins in desiccation-sensitive (recalcitrant) seeds of temperate trees. Planta. 1994; 193:478–485.

[42] Farrant JM, Pammenter NW, Berjak P, Farnsworth EJ, Vertucci CW. Presence of dehydrin-like proteins and level of abscisic acid in recalcitrant (desiccation sensitive) seeds may be related to habitat. Seed Sci. Res. 1996; 6:175–182.

[43] Han B, Berjak P, Pammenter N, Farran tJ, Kermode AR. The recalcitrant plant species, *Castanospermum australe* and *Trichilia dregeana,* differ in their ability to produce dehydrin-related polypeptides during seed maturation and in response to ABA or water-deficit-related stresses.Journal of Experimental Botany.1997;48:1717–172.

[44] Šunderlíková V, Salaj J, Kopecky D, Salaj T, Wilhem E, Marušíková I. Dehydrin genes and their expression in recalcitrant oak (*Quercus robur*) embryos. Plant Cell Rep. 2009; 28:1011–1021, doi: 10.1007/s00299-009-0710-6

[45] Hinninger C, Caillet V, Michoux F, Amor MB, Tanksley S, Lin C, McCarthy J. Isolation and characterization of cDNA encoding three dehydrins expressed during *Coffea canephora* (Robusta) grain development. Annals of Botany. 2006; 97: 755–765, doi: 10.1093/aob/mcl032

[46] Delahaie J, Hundertmark M, Bove J, Leprince O, Rogniaux H, Buitink J. LEA polypeptide profiling of recalcitrant and orthodox legume seeds reveals ABI3-regulated LEA protein abundance linked to desiccation tolerance. Journal of Experimental Botany. 2013; 64:4559–4573, doi:10.1093/jxb/ert274

[47] Kalemba EM, Pukacka S. Association of protective proteins with dehydration and desiccation of orthodox and recalcitrant category seeds of three Acer genus species. J Plant Growth Regul 2012; 31:351–362, doi: 10.1007/s00344-011-9246-4

[48] Radwan A, Hara M, Kleinwächter M, Selmar D. Dehydrin expression in seeds and maturation drying: a paradigm change. Plant Biology. 2014; 16:853–855, doi: 10.1111/plb.12228

[49] Kleinwächter M, Radwan A, Hara M, Selmar D. Dehydrin expression in seeds: an issue of maturation drying. Frontiers in Plant Science. 2014; 5.Article 402. P. 1–3.

[50] Gumilevskaya NA, Azarkovich MI, Komarova ME, Obroucheva NV. Proteins of axial organs of dormant and germinating horse chestnut seeds: 1. General characterization. Russ. J. Plant Physiol. 2001; 48:1–11.

[51] Azarkovich MI, Gumilevskaya NA. Proteins of cotyledons of mature horse chestnut seeds. Russ. J. Plant Physiol. 2006; 53:629–637, doi: 10.1134/S1021443706050062

[52] Gumilevskaya NA, Azarkovich MI. Identification and characterization of dehydrins in horse chestnut recalcitrant seeds. Russ. J. Plant Physiol. 2010; 57:859–864, doi: 10.1134/S1021443710060154

[53] Gumilevskaya NA, Azarkovich MI. Growth capacity of embryo axes excised from dormant and germinating horse chestnut seeds and their response to exogenous abscisic acid. Russ. J. Plant Physiol. 2004; 51: 75–85.

[54] Vierling E. The roles of heat shock proteins in plants. Ann. Rev. Plant Physiology and Plant Mol. Biol. 1991; 42:579 – 620.

[55] Vierling E. The small heat shock proteins in plants are members of an ancient family of heat induced proteins. Acta Physiol. Plantarum.1997; 19: 539–547.

[56] Kimpel JA, Key JL. Heat shock in plants. Trends in Biochem. Sci. 1985; 10: 353 – 357.

[57] Walters ER, Lee CJL, Vierling E. Evolution, structure and function of small heat shock proteins in plants. Journal of Exp. Botany. 1996; 47:325 – 338.

[58] Collada C, Gomez L, Casado R, Aragoncill C. Purification and *in vitro* chaperone activity of a class I small heat-shock protein abundant in recalcitrant chestnut seeds. Plant Physiol. 1997; 115: 71–77

[59] Queitsch C, Hong S-W, Vierling E, and Lindquist S. Heat shock protein 101 plays a crucial role in thermotolerance in Arabidopsis. The Plant Cell. 2000; 12:479–492.

[60] Sabehat A, Weiss D, Lurie S. The correlation between heat-shock proteins accumulation and persistence and chilling tolerance in tomato fruit. Plant Physiol. 1996; 110:510–527.

[61] Wehmeyer N, Hernandes LD, Finkelstein RR, Vierling E. Synthesis of small heat shock proteins is a part of the developmental program of a late seed maturation. Plant Physiol. 1996; 112: 747–757.

[62] Wehmeyer N, Vierling E. The expression of small heat shock proteins in seeds responds to discrete developmental signals and suggests a general protective role in desiccation tolerance. Plant Physiol. 2000; 122: 1099–1108.

[63] Wang W, Vinocur B, Shoseyov O, Altman A. Role of plant heat-shock proteins and molecular shaperones in the abiotic stress response. Trends in Plant Science. 2004; 9:244–252.

[64] Mascarenhas JP, Altshuler M. Responses to environmental heat shock in the plant embryos. In: Atkinson BG, Walden DB, editors. Changes in eukaryotic gene expression in response to environmental stress. Orlando: Acad. Press; 1985. P. 315–326.

[65] Chrispeels MJ, Greenwood JS. Heat stress enhances phytogemagglutinin synthesis but inhibits its transport of the endoplasmic reticulum. Plant Physiol. 1987; 83:778–784.

[66] Helm K, Peterson N, Abernethy R. Heat shock response of germinating embryos of wheat. Plant Physiol. 1989; 90:598–605.

[67] Howarth CJ. Molecular responses of plants to an increased incidence of heat shock. Plant Cell Environ. (1991) 14:831–841.

[68] Riley GJP. Effects of high temperature on the germination of maize (*Zea mays* L.). Planta. 1981; 151:68–74.

[69] Gumilevskaya NA, Chumikina LV, Arabova LI, Zimin MV, Shatilov VR. Effect of high temperature on protein synthesis in the axes of imbibing pea embryos. Russian Journal of Plant Physiology. 1996; 43: 215–222.

[70] DeRocher AE, Vierling E. Developmental control of small heat shock protein expression during pea seed maturation. Plant J. 1994; 5:93–102.

[71] Reddy PS, Kavi Kishor PB, Seiler C, Kuhlmann M, Eschen-Lippold L, Lee J, Reddy MK, Sreenivasulu N. Unraveling regulation of the small heat shock proteins by the heat shock factor *HvHsfB2c* in barley: its implications in drought stress response and seed development. PLoS ONE. 2014; 9(3): e89125, doi:10.1371/journal.pone.0089125

[72] Azarkovich MI, Gumilevskaya NA. Response of dormant recalcitrant horse chestnut (*Aesculus hippocastanum* L.) seeds to heat shock. Plant Stress.2012; 6:14–19.

[73] Simon WJ, Slabas AR, and Murata N. The histidine kinase Hik34 is involved in thermotolerance by regulating the expression of heat shock genes in *Synechocystis*. Plant Physiol. 2005; 138:1409–1421.

[74] Suzuki I, Simon WJ, Slabas AR. The heat shock response of *Synechocystis* sp. PCC 6803 analyzed by transcriptomics and proteomics. J. Exp. Bot. 2006; 57:1573–1578.

[75] Inaba M, Suzuki I, Szalontai B, Kanesaki Y, Los DA, Hayashi H, Murata N. Gene-engineered rignification of membrane lipids enhances the cold inducibility of gene expression in *Synechocystis*. J. Biol. Chem. 2003; 278:12191–12198.

[76] Zorina AA, Mironov KS, Stepanchenko NS, Sinetova MA, Koroban NV, Zinchenko VV, Kupriyanova EV, Allakhverdiev SI, Los DA. Regulation systems for stress responses in cyanobacteria. Russ. J. Plant Physiol. 2011; 58:749–767, doi: 10.1134/S1021443711050281

[77] Kreslavski VD, Zorina AA, Los DA, Fomina IR, Allakhverdiev SI. Molecular mechanisms of stress resistance of photosynthetic machinery. In: Rout GR, Das AB, editors. Molecular Stress Physiology of Plants. Springer India; 2013. p. 21–51, doi: 10.1007/978-81-322-0807-5_2

11

Plant Evolution in Response to Abiotic and Biotic Stressors at "Rear-edge" Range Boundaries

Gunbharpur S. Gill, Riston Haugen, Jesse Larson, Jason Olsen and David H. Siemens

Abstract

Most species in the mustard family are restricted to higher elevations and latitudes where they also have restricted local spatial distributions. In this chapter, we describe a novel hypothesis for the development of low-elevation range limits in upland mustard species. The hypothesis suggests that defense regulation of glucosinolates could underlie the evolution of the spatially restricted distributions. A list of testable predictions is presented to evaluate the hypothesis. An interdisciplinary Ecological Genomics approach is needed to test the predictions; therefore, we also describe the field of Ecological Genomics. Although there is already support for some of the predictions, which we discuss, most of the predictions remain untested. Therefore, we also describe several tests that help evaluate each of the predictions.

Keywords: Range limits, signaling pathways, crosstalk, evolutionary constraints, ecological genomics

1. Introduction

Mustard plants (Brassicaceae) include approximately 3,700 species, several crop species (cabbage, radish, canola, etc.), and the model for molecular plant biology, *Arabidopsis thaliana*. Despite this diversity, mustard species generally inhabit high-altitude temperate regions where populations have patchy distributions (Al-Shehbaz personal communication). At lower elevations and latitudes, species often face both abiotic and biotic stressors, which populations must adapt to for range expansion or to survive climate shifts. Mustard species are also characterized by the production of glucosinolate defense toxins [1]. In this chapter, we evaluate a recent hypothesis [2] that regulation of glucosinolates could underlie the evolution of the spatially restricted distributions. This hypothesis is in contrast to previous hypotheses on defense evolution that argue the opposite that variation in defensive chemistry is the consequence of spatial distributions, life history patterns, inherent growth rates, etc. [3, 4]. As such, we begin this chapter with a description of the central hypothesis, followed by a set of predictions and then a description of the interdisciplinary approach needed to evaluate the hypothesis.

2. Hypothesis

Understanding the causes and dynamics of naturally occurring range limits in plants has become a central issue in both basic (evolutionary ecology) and applied (conservation and agriculture) areas of biology because of climate change and land use concerns [5-8]. Most transplant studies show decreased performance just across geographic range boundaries ([6] for review); therefore, it is generally assumed that many range boundaries are spatial manifestations of niche limits, requiring adaptation for local range expansion or for the persistence of populations at range edges as climate changes. However, the existence of range limits suggests that adaptation to stressful environments just outside the range is often prevented. What prevents this adaptation from occurring?

Many factors and processes can contribute to range limit development. These factors include lack of genetic variation in range margin populations, barriers to dispersal, gene flow from elsewhere within the range, and various kinds of tradeoffs [6]. Any of these factors, alone or in combination with other factors, could prevent adaptation to stressful environments outside the range. However, there is often sufficient genetic variation within and among range margin populations [9] for natural selection to presumably act upon, and often there are no obvious barriers to dispersal at range boundaries. In these cases, possible constraints on the process of adaptation to stressful environments just outside the range would include gene flow and tradeoffs. But because many range margin populations are geographically and genetically isolated (e.g., [10]), it is thought that the study of range limit development should often focus on molecular, physiological, or developmental tradeoffs [11]. What kind of tradeoffs might be important to mustards at low elevation or low latitudinal range limits?

The process of adaptation often proceeds by modifying existing structures and pathways. Within ranges, stress response signal transduction pathways help plants to survive temporary challenges from abiotic and biotic stressors [12]. Just across range boundaries, some of these same stressors increase in frequency; therefore, one would predict that adaptation to stressful environments across range limits would involve the upregulation of stress response pathways such that the pathways and the traits that they regulate were expressed more frequently or stably.

However, evolutionary models predict that a problem may arise when antagonistic response pathways are co-opted simultaneously for evolutionary change [13]. The problem is with negative pleiotropic and epistatic effects. Multiple signaling pathways often form networks of regulatory genes (transcription factors) that may interact for multiple positive and negative integrative effects. An excellent example is the flowering time signaling network in Arabidopsis, which involves many positive and negative interactions among photoperiod, circadian clock, vernalization, autonomous and Gibberellic acid pathways. In general, quantitative geneticists predict that the evolution of complex traits may involve many genetic correlations, which is why there has been such an interest in the analysis of quantitative genetic variance–covariance G-matrices [14]. Indeed, early theoreticians such as Fisher predicted that the evolution of complex traits would only involve many genes of small effects to avoid such pleiotropic effects [15]. Epistatic interactions between major flowering time network genes FRI and FLC, for example, were one of the contributing factors in the maintenance of genetic variation in Arabidopsis flowering time [16] and therefore may represent an evolutionary constraint. FRI and FLC are major transcription factors (TFs) in the flowering time signaling network that allow large behavioral shifts involving many genes, but these major effects might impede evolution through multiple epistatic and pleiotropic effects.

In another example, it is generally assumed that plants face both abiotic and biotic stressors across range boundaries, especially at low elevation "trailing edges" of species ranges [17], yet it is well known that stress response pathways, such as abscisic acid (ABA) signaling for coping with abiotic stressors (e.g., drought) and jasmonic acid (JA) signaling for coping with biotic stressors (e.g., herbivores) may negatively interfere with one another ([18 - 20] for reviews). Thus, the simultaneous co-option of these antagonistic pathways for low-elevation range expansion where organisms face both increased abiotic and biotic stressors may be problematic because of the crosstalk. Although limited phylogenetic evidence suggests that ancient antagonistic crosstalk between signaling pathways may not constrain evolution [21], more thorough and experimental work is needed to address this issue.

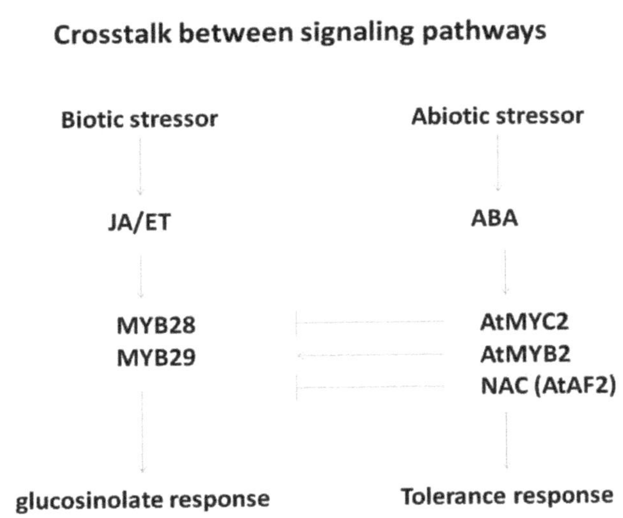

Figure 1. Crosstalk between JA/ET and ABA signaling. MYBs, MYCs, and NAC are transcription factors. The key candidate TFs in *Arabidopsis thaliana* either regulate aliphatic GS toxins (MYB28 [At5g61420] and MYB29 [At5g07690]) (44-48) or regulate both defense and drought tolerance pathways (AtMYC2 At1g32640, AtMYB2 At2g47190, and a NAC TF AtAF2 At5g08790) (18, 19, 20 for reviews).

Specifically, our hypothesis states that components of defense (e.g., JA – jasmonic acid) and stress tolerance (e.g., ABA – abscisic acid) signaling pathways are not genetically independent of one another, which may constrain the simultaneous evolution of defense and stress tolerance. Candidate genes in Arabidopsis for the evolutionary constraint include specific TFs that are involved in gene regulation within and among signaling pathways (Figure 1). For example, if there exists genetic variation in TFs MYC2 or AF2 that natural or artificial selection could act upon to increase drought stress tolerance traits for more stable expression, we predict that [1] a glucosinolate defense response would also change because these TFs help to regulate both pathways, and [2] other pleiotropic or epistatic effects would reduce fitness.

3. Predictions

1. Defense and stress tolerance phenotypes will be genetically correlated in family structured common garden experiments.
2. Genetically diverged populations from different elevations will also be diverged for defense and abiotic stress tolerance traits; neither population will have high values of both traits simultaneously.
3. Defense allocation and stress tolerance phenotypes will not segregate independently from one another in extended generation crosses between diverged populations.
4. Defense and stress tolerance genetic covariation will be associated with markers linked to candidate transcription factors (TFs) that regulate both defense and drought tolerance pathways (AtMYC2 [At1g32640], AtMYB2 [At2g47190], and a NAC TF AtAF2 [At5g08790]).
5. DNA sequence variation of candidate regulatory genes (TFs) implicated in prediction #3 will also correlate with the defense and stress tolerance tradeoff in (a) the segregating crosses and in (b) unrelated individuals (linkage disequilibrium association analysis). Comparison of DNA sequence of coding regions will also show molecular evidence for evolution (McDonald-Kreitman Test and dS/dN ratios).
6. There may be other regions of the genome besides those containing the candidate TFs that also simultaneously affect defense and stress tolerance phenotypes and thus would also provide molecular evidence for the tradeoff. For whole genome marker analysis away from candidate TFs, defense and stress tolerance phenotypes will either co-locate in linkage mapping or their QTL will show negative epistasis.
7. Gene expression of candidate genes in genetically diverged populations will reflect the evolution of more stable expression of defense and stress tolerance response pathways.

4. Approach and Discussion

An Ecological Genomics approach would satisfy the need for experimental and molecular genetics to evaluate the hypothesis in several of the predictions. Ecological Genomics (EG) is an interdisciplinary approach in the biological sciences that seeks to find the genes underlying species interactions in natural habitats and to study the evolutionary forces that have shaped these genes, their expression patterns, and the phenotypes that they encode [22-26]. As such, nucleic acid sequencing, forward and reverse genetic tools, comparative methods, and other molecular techniques are required to find the relevant genes, to establish databases of candidate genes, and to study their expression patterns and allelic variants.

Although many model species (species whose genomes have been completely sequenced and that have broad interest from molecular biologists) meet and exceed these molecular criteria, often they lack the attributes for ecological studies. For example, the natural distribution of *A. thaliana* is in Eurasia, making it difficult for study by North American researchers. Furthermore, there are still relatively few model species. With the advent of affordable high-throughput next-generation sequencing (NGS), however, sequence information is becoming available for more species and populations. As of June 2015, for example, there were 6,653 completely sequenced genomes, but only a small fraction of these are eukaryotes (http://www.genomesonline.org). The GOLD website also listed 60,631 genome sequencing projects (June, 2015), but only 9,059 or 15% were eukaryotes. Thus, although NGS is helping the field of EG to move away from model species, especially, for example, in evolutionary studies [27], there is still a need to understand the attributes that constitute an ideal model organism for EG studies.

Feder and Mitchell-Olds [22] listed the criteria for an ideal model species in Evolutionary and Ecological Functional Genomics (EEFG), which is synonymous with EG. The criteria stated that there needs to be: (1) a co-operative community of researchers from different disciplines that share resources and information; (2) the tools to find genes and study their variation within and among species; (3) natural, undisturbed habitats such that genetically diverged populations can be studied for local adaptation; (4) molecular data on sequences and chromosomal maps for marker development and mapping, cis and trans regulatory regions identified, and gene function and its fitness consequences known under natural conditions; and, finally, (5) the ability to study the ecological consequences of natural genetic sequence variation in the genes for evolutionary inferences. Thus, access to NGS alone does not necessarily make a species ideally suited for EG study.

Boechera stricta, a close wild relative of Arabidopsis, satisfies many of these criteria and is an emerging ecological model species that inhabits environments that differ substantially in drought stress, herbivore community, and other abiotic and biotic conditions [28]. The selfing rate of *B. stricta* in the northern portion of its geographic range is 0.95 [29], enabling the creation of experimental advanced generation hybrids for forward ecological genetic studies (e.g., [30]). The genome of *B. stricta* has also been recently sequenced (http://phytozome.jgi.doe.gov/pz/portal.html#!info?alias=Org_Bstricta).

Forward genetics for finding candidate genes in ecology include population genomics, association mapping, linkage mapping, and transcriptomics [31, 32]. Population genomics identifies outlier marker loci in the statistical analysis of population genetic parameters, but there is no knowledge of associated phenotypes. Association and linkage mapping both include measurement of phenotypes, but a distinct advantage of association mapping is that no pedigree is required and that allelic variants representative of naturally occurring populations can be used in analyses [26, 33]. Association mapping can be conducted on unrelated individuals because it is based on general inherent linkage disequilibrium (LD). When markers for association analysis are developed from candidate genes, significant associations may actually identify relevant genes. We assume that candidate genes in Arabidopsis could often be studied with success under natural conditions in close relatives in the genus *Boechera*. Unidentified causal genes in LD with significant candidate markers are unlikely in *B. stricta* where LD decays rapidly, within 10 kb [34]. If present, population substructure (i.e., stratification, admixture, or inbreeding) must be controlled for genetic association analyses because of confounding effects on LD that can lead to false positives [33].

4.1. Tests of prediction #1

Using family structured quantitative genetic analyses, previous studies have examined the genetic correlation between defense and stress tolerance of prediction #1. This work has been conducted on a close wild relative of Arabidopsis in the genus *Boechera*. *Boechera stricta* is a genetically diverse, diploid, predominantly self-fertilizing species that occurs at higher elevations throughout western North America in natural habitats [29, 34]. The phenotype of an individual of any species is determined by genetic and environmental factors (P = G + E), and these factors and thus the phenotypes vary among individuals within populations ($V_P = V_G + V_E$). If the phenotypes are measured from in a common garden experiment, environmental variation is eliminated so that the phenotypic and genetic variations are equal ($V_P = V_G$). Among full-siblings of clonal or self-fertilizing species such as *B. stricta*, the genetic variation (V_G) measured from a common environment can be used for evolutionary inferences because all allele combinations within and among loci are inherited without change. In accordance with prediction #1, the negative genetic correlation between glucosinolate (GS) toxin defense allocation and stress tolerance associated with range limits of *B. stricta* has been observed using these methods in the field and lab five times previously [2, 35-37]. Stressors involved in the tradeoff have included drought, nutrient deficiency, and change in plant community structure across the range boundary (suggesting competition) and the multivariate stress of the range boundary itself. In these studies, it was hypothesized that these tradeoffs occurred because of antagonistic crosstalk between abscisic acid (ABA) stress tolerance and jasmonic acid/ethylene (JA/ET) or salicylic acid (SA) defense-signaling pathways. Circumstantial evidence implicating the pathways in the tradeoff comes from experimental ABA soil inoculations that depended on endogenous GS level [35]. These conclusions are based mainly on correlative methods and circumstantial evidence. Experimental and molecular genetics and more direct measures of pathway components are needed.

4.2. Tests of prediction #2

For gene mapping in *B. stricta*, a replicated cross has been conducted between populations from the Big Horn Mountains, Wyoming, and the Black Hills, South Dakota. These are geographically isolated and genetically diverged populations [36, 38]. The populations are located at different ends of the altitudinal range of *B. stricta* (Big Horns 3,000 m, Black Hills 1,700m) and thus the sites differ by several environmental factors. The populations have diverged for glucosinolate content and stress tolerance traits such as root:shoot ratio as predicted; neither population had high values of both traits (Figure 2).

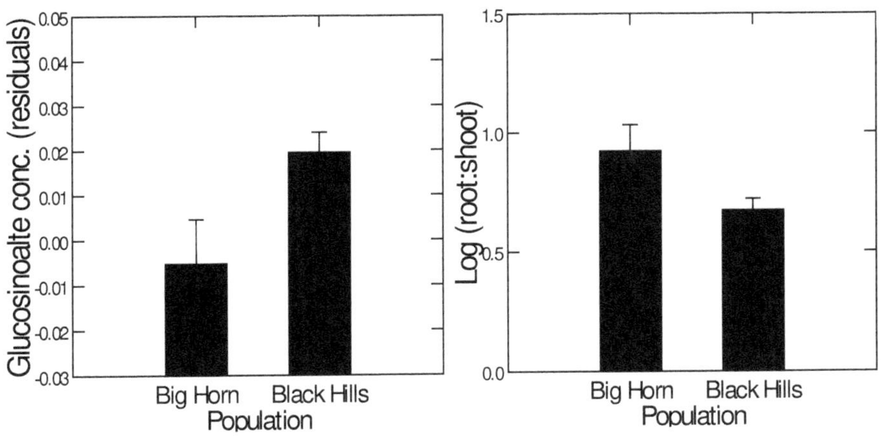

Figure 2. Genetic variation in GS levels ($F_{18, 701}$ = 7.101, P < 0.001) (left) and root:shoot ratios ($F_{5, 251}$ = 3.576, P = 0.004) (right) between genetically diverged populations. Data are from a growth chamber experiment. Residuals controlled for plant size, thus GS levels reflect content.

4.3. Tests of prediction #3

The segregating F2 populations from the crosses can be used to test prediction #3 that the defense and stress tolerance traits will not be inherited independently of one another. F2s from more than one cross allow for broader inference and increased statistical power [39]. Because each genotype in the F2 segregating generation cannot be replicated, drought treatments in the lab would need to be imposed on all plants after the plants were first monitored for performance (e.g., growth) under controlled watering conditions. Measures of drought tolerance could include relative growth (before and after drought treatments), leaf mass area (LMA), water use efficiency (carbon isotope ratio), and root:shoot ratio. Several measures of stress tolerance increase the probability of detecting the tradeoff, and provide a general assessment of stress tolerance [40]. Glucosinolate analysis should be conducted on leaf tissue at the time of drought stress.

Of course, the goal should be to test for the tradeoff in the field across the low-elevation range boundary using the segregating generation of the experimental crosses. However, this requires replication of each F2 genotype that could be compared within and just outside the range boundary. In the F3 generation, there is replication of each F2 lineage. Common garden experiments of F3 families could be established within and across low-elevation range boundaries. Previous field transplant experiments [2, 35, 36] have established the areas just across the low-elevation range boundary as stressful in terms of several correlated abiotic and biotic stressors, which manifest as slower growth and survivorship. Large sample sizes would guarantee the detection of effects if they exist, and would also allow for assessment across multiple years and the possibility of ecological gradient manipulations. For example, removal treatments of candidate competitors, such as *Lithophragma* parviflorum [36], could be included within and outside the range. Competitive interactions also induces ABA [36] and therefore might also induce the tradeoff.

4.4. Tests of predictions #4 and #5

Here the goal is to test whether TFs can be implicated in the tradeoff using TF-linked markers in the crosses, and then to conduct further genetic association analyses using markers within the implicated candidate TFs. Previous molecular work allows one to locate markers on a *B. stricta* linkage map that are linked to the candidate TFs in the crosses. Chromosomal painting and end sequencing have shown that there are large syntenic blocks that align between the *A. thaliana* and *B. stricta* genomes [41]. These Arabidopsis blocks have been located in the *B. stricta* genome. This can now also be verified with the recently sequenced *B. stricta* genome.

Analyses within the candidate genes could be conducted in populations of related and unrelated individuals. The analyses of related individuals could use F2 and F3 mapping populations, but there would be confounding effects of linkage and linkage disequilibrium (LD) from other unknown genes and alleles. Genotyping in the F2 generation also allows for genetic association analyses in the F3 field experiments. Significant associations using unrelated individuals would help to eliminate confounding effects of linkage and linkage disequilibrium (LD) from other unknown genes and alleles. Another set of markers such as microsatellite markers (e.g., [29, 36]) would need to be used to identify and control population genetic structure among the unrelated individuals [33].

For marker development within genes of *B. stricta*, identification of genetic variation (e.g., single nucleotide polymorphisms – SNPs) in marker-implicated candidate TFs can be performed by first sequencing the TFs and their promoter regions from the diverged populations. Alignment of the diverged sequences would identify any polymorphisms. The recently sequenced *B. stricta* genome allows one to readily design primers to amplify and then sequence the genes.

To test whether polymorphic genes are under selection, statistical tests of synonymous to nonsynonymous substitution ratios in coding regions would be performed. Comparing dN/dS ratios allows for detection of positive or negative selection, albeit conservatively. The MacDonald-Kreitman test is more liberal for detecting deviation from the neutral model of molecular evolution [42].

4.5. Test of prediction #6

To determine whether drought tolerance and defensive QTL co-localize on the *B. stricta* genome or whether there are epistatic interactions between QTL from different traits, linkage mapping would need to be conducted in F2 lab and F3 field experiments. Genotyping conducted in the F2 generation can also be used for successful mapping among F3 sib families [43]. Linkage analysis is often extended to incorporate information from several markers, called multi-point or interval mapping; therefore, composite interval mapping followed by multiple interval mapping using QTL-Cartographer will be used.

4.6. Test of prediction #7

If natural selection has acted upon genetic variation in signaling pathways for more stable or frequent expression of traits, this should be detectable in a comparison of the high- and low-elevation populations for trait responses and gene expression. For example, the populations could be compared in a double challenge experiment in which plants are first drought-stressed and then fed upon by generalist insect herbivores. One would predict that drought stress responses would take precedence over herbivore-induced defense responses in close wild relatives of Arabidopsis, reflecting the crosstalk between ABA and JA signaling that has been well documented in Arabidopsis. But we would predict that this plastic tradeoff would vary between high and low diverged populations and that this would be reflected in differential gene expression involving the candidate signaling pathways. To examine the expression levels of candidate TFs, one could use qPCR, but genome-wide gene expression using RNAseq would also allow the assessment of other genes as well.

5. Conclusion

An Ecological Genomics approach is needed to evaluate most of the listed predictions that remain untested. We mainly advocate a candidate gene approach that leverages the vast functional genomics literature of Arabidopsis for functional ecological genetics studies on close wild relatives. However, next-generation sequencing is rapidly increasing the potential of many ecological systems. The suggested studies outlined here are important because if our hypothesis continues to be supported, there may be important implications for understanding range limits, defense evolution, canalization, conservation, and crop improvement. If defense regulation can be used to help predict population sensitivity to environmental stress, then there would be several important applied implications.

Author details

Gunbharpur S. Gill, Riston Haugen, Jesse Larson, Jason Olsen and David H. Siemens*

*Address all correspondence to: David.Siemens@bhsu.edu

Integrative Genomics Program, Black Hills State University, Spearfish, SD, USA

References

[1] Fahey JW, Zalcmann AT, Talalay P. The chemical diversity and distribution of glucosinolates and isothiocyanates among plants. Phytochemistry. 2001;56(1):5-51.

[2] Siemens DH, Haugen R, Matzner S, VanAsma N. Plant chemical defense allocation constrains evolution of local range. Molecular Ecology 2009;18:4974-83.

[3] Stamp N. Out of the quagmire of plant defense hypotheses. Quarterly Review of Biology. 2003;78(1):23-55.

[4] Hamilton JG, Zangerl AR, DeLucia EH, Berenbaum MR. The carbon-nutrient balance hypothesis: its rise and fall. Ecology Letters. 2001;4:86-95.

[5] Parmesan C, Gaines S, Gonzalez L, Kaufman D, Kingsolver J, Townsend Peterson A, et al. Empirical perspectives on species borders: from traditional biogeography to global change. Oikos. 2005;108:58-75.

[6] Sexton JP, McIntyre PJ, Angert AL, Rice KJ. Evolution and Ecology of Species Range Limits. Annual Review of Ecology Evolution and Systematics. 2009;40:415-36.

[7] Gaston KJ. Geographic range limits: achieving synthesis. Proceedings of the Royal Society B-Biological Sciences. 2009;276(1661):1395-406.

[8] Wiens JJ. The niche, biogeography and species interactions. Philosophical Transactions of the Royal Society B-Biological Sciences. 2011;366(1576):2336-50.

[9] Eckert CG, Samis KE, Lougheed SC. Genetic variation across species' geographical ranges: the central-marginal hypothesis and beyond. Molecular Ecology. 2008;17(5): 1170-88.

[10] Vergeer P, Kunin WE. Adaptation at range margins: common garden trials and the performance of Arabidopsis lyrata across its northwestern European range. New Phytologist. 2013;197(3):989-1001.

[11] Kawecki TJ. Adaptation to Marginal Habitats. Annual Review of Ecology Evolution and Systematics. 2008;39:321-42.

[12] van Straalen NM, Roelofs D. An Introduction to Ecological Genomics. second ed: Oxford University Press; 2012. 376 p.

[13] Des Marais DL, Juenger TE. Pleiotropy, plasticity, and the evolution of plant abiotic stress tolerance. Ann N Y Acad Sci. 2010;1206:56-79.

[14] Conner JK, Hartl DL. A primer of ecological genetics. Sunderland, MA: Sinauer; 2004. 304 p.

[15] Rockman MV. The QTN program and the alleles that matter for evolution: all that's gold does not glitter. Evolution. 2012;66(1):1-17.

[16] Korves TM, Schmid KJ, Caicedo AL, Mays C, Stinchcombe JR, Purugganan MD, et al. Fitness effects associated with the major flowering time gene FRIGIDA in Arabidopsis thaliana in the field. American Naturalist. 2007;169(5):E141-E57.

[17] Ettinger AK, Ford KR, Hillerislambers J. Climate determines upper, but not lower, altitudinal range limits of Pacific Northwest conifers. Ecology. 2011;92(6):1323-31.

[18] Fujita M, Fujita Y, Noutoshi Y, Takahashi F, Narusaka Y, Yamaguchi-Shinozaki K, et al. Crosstalk between abiotic and biotic stress responses: a current view from the

points of convergence in the stress signaling networks. Current Opinion in Plant Biology. 2006;9(4):436-42.

[19] Asselbergh B, De Vieesschauwer D, Hofte M. Global switches and fine-tuning - ABA modulates plant pathogen defense. Mol Plant-Microbe Interact. 2008;21(6):709-19.

[20] Ton J, Flors V, Mauch-Mani B. The multifaceted role of ABA in disease resistance. Trends in plant science. 2009;14(6):310-7.

[21] Thaler JS, Humphrey PT, Whiteman NK. Evolution of jasmonate and salicylate signal crosstalk. Trends Plant Sci. 2012;17(5):260-70.

[22] Feder ME, Mitchell-Olds T. Evolutionary and ecological functional genomics. Nature-Reviews-Genetics. 2003;4(8):651-7.

[23] Ungerer MC, Johnson LC, Herman MA. Ecological genomics: understanding gene and genome function in the natural environment. Heredity. 2008;100(2):178-83.

[24] Mitchell-Olds T, Feder M, Wray G. Evolutionary and ecological functional genomics. Heredity. 2008;100(2):101-2.

[25] Renn SCP, Siemens DH. Meeting Review: Ecological Genomics -- Changing perspectives on Darwin's basic concerns. Molecular Ecology. 2010;19:3025-30.

[26] Anderson JT, Mitchell-Olds T. Ecological genetics and genomics of plant defences: evidence and approaches. Functional Ecology. 2011;25:312-24.

[27] Tautz D, Ellegren H, Weigel D. Next Generation Molecular Ecology. Molecular Ecology. 2010;19:1-3.

[28] Rushworth CA, Song BH, Lee CR, Mitchell-Olds T. Boechera, a model system for ecological genomics. Molecular Ecology. 2011;20(23):4843-57.

[29] Song B, Clauss MJ, Pepper A, Mitchell-Olds T. Geographic patterns of microsatellite variation in *Boechera stricta*, a close relative of *Arabidopsis*. Molecular Ecology. 2006;15:357-69.

[30] Prasad KV, Song BH, Olson-Manning C, Anderson JT, Lee CR, Schranz ME, et al. A gain-of-function polymorphism controlling complex traits and fitness in nature. Science. 2012;337(6098)(Aug 31):1081-4.

[31] Stinchcombe JR, Hoekstra HE. Combining population genomics and quantitative genetics: finding the genes underlying ecologically important traits. Heredity. 2008;100(2):158-70.

[32] Lee YW, Gould BA, Stinchcombe JR. Identifying the genes underlying quantitative traits: a rationale for the QTN programme. AoB Plants. 2014;6:1-14.

[33] Laird NM, Lange C. The fundamentals of modern statistical genetics. Gail M, Krckeberg K, Samet JM, Tsiatis A, Wong W, editors. New York: Springer Science+Business Media, LLC; 2011. 223 p.

[34] Song BH, Windsor AJ, Schmid KJ, Ramos-Onsins SE, Schranz ME, Heidel AJ, et al. Multilocus patterns of nucleotide diversity, population structure and linkage disequilibrium in *Boechera stricta*, a wild relative of *Arabidopsis*. Genetics. 2009;181(3): 1021-33.

[35] Siemens DH, Duvall-Jisha J, Jacobs J, Manthey J, Haugen R, Matzner S. Water deficiency induces stress tolerance-chemical defense evolutionary tradeoff that may help explain restricted range in plants. Oikos. 2012;121:790-80.

[36] Siemens DH, Haugen R. Plant chemical defense allocation constrains evolution of tolerance to community change across a range boundary. Ecology and Evolution. 2013;3(13):4339-47.

[37] Alsdurf JD, Ripley TJ, Matzner SL, Siemens DH. Drought-induced trans-generational tradeoff between stress tolerance and defence: consequences for range limits? AoB Plants. 2013;5:plt038.

[38] Lee CR, Mitchell-Olds T. Quantifying effects of environmental and geographical factors on patterns of genetic differentiation. Mol Ecol. 2011;20(22):4631-42.

[39] Lynch M, Walsh B. Genetics and Analysis of Quantitative Traits. Sunderland, Massachusetts: Sinauer; 1998.

[40] Donovan LA, Maherali H, Caruso CM, Huber H, de Kroon H. The evolution of the worldwide leaf economics spectrum. Trends in Ecology and Evolution 2011;26:88-95.

[41] Schranz ME, Windsor AJ, Song BH, Lawton-Rauh A, Mitchell-Olds T. Comparative genetic mapping in Boechera stricta, a close relative of Arabidopsis. Plant Physiol. 2007;144(1):286-98.

[42] Tiffin P, Moeller DA. Molecular evolution of plant immune system genes. Trends in Genetics. 2006;22(12):662-70.

[43] Schranz ME, Manzaneda AJ, Windsor AJ, Clauss MJ, Mitchell-Olds T. Ecological genomics of Boechera stricta: identification of a QTL controlling the allocation of methionine- vs branched-chain amino acid-derived glucosinolates and levels of insect herbivory. Heredity. 2009;102(5):465-74.

[44] Gigolashvili T, Yatusevich R, Berger B, Müller C, Flügge U-I. The R2R3-MYB transcription factor HAG1/MYB28 is a regulator of methionine-derived glucosinolate biosynthesis in Arabidopsis thaliana. The plant journal. 2007;51(2):247-61.

[45] Hirai MY, Sugiyama K, Sawada Y, Tohge T, Obayashi T, Suzuki A, et al. Omics-based identification of Arabidopsis Myb transcription factors regulating aliphatic glucosinolate biosynthesis. Proceedings of the National Academy of Sciences of the United States of America. 2007;104(15):6478-6483.

[46] Sonderby IE, Hansen BG, Bjarnholt N, Ticconi C, Halkier BA, Kliebenstein DJ. A Systems Biology Approach Identifies a R2R3 MYB Gene Subfamily with Distinct and

Overlapping Functions in Regulation of Aliphatic Glucosinolates. PLoS ONE. 2007;2(12): e1322.

[47] Beekwilder J, van Leeuwen W, van Dam NM, Bertossi M, Grandi V, Mizzi L, et al. The Impact of the Absence of Aliphatic Glucosinolates on Insect Herbivory in Arabidopsis. PLoS ONE. 2008;3(4).

[48] Kliebenstein DJ. Advancing Genetic Theory and Application by Metabolic Quantitative Trait Loci Analysis. The plant cell. 2009;21(6):1637-46.

Permissions

All chapters in this book were first published in ABSPRAFP, by InTech Open; hereby published with permission under the Creative Commons Attribution License or equivalent. Every chapter published in this book has been scrutinized by our experts. Their significance has been extensively debated. The topics covered herein carry significant findings which will fuel the growth of the discipline. They may even be implemented as practical applications or may be referred to as a beginning point for another development.

The contributors of this book come from diverse backgrounds, making this book a truly international effort. This book will bring forth new frontiers with its revolutionizing research information and detailed analysis of the nascent developments around the world.

We would like to thank all the contributing authors for lending their expertise to make the book truly unique. They have played a crucial role in the development of this book. Without their invaluable contributions this book wouldn't have been possible. They have made vital efforts to compile up to date information on the varied aspects of this subject to make this book a valuable addition to the collection of many professionals and students.

This book was conceptualized with the vision of imparting up-to-date information and advanced data in this field. To ensure the same, a matchless editorial board was set up. Every individual on the board went through rigorous rounds of assessment to prove their worth. After which they invested a large part of their time researching and compiling the most relevant data for our readers.

The editorial board has been involved in producing this book since its inception. They have spent rigorous hours researching and exploring the diverse topics which have resulted in the successful publishing of this book. They have passed on their knowledge of decades through this book. To expedite this challenging task, the publisher supported the team at every step. A small team of assistant editors was also appointed to further simplify the editing procedure and attain best results for the readers.

Apart from the editorial board, the designing team has also invested a significant amount of their time in understanding the subject and creating the most relevant covers. They scrutinized every image to scout for the most suitable representation of the subject and create an appropriate cover for the book.

The publishing team has been an ardent support to the editorial, designing and production team. Their endless efforts to recruit the best for this project, has resulted in the accomplishment of this book. They are a veteran in the field of academics and their pool of knowledge is as vast as their experience in printing. Their expertise and guidance has proved useful at every step. Their uncompromising quality standards have made this book an exceptional effort. Their encouragement from time to time has been an inspiration for everyone.

The publisher and the editorial board hope that this book will prove to be a valuable piece of knowledge for researchers, students, practitioners and scholars across the globe.

List of Contributors

V. C. Dilukshi Fernando and Dana F. Schroeder
University of Manitoba, Winnipeg, MB, Canada

Allan Klynger da Silva Lobato and Gélia Dinah Monteiro Viana
Núcleo de Pesquisa Vegetal Básica e Aplicada, Universidade Federal Rural da Amazônia, Paragominas, Brazil

Gleberson Guillen Piccinin
Campus de Parauapebas, Universidade Federal Rural da Amazônia, Belém, Brazil

Milton Hélio Lima da Silva
Coordenação de Botânica, Museu Paraense Emílio Goeldi, Belém, Brazil

Gabriel Mascarenhas Maciel
Instituto de Ciências Agrárias, Universidade Federal de Uberlândia, Monte Carmelo, Brazil

Douglas José Marques
Setor de Olericultura e Experimentação, Universidade José do Rosário Vellano (UNIFENAS), Alfenas, Brazil

Zoltán Á. Nagy, György Kátay, Gábor Gullner and Attila L. Ádám
Plant Protection Institute, Centre for Agricultural Research, Hungarian Academy of Sciences, Budapest, Hungary

Poornananda M. Naik and Jameel M. Al-Khayri
Department of Agricultural Biotechnology, College of Agriculture and Food Sciences, King Faisal University, Al-Hassa, Saudi Arabia

Sabina Berne and Branka Javornik
Department of Agronomy, Biotechnical Faculty, University of Ljubljana, Ljubljana, Slovenia

Santosh Kumar Upadhyay
Department of Botany, Panjab University, Chandigarh, India

Sudhir P. Singh
Center of Innovative and Applied Bioprocessing, Mohali, India

Zerihun Tadele
Institute of Plant Sciences, University of Bern, Bern, Switzerland
Addis Ababa University, Institute of Biotechnology, Addis Ababa, Ethiopia

Sebnem Kusvuran
Cankiri Karatekin University, Kizilirmak Vocational High School, Cankiri, Turkey

Sevinç Kiran
Soil Fertilizer and Water Resources Central Research Institute, Ankara, Turkey

S. Sebnem Ellialtioglu
Department of Horticulture, Faculty of Agriculture, Ankara University, Ankara, Turkey

Sevgi Marakli and Nermin Gozukirmizi
Istanbul University, Faculty of Science, Department of Molecular Biology and Genetics, Vezneciler, Istanbul, Turkey

Marina I. Azarkovich
Timiryazev Institute of Plant Physiology, Russian Academy of Sciences, Moscow, Russia

Gunbharpur S. Gill, Riston Haugen, Jesse Larson, Jason Olsen and David H. Siemens
Integrative Genomics Program, Black Hills State University, Spearfish, SD, USA

Index

A

Abiotic Stress Tolerance, 1-2, 7, 12, 17 69, 104-105, 152, 180, 222, 228

Abiotic Stresses, 20, 75, 79, 83, 88-89 133, 139-140, 152-153, 155, 157 167-168, 176, 178, 180, 187, 192, 200

Aesculus Hippocastanum, 205, 209, 218

Agronomic Benefits, 130, 132

Antioxidant Enzymes, 63, 148, 152-159 161-162, 164, 166-171, 173-174, 181 183, 189, 193

Antioxidant Systems, 55, 68, 153, 158 174

Arabidopsis Thaliana, 1-4, 7, 13, 15-18 59, 78, 86, 93, 95-97, 103, 113, 126 188, 192, 196, 198, 213-215, 221, 230

Arachis Hypogaea, 24, 175, 192

Ascorbate Peroxidase, 137-138, 142 153, 175, 181

B

Brassica Oleraceae, 198

C

Carotenoids, 138, 155, 164-165

Cellular Dehydration Tolerance, 8

Chaperones, 9-10, 86, 101, 207-209, 212

Crop Management, 141

Cyanobacteria, 2, 6, 202, 214, 218

Cytotoxic Effects, 152-153

D

Deep Dormancy, 200, 205, 210-211

Dehydrins, 7, 10, 200-207, 212-216

Desiccation Tolerance, 1, 8, 10-11, 17 203-204, 209, 215-217

Downstream Genes, 3, 5

Drought Avoidance, 128, 136, 138, 187

Drought Insurance, 141, 150-151

Drought Stress, 6-10, 12, 16, 29-31, 33 51, 63-64, 138, 146, 149, 157-160 171-173, 180-181, 191-192, 203-204 213, 218, 222-223, 225, 227

E

Element Binding Factors, 3, 7, 15

Environmental Stresses, 1, 50, 52, 138 152-153, 195, 201

G

Gas Exchange, 19-20, 22, 28, 191

Gene Expression, 2-5, 7-10, 15, 20, 29 66-67, 71, 73, 82, 98, 119-120, 181 184, 190, 194, 200-201, 207-208 210-211, 222, 227

Germplasm, 45, 140, 145

Global Agriculture, 130, 140

Glutathione Reductase, 20, 137, 142 153, 165, 174, 181

Glycine Max, 21-22, 25, 32, 192

Gossypium Hirsutum, 20, 169

H

Heat Shock Proteins, 188, 200, 204 207-212, 217-218

Heat-stable Proteins, 204-207, 210

Herbicide Treatment, 153

Hydrophilic Proteins, 202, 207

I

Ionic Stress, 8

L

Late Embryogenesis Abundant, 8, 17 142-143, 200-201, 213

Leguminous Crops, 19, 180

Leucine Zipper Transcription Factor, 16

M

Marker-assisted Selection, 140

Metal Toxicity, 153, 175

Moisture Scarcity, 139

N

Nicotiana Rustica, 156, 170

Non-grafted Seedlings, 156

Non-inoculated Plants, 22-23

Non-stratified Seeds, 210-211

O

Okra Genotypes, 155-156, 158, 172

Osmolytes, 9-10, 201

Osmotic Adjustment, 10, 19-20, 25, 27 29-30, 33, 50, 133, 136, 138, 161, 201

Osmotic Stress, 3, 7-11, 13, 15-16 49-50, 63, 139, 155-156, 180, 192-193 202, 204, 214

Osmotic Stress Tolerance, 3, 7, 16

Oxidative Stress, 30, 146, 148, 152 154-155, 157-159, 161-165, 167-169 171, 173-174, 176, 179, 181, 184, 189 193-194, 196-197, 199, 212

P
Panicum Miliaceum, 128, 130, 144
Paraheliotropic Leaf Movement, 21, 28
Phospholipid Signaling, 7
Photosynthetic Pigments, 20, 166, 175
Phytohormone Abscisic Acid, 2 202-203
Protein Kinase, 3-4, 8, 11, 14, 18, 80-82 84, 91, 94-95

Q
Quenching Activity, 152-153

R
Reactive Oxygen Species, 7, 9, 11, 20 30, 55, 69, 76, 78, 81, 94-95, 138, 143 152-153, 175, 179
Recalcitrant Seeds, 200-201, 203-205 207, 209, 216
Root Dry Matter, 21-22, 24, 28

S
Salinity Stress, 2, 8, 63, 68, 155-157 167-169, 181
Seed Dormancy, 2, 205
Seed Maturation, 2, 7-8, 10, 204, 216-218
Sesquiterpenoid, 2, 49, 52, 66
Shoot Dry Matter, 21-22, 28
Signaling Mechanisms, 1, 6, 89, 94
Signalization Pathway, 25, 27
Solanum Lycopersicum, 156, 196
Sorghum Bicolor, 20, 214
Stomatal Aperture, 1, 11
Stomatal Conductance, 19, 22, 24, 28, 32
Stratification, 205-207, 210-211
Stress Tolerance, 1-3, 7-10, 12, 16-17 30, 51, 69, 104-105, 113, 127, 149, 152 154, 159, 168, 171-172, 178, 180, 194 203, 214, 221-222, 224-225, 228, 230
Stress-induced Proteins, 200-201, 207
Stress-responsive Genes, 2, 9, 11
Superoxide Dismutase, 138, 143, 175 181, 212

T
Transcription Factors, 3-10, 14-15, 75 79, 82-83, 96-98, 103, 125, 220-222, 230
Transgenic Tobacco Plants, 31, 107 110, 113, 115, 127
Turgor Loss, 201

V
Vigna Unguiculata, 19-20, 23, 26-27 31-33

W
Water Relations, 20, 22, 31, 133, 191 201
Water Requirement Satisfaction Index 134, 143

www.ingramcontent.com/pod-product-compliance
Lightning Source LLC
LaVergne TN
LVHW080733210926
848485LV00003B/545